Mechanics of Materials FOR DUMMIES®

by James H. Allen III, PE, PhD

Wiley Publishing, Inc.

Mechanics of Materials For Dummies®

Published by
Wiley Publishing, Inc.
111 River St.
Hoboken, NJ 07030-5774
www.wiley.com

Published by Wiley Publishing, Inc., Indianapolis, Indiana

Published simultaneously in Canada

For general information on our other products and services, please contact our Customer Care Department within the U.S. at 877-762-2974, outside the U.S. at 317-572-3993, or fax 317-572-4002.

For technical support, please visit www.wiley.com/techsupport.

Wiley also publishes its books in a variety of electronic formats and by print-on-demand. Some content that appears in standard print versions of this book may not be available in other formats. For more information about Wiley products, visit us at www.wiley.com.

Library of Congress Control Number: 2011926319

ISBN: 978-0-470-94273-4 (pbk); 978-1-118-08899-9 (ebk); 978-1-118-08900-2 (ebk); 978-1-118-08901-9 (ebk)

Manufactured in the United States of America

10 9 8 7 6 5 4 3 2 1

About the Author

James H. Allen III, PE, PhD, serves on the civil engineering faculty at the University of Evansville, where he teaches statics, mechanics of materials, structural analysis, and structural design courses. Dr. Allen received his Ph.D. from the University of Cincinnati in structural engineering and performed his undergraduate work at the University of Missouri-Rolla (now the Missouri University of Science and Technology).

Dedication

For my loving wife, Miranda.

Author's Acknowledgments

I wish to thank the many people associated with the creation of this book, including the dedicated staff at Wiley. My continued appreciation is extended to my senior project editor, Alissa Schwipps, and my copy editor, Megan Knoll, for their continued guidance in this project. I'd also like to thank Tracy Boggier and the folks at Wiley's Composition Services department for their help in the completion of this endeavor.

Publisher's Acknowledgments

We're proud of this book; please send us your comments at http://dummies.custhelp.com. For other comments, please contact our Customer Care Department within the U.S. at 877-762-2974, outside the U.S. at 317-572-3993, or fax 317-572-4002.

Some of the people who helped bring this book to market include the following:

Acquisitions, Editorial, and Media Development

Senior Project Editor: Alissa Schwipps

Acquisitions Editor: Tracy Boggier

Copy Editor: Megan Knoll

Assistant Editor: David Lutton

Technical Editors: Guillermo Ramirez, PE, PhD; Dr. Thomas Siegmund

Editorial Manager: Christine Meloy Beck

Editorial Assistants: Rachelle Amick, Jennette ElNaggar, Alexa Koschier

Cover Photo: © iStockphoto.com/Nikada

Cartoons: Rich Tennant (www.the5thwave.com)

Composition Services

Project Coordinator: Katherine Crocker

Layout and Graphics: Carl Byers, Carrie A. Cesavice, Cheryl Grubbs, Corrie Socolovitch, Laura Westhuis

Proofreader: Melissa D. Buddendeck

Indexer: Becky Hornyak

Special Help

Danielle Voirol

Publishing and Editorial for Consumer Dummies

Diane Graves Steele, Vice President and Publisher, Consumer Dummies

Kristin Ferguson-Wagstaffe, Product Development Director, Consumer Dummies

Ensley Eikenburg, Associate Publisher, Travel

Kelly Regan, Editorial Director, Travel

Publishing for Technology Dummies

Andy Cummings, Vice President and Publisher, Dummies Technology/General User

Composition Services

Debbie Stailey, Director of Composition Services

Contents at a Glance

Table of Contents

Introduction

Students undertaking a mechanics of materials class often find themselves facing a common dilemma: In their basic statics and dynamics classes, they focused on dealing exclusively with a key set of assumptions — namely, that objects subjected to load don't deform — but mechanics of materials throws many of those assumptions out the window.

Mechanics of materials is often your first foray into the real world from the land of theory in mechanics and physics. This class is where you start to take your basic understanding of the world around you and shape your surroundings to perform specific tasks; that is, you design stuff. This point is where I tell students that with a bit of knowledge, you can become quite dangerous.

Mechanics of materials at its core is still a very theoretical class, but it quickly takes these basic theories and applies them in new and unfamiliar ways. That's why I've written *Mechanics of Materials For Dummies:* to help make your transition from theoretical to practical as smooth and simple as possible. My goal in this text is to illustrate the basic theory while showing you how to actually apply these theories to real-world applications.

About This Book

No mechanics of materials book can possibly show you how to analyze every type of problem you may come across. Most mechanics of materials textbooks focus on complex derivations and variables that result in several relatively simple formulas without providing a whole lot of explanation along the way.

Mechanics of Materials For Dummies gives you the basic rundown of the theory but focuses more on why you need to know the formulas and how to apply them rather than where exactly they came from. I intend this book to serve more as an application-oriented text that utilizes the basic theories. What exactly is a stress, and how do you relate it to the load-carrying capability of a material? How do you determine the capacity of a long, slender column? How do you compute the angle of twist of a shaft under torsion loads? All these topics (and many, many more) are common application problems in engineering, and they provide a basis for the core of discussion covered in this text.

Tip: For even more background on the topics in this book, check out my *Statics For Dummies* (Wiley); it can help you refresh the statics vital to mechanics of materials.

I've broken each chapter into several sections, and each section deals with a specific concept relevant to the major chapter topic, such as

- How is normal stress different from shear stress?
- How do you determine cross-sectional dimensions for a beam subjected to flexural loads?
- What techniques can you use to solve statically indeterminate problems?

Because methodical analysis is key in mechanics of materials, I present analysis and design techniques in a step-by-step format whenever possible.

As with any *For Dummies* book, you can control where you want to start. For example, if all you need is information on analyzing stress, turn to Part II. If you already have a firm grasp of stress and strain, but need help applying these topics, turn to Part IV.

Conventions Used in This Book

I use the following conventions throughout the text to make things consistent and easy to understand:

- I format new terms in *italics* and follow them closely with an easy-to-understand definition.
- I also use italics to denote a variable (and its magnitude value) in text.
- **Bold** highlights the action parts of numbered steps, as well as the keywords in bulleted lists.

I also utilize other, mechanics-specific conventions that I may not explain every time they appear:

- **Origin:** The origin used in mechanics of materials calculations is a reference point that is typically located at a special location known as the centroid of an area or region. In this book, unless I state otherwise, this is the location I also use.
- **Significant digits:** I usually try to carry at least three significant digits in all my calculations to help ensure enough precision to demonstrate the fundamental principles.
- **Internal force variables**: Because the calculation of stress is entirely dependent on the internal forces, being consistent with notation can alleviate a lot of potential headaches. For internal forces in this text, I use N to denote an axial (or normal) force, V to indicate a shear force, and M to

represent a moment. If any of these internal forces acts in a specific direction or about a specific axis, I include subscripts related to the Cartesian axes or specific locations on a member to help distinguish them.

- **Plus signs (+) with magnitude values:** Although it's optional, I use the plus symbol before positive numbers in some calculations to remind myself (and you) that I've considered the *sense* (direction) of the vector on the Cartesian plane.

What You're Not to Read

I readily admit that you can skip over a few items in this text if you're short on time or just after the most important and practical stuff:

- **Text in sidebars:** *Sidebars* are the shaded boxes that provide extra information that goes into more detail about the topic at hand than is necessary.
- **Anything with a Technical Stuff icon:** The in-depth info associated with this icon is useful but may not be necessary for solving everyday problems.
- **The stuff on the copyright page:** The copyright page provides some of the best information in the book. Too bad none of it applies to mechanics of materials!

Foolish Assumptions

As I wrote this book, I made a few assumptions about you, the reader.

- You're a college student taking an engineering mechanics of materials (or strength of materials) class who has successfully completed a basic engineering statics class. Or if you're not a student currently, you're at least familiar with basic statics and computation of internal forces. Just in case though, I provide a bit of a review in Chapter 3.
- You remember some basic math skills, including basic algebra and trigonometry, as well as some basic calculus topics (such as differentiation, simple integration, and how to find maximum and minimum values of functions).
- You're proficient in geometry and trigonometry. Being familiar with the Cartesian coordinate system and its terminology as well as knowing the basic rules governing sines, cosines, and tangents of angles (both in degrees and radians) is invaluable as you work mechanics of materials problems.

How This Book Is Organized

This book is organized into parts and chapters, starting with a basic review of math and static equilibrium concepts and going through section property calculations, analysis of stress and strain, and practical mechanics of materials applications.

Part I: Setting the Stage for Mechanics of Materials

In Part I, you get a brief rundown of basic information you need in mechanics of materials, such as a quick refresher on math and units, a brief review of essential statics topics, and fundamentals for computing basic section properties. Chapter 1 introduces the basic concept of mechanics of materials; explains the basic differences among statics, dynamics, and mechanics of materials; and touches on basic terminology that you need. Chapter 2 provides you with a brief refresher about a wide range of mathematics topics, including basic trigonometric relationships and calculus computations such as differentiation and integration. It also reviews systems of units and the base units you need in mechanics of materials.

Chapter 3 highlights essential statics skills you need, including equilibrium calculations and internal force diagrams. Chapter 4 gives a quick description of cross-sectional properties (including area calculations) and shows how to locate the centroid of a region. Chapter 5 introduces the first moment of area, different variations of the second moment of area (also known as the area moments of inertia), and the radius of gyration — some of the more complex section properties that you need.

Part II: Analyzing Stress

Part II introduces you to the concept of intensity of load, also known as stress. Chapter 6 leads off by explaining the basic types of stress and highlighting the difference between average stress and stress at a point. In Chapter 7, I show you how to determine the maximum and minimum (or principal) values and their orientation angles by using transformation equations and the graphical technique known as Mohr's circle for stress.

Next, I delve into the different types of stress that can be developed from various loading situations that you may encounter. In Chapter 8, I explain the different types of axial stress calculations, such as bearing stress, pressure vessels, and maximum stresses concentrations. Chapter 9 focuses on flexural bending effects; I show you how to determine the normal stress at a point within the cross section due to applied bending moment. In Chapter 10, I discuss different types of shear stresses, including direct shear of bolts and shafts as well as shear stresses that arise from flexural effects. Finally, Chapter 11 demonstrates how to compute shear stresses that result when you twist an object.

Part III: Investigating Strain

In Part III, I explore how objects deform in response to applied load, known as strain. Chapter 12 covers the different types of strain, including normal and shear strains, and shows how thermal strains can result in deformation without applied physical forces. In Chapter 13, I demonstrate how to compute maximum and minimum strain values (known as principal strains) and how to determine their orientation within an object. I explain strain transformation by using both equations and another form of Mohr's circle for strain. Chapter 14 discusses several important material properties, such as Young's modulus of elasticity and the Poisson ratio, and shows how you can use these properties to correlate stresses to strains in a material through the fundamental relationship, Hooke's law.

Part IV: Applying Stress and Strain

Part IV shows you how to take the principles from Parts I, II, and III and apply them to a wide array of important engineering applications. In Chapter 15, I show you how to combine different types of stresses into a single net effect. Chapter 16 turns your attention to computing deformations, deflections, and angles of twist for different objects. In Chapter 17, you discover how you can use mechanics of materials to solve indeterminate statics problems. Chapter 18 covers columns and compression members; in this chapter, I discuss how compression members can fail at loads less than the failure stress of the material from which they're made. Chapter 19 provides examples illustrating how you can use mechanics of materials to design members to support known loads. Finally, in Chapter 20, you find out how you can apply the physics concept of energy to analyze the effects of load on an object.

Part V: The Part of Tens

Part V includes a couple of top-ten lists on interesting mechanics of materials topics. Chapter 21 shows you ten things to remember when working with mechanics of materials. Chapter 22 gives you ten tips for solving a mechanics of materials problem.

Icons Used in This Book

To make this book easier to read and simpler to use, I include some icons that can help you quickly find and identify key ideas and information.

I use this icon to highlight an idea that contains a shortcut procedure or a method for remembering an idea or equation.

The information with this icon draws your attention to facts and ideas that are important for the proper application of the topic at hand.

This icon flags information that you need to be careful about. I use this icon to highlight common missteps that I've seen (or taken myself) in applying the theory or equations of mechanics of materials.

This icon gives additional information that, although handy and interesting, may not be totally necessary for your everyday survival in mechanics of materials. But you may be able to use this information to impress your friends or professor!

Where to Go from Here

You can use *Mechanics of Materials For Dummies* to supplement a course you're currently taking or on its own as a text for understanding the basic principles of mechanics of materials. I wrote this book to allow you to move freely among chapters, with each chapter being a self-contained topic; unlike a classical mechanics textbook, you don't necessarily need to move through the book in order.

However, if you're new to the subject of mechanics of materials, I strongly suggest you start at the beginning with Chapter 1 and proceed through the chapters in order. Topics later in the text use principles that are developed early on (although I do provide cross references to those discussions so you don't feel like you're out of luck if you've been skipping around). On the other hand, if you're simply brushing up on your skills; feel free to use the table of contents or index to jump to the material you need.

Part I

Setting the Stage for Mechanics of Materials

"I looked over your neutral axis, Mrs. Dundt. Your polar and product moments are clean and nothing's wrong with your centroid and cross-sectional area. It may be your gyradius, but I won't be able to look at it until Thursday."

In this part . . .

This part introduces you to the basic concepts of mechanics of materials and its relationship to and differences from basic statics and dynamics (known simply as mechanics). You get a short refresher in several mathematics areas, including geometry, trigonometry, and basic calculus, that you may need along the way, and I discuss the basic unit systems while showing you the base units mechanics of materials uses from each system.

But that's not all! I also provide a short review of basic statics skills and of computing internal forces of structural members, which are critical to your continued analysis of mechanics of materials. I round out the part with chapters on computing section properties such as the cross-sectional area, centroid location, and the first and second moments of area, all of which are integral to mechanics of materials.

Chapter 1

Predicting Behavior with Mechanics of Materials

In This Chapter

- Defining mechanics of materials
- Introducing stresses and strains
- Using mechanics of materials to aid in design

Mechanics of materials is one of the first application-based engineering classes you face in your educational career. It's part of the branch of physics known as *mechanics,* which includes other fields of study such as rigid body statics and dynamics. Mechanics is an area of physics that allows you to study the behavior and motion of objects in the world around you.

Mechanics of materials uses basic statics and dynamics principles but allows you to look even more closely at an object to see how it deforms under load. It's the area of mechanics and physics that can help you decide whether you really should reconsider knocking that wall down between your kitchen and living room as you remodel your house (unless, of course, you like your upstairs bedroom on the first floor in the kitchen).

Although statics can tell you about the loads and forces that exist when an object is loaded, it doesn't tell you how the object behaves in response to those loads. That's where mechanics of materials comes in.

Tying Statics and Mechanics Together

Since the early days, humans have looked to improve their surroundings by using tools or shaping the materials around them. At first, these improvements were based on an empirical set of needs and developed mostly through a trial-and-error process. Structures such as the Great Pyramids in Egypt or the Great Wall of China were constructed without the help of fancy materials or formulas. Not until many centuries later were mathematicians such as Sir Isaac Newton able to formulate these ideas into actual numeric equations (and in many cases, to remedy misconceptions) that helped usher in the area of physics known as mechanics.

Mechanics, and more specifically the core areas of statics and dynamics, are based on the studies and foundations established by Newton and his laws of motion. Both statics and dynamics establish simple concepts that prove to be quite powerful in the world of analysis. You can use *statics* to study the behavior of objects at rest (known as *equilibrium*), such as the weight of snow on your deck or the behavior of this book as it lies on your desk. *Dynamics*, on the other hand, explains the behavior of objects in motion, from the velocity of a downhill skier to the trajectory of a basketball heading for a winning shot.

What statics and dynamics both have in common is that at their fundamental level, they focus on the behavior of *rigid bodies* (or objects that don't deform under load). In reality, all objects deform to some degree (hence why they're called *deformable bodies*), but the degree to which they deform depends entirely on the mechanics of the materials themselves. *Mechanics of materials* (which is sometimes referred to as *strength of materials* or *mechanics of deformable bodies*) is another branch of mechanics that attempts to explain the effect of loads on objects.

The development of mechanics of materials over the centuries has been based on a combination of experiment and observation in conjunction with the development of equation-based theory. Famous individuals such as Leonardo da Vinci (1452–1519) and Galileo Galilei (1564–1642) conducted experiments on the behavior of a wide array of structural objects (such as beams and bars) under load. And mathematicians and scientists such as Leonhard Euler (1707–1783) developed the equations used to provide the basics for column theory.

Mechanics of materials is often the follow-up course to statics and dynamics in the engineering curriculum because it builds directly on the tools and concepts you learn in a statics and dynamics course, and it opens the door to engineering design. And that's where things get interesting.

Defining Behavior in Mechanics of Materials

The fact that all objects deform under load is a given. Mechanics of materials helps you determine how much the object actually deforms. Like statics, mechanics of materials can be very methodical, allowing you to establish a few simple, guiding steps to define the behavior of objects in the world around you. You can initially divide your analysis of the behavior of objects under load into the study and application of two basic interactions: stress and strain.

With the basic concepts of stress and strain, you have two mechanisms for determining the maximum values of stress and strain, which allow you to investigate whether a material (and the object it creates) is sufficiently strong while also considering how much it deforms. You can then turn your attention to specific sources of stress, which I introduce a little later in this chapter.

Stress

Stress is the measure of the intensity of an internal load acting on a cross section of an object. Although you know a bigger object is capable of supporting a bigger load, stress is what actually tells you whether that object is big enough. This intensity calculation allows you to compare the intensity of the applied loads to the actual strength (or *capacity*) of the material itself. I introduce the basic concept of stress in Chapter 6, where I explain the difference between the two types of stress, *normal stresses* and *shear stresses.*

With this basic understanding of stress and how these normal and shear stresses can exist simultaneously within an object, you can use *stress transformation* calculations (see Chapter 7) to determine maximum stresses (known as *principal stresses*) and their orientations within the object.

Strain

Strain is a measure of the deformation of an object with respect to its initial length, or a measure of the intensity of change in the shape of a body. Although stress is a function of the load acting inside an object, strain can occur even without load. Influences such as thermal effects can cause an object to elongate or contract due to changes in temperature even without a physical load being applied. For more on strain, turn to Chapter 12.

As with stresses, strains have maximum and minimum values (known as *principal strains*), and they occur at a unique orientation within an object. I show you how to perform these *strain transformations* in Chapter 13.

Using Stresses to Study Behavior

Stresses are what relate loads to the objects they act on and can come from a wide range of internal forces. The following list previews several of the different categories of stress that you encounter as an engineer:

- **Axial**: *Axial stresses* arise from internal axial loads (or loads that act along the longitudinal axis of a member). Some examples of axial stresses include tension in a rope or compression in a short column. For more on axial stress examples, turn to Chapter 8.
- **Bending:** *Bending stresses* develop in an object when internal bending moments are present. Examples of members subject to bending are the beams of your favorite highway overpass or the joists in the roof of your house. I explain more about bending stresses in Chapter 9.
- **Shear:** *Shear stresses* are actually a bit more complex because they can have several different sources. *Direct shear* is what appears when you try to cut a piece of paper with a pair of scissors by applying two forces in opposite direction across the cut line. *Flexural shear* is the result of bending moments. I discuss both of these shear types in Chapter 10. *Torsion* (or *torque*) is another type of loading that creates shear stresses on objects through twisting and occurs in rotating machinery and shafts. For all things torsion, flip to Chapter 11.

Studying Behavior through Strains

You can actually use strains to help with your analysis in a couple of circumstances:

- **Experimental analysis:** Strains become very important in experiments because, unlike stresses, they're quantities that you can physically measure with instruments such as electromechanical strain gauges. You can then correlate these strains to the actual stresses in a material using the material's properties.
- **Deformation without load:** Strain concepts can also help you analyze situations in which objects deform without being subjected to a load such as a force or a moment. For example, some objects experience changes in shape due to temperature changes. To measure the effects of temperature change, you must use the concepts of strain.

Incorporating the "Material" into Mechanics of Materials

After you understand the calculations behind stress and strains, you're ready to turn your attention to exploring the actual behavior of materials. All materials have a unique relationship between load (or stress) and deformation (or strain), and these unique material properties are critical in performing design.

One of the most vital considerations for the stress-strain relationship is Hooke's law (see Chapter 14). In fact, it's probably the single most important concept in mechanics of materials because it's the rule that actually relates stresses directly to strain, which is the first step in developing the theory that can tell you how much that tree limb deflects when you're sitting on it. This relationship also serves as the basis for design and the some of the advanced calculations that I show you in Part IV.

Putting Mechanics to Work

When you have the tools to analyze objects in the world around them, you can put them to work for you in specific applications. Here are some common mechanics of materials applications:

- **Combined stresses:** In some cases, you want to combine all those single and simple stress effects from Part II into one net action. You can analyze complex systems such as objects that bend in multiple directions simultaneously (known as *biaxial bending*) and bars with combined shear and torsion effects. Flip to Chapter 15 for more.
- **Displacements and deformations:** *Deformations* are a measure of the response of a structure under stress. You can use basic principles based on Hooke's law to calculate deflections and rotations for a wide array of scenarios. (See Chapter 16.)
- **Indeterminate structures:** For simple structures, the basic equilibrium equations you learn in statics can give you all the information you need for your analysis. However, the vast majority of objects are much more complex. When the equilibrium equations from statics become insufficient to analyze an object, the object is said to be *statically indeterminate*. In Chapter 17, I show you how to handle different types of these indeterminate systems by using mechanics of materials principles.

- **Columns:** Unlike most objects that fail when applied stresses reach the limiting strength of the material, columns can experience a geometric instability known as *buckling,* where a column begins to bow or flex under compression loads. Chapter 18 gives you the lowdown on columns.
- **Design:** *Design* is the ability to determine the minimum member size that can safely support the stresses or deflection criteria. This step requires you to account for factors of safety to provide a safe and functional design against the real world. Head to Chapter 19 for more.
- **Energy methods:** *Energy methods* are another area of study that relates the principles of energy that you learned in physics to concepts involving stresses and strain. In Chapter 20, I introduce you to energy method concepts such as strain energy and impact.

Chapter 2

Reviewing Mathematics and Units Used in Mechanics of Materials

In This Chapter

- Refreshing basic trigonometry and geometry
- Applying some basic calculus
- Dealing with SI and U.S. customary units

As with other areas of engineering and the sciences, mathematics plays a significant role in mechanics of materials. The math is what takes advantage of all those awesome statics equations you created and gives mechanics of materials its basic punch in design and analysis of stress and strain.

In the beginning, basic mathematics skills such as algebra, geometry, and trigonometry can carry you a long way in your mechanics of materials endeavors. Later, the calculus — particularly integration and differentiation — helps you estimate such things as deflections in beams and relationships between internal forces.

In this chapter, I provide a refresher on some important math foundations for mechanics of materials. I also address this field's unique units as well as its systems of units.

Grasping Important Geometry Concepts

You encounter several geometric principles in mechanics of materials, including angle units and the famous Pythagorean theorem. The following sections fill you in on how those issues play into your mechanics work.

One of the most common geometric relationships involves the relationship between the sides of a right triangle (or a triangle with exactly one angle of 90 degrees). This relationship is known as the *Pythagorean theorem,* and it's a crucial piece of the transformation calculations in Chapters 7 and 13. The triangle in Figure 2-1 illustrates this theorem.

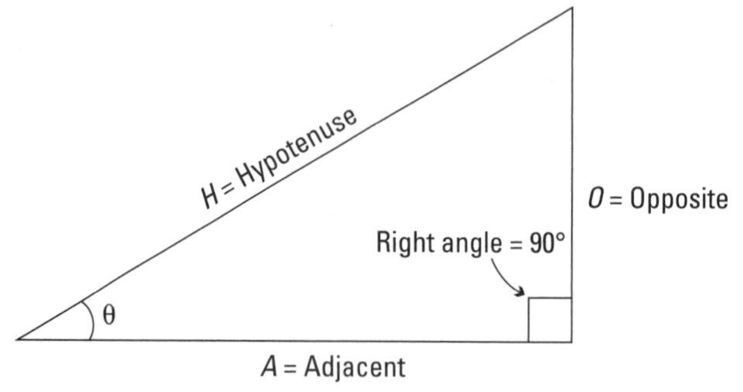

Figure 2-1: Trigonometric functions and the Pythagorean theorem.

The basic equation for the Pythagorean theorem can be given by the following relationship:

$$H^2 = A^2 + O^2$$

where H is the hypotenuse (or the side opposite of the right angle), O is the side opposite of the reference angle θ, and A is the side adjacent to the angle θ.

Some textbooks write the Pythagorean theorem as $A^2 + B^2 = C^2$. This formula simply substitutes C for H and B for O (A stays the same).

Tackling Simultaneous Algebraic Equations

An aspect of algebra that appears repeatedly in mechanics of materials is the solution of *simultaneous* algebraic equations — those pesky equations with several different variables. These equations appear frequently in mechanics of materials when you work with strain rosettes (which I discuss in Chapter 13) or solve indeterminate mechanics problems (see Chapter 17).

To tackle these equations, you employ a bit of basic algebra. Consider a linear system of two equations with two different variables, x and y, such that

$$3x+2y=11$$
$$9x-3y=-3$$

Because you have two different equations with the same two unknown variables, you can solve these equations simultaneously to find the values of the variables using a few simple steps:

1. **Solve one of the equations for one of the unknown variables.**

   For example, you can solve for x in the first equation by using basic algebra.

   $$x=\frac{1}{3}(11-2y)$$

2. **Substitute the expression for the variable of Step 1 into the remaining equations and solve for the other unknown variable.**

   $$9\left(\frac{1}{3}(11-2y)\right)-3y=-3$$
   $$\Rightarrow 33-9y=-3$$
   $$y=+4$$

3. **Substitute the result from Step 2 into the equation of Step 1.**

   $$x=\frac{1}{3}(11-2(4))$$
   $$\Rightarrow x=+1$$

4. **Check your answers with one of the original equations.**

   $$3(1)+2(4)=11$$
   $$\Rightarrow 11=11 \qquad \therefore OK$$

You can use these same principles to solve multiple simultaneous equations as well; you just need to repeat Steps 1 and 2 additional times, solving for the different unknown variables. Just remember that the number of variables you are solving for must be the same as the number of different equations you have.

Taking On Basic Trig Identities

Trigonometry (or trig) is the branch of mathematics that deals with triangles. Three of the most important functions in all of engineering arise from the sine, cosine, and tangent functions that define the relationships among the sides of a right triangle. Referring to Figure 2-1, you can express the relationships among the sides as follows:

$$SOH \Rightarrow \boxed{s}\text{in}(\theta) = \boxed{\frac{O}{H}} = \frac{\text{opposite}}{\text{hypotenuse}}$$

$$CAH \Rightarrow \boxed{c}\text{os}(\theta) = \boxed{\frac{A}{H}} = \frac{\text{adjacent}}{\text{hypotenuse}}$$

$$TOA \Rightarrow \boxed{t}\text{an}(\theta) = \boxed{\frac{O}{A}} = \frac{\text{opposite}}{\text{adjacent}}$$

In these relationships, I've boxed a couple of the letters to illustrate a simple anagram — SOHCAHTOA — that can help you remember the relationships between the sides. *SOH* refers to the sine (*S*) relationship and is expressed as the opposite (*O*) over the hypotenuse (*H*). Similarly, the *CAH* relates the cosine (*C*) function to the adjacent (*A*) over the hypotenuse (*H*), and TOA relates the tangent (*T*) function to the opposite (*O*) over the adjacent (*A*). Just remember, it's spelled S-O-H-C-A-H-T-O-A.

Where you assign the opposite and adjacent sides is completely dependent on which angle you choose as your reference angle (θ). So be cautious!

Covering Basic Calculus

As you work with mechanics and materials concepts, you quickly discover that you can express many of the expressions you use as polynomials. Therefore, you can use the tools of calculus (such as differentiation and integration) to find the locations and magnitudes of the minimum and maximum values. I cover these topics in the following sections.

Integration and differentiation of polynomials

In a basic mechanics of materials class, certain fundamental calculus skills become very handy, including simple integration and differentiation of polynomial functions — functions where you can apply the power rule. Of course, these basic skills entail significantly more (read: tons) than what I cover here. However, for the purposes of this book, understanding how to apply the power rule is usually sufficient for the type of functions you end up creating.

Basic differentiation and tangents to functions

The *derivative* of a function represents the slope of the tangent line to the function at a particular location (x). For a simple function $f(x)$, you denote the derivative as either $f'(x)$ or $\frac{df(x)}{dx}$.

The *power rule* states that for a smooth and continuous polynomial (meaning no gaps or kinks in the function) of order n, you can express the derivative of a function $f(x)$ as

$$f(x)=x^n \text{ then } f'(x)=\frac{df(x)}{dx}=n\cdot x^{n-1}$$

For example, for the function $f(x) = 3x^6 + 7x^3 - 9$, you can compute the derivative of the function $f(x)$ as

$$f'(x)=(6)(3)x^{(6-1)}+(3)(7)x^{(3-1)}+0=18x^5+21x^2$$

The terms inside the parentheses indicate the powers of the original term being differentiated. Because the derivative of a constant is always zero, the -9 in the original function has disappeared.

This particular example demonstrates how to calculate a simple first derivative. But you can actually have higher-order derivatives as well. If you want to calculate a second derivative, you differentiate the function $f(x)$ and then differentiate that differentiation. The higher the order of derivative you want to compute, the more derivatives you have to take. In mechanics of materials, a third- or fourth-order derivative usually does the job.

Basic integration

If you evaluate an integral between an upper limit b and a lower limit a, you're actually computing a special type of integral known as a definite integral. A definite integral for the function $f(x)$ can be evaluated as follows:

$$f(x)=\int_a^b f'(x)\cdot dx=f(b)-f(a)$$

When you perform an integration, you're actually calculating the area under the curve (or function) between the limits of a and b. This area can be quite helpful when you calculate centroids and section properties (flip to Chapters 4 and 5). The definite integral for a smooth and continuous polynomial of order n such that $f'(x)=x^n$ becomes

$$f(x)=\int_a^b x^n\cdot dx=\left.\frac{x^{n+1}}{n+1}\right|_a^b=\left(\frac{1}{n+1}\right)\left(b^{n+1}-a^{n+1}\right)$$

If you perform the reverse process of the power rule (see the preceding section), you're actually performing a basic integration known as an *indefinite integral,* which is crucial to the deflection calculations in Chapter 16. When you calculate an indefinite integral, a constant C_i shows up each time you integrate. To integrate a smooth and continuous polynomial of order n such that $f'(x)=x^n$, the integral becomes

$$f(x)=\int x^n\cdot dx=\frac{x^{n+1}}{n+1}+C_1$$

Integrating the function $f''(x)=3x^6+7x^3-9$ twice produces the following expressions:

$$f'(x)=\frac{3}{7}x^7+\frac{7}{4}x^4-9x+C_1$$

$$f(x)=\frac{3}{56}x^8+\frac{7}{20}x^5-\frac{9}{2}x^2+C_1x+C_2$$

where C_1 and C_2 are numerical constants of integration that are determined by *boundary conditions* (known specific values of the function). I explain more about boundary conditions in Chapter 16.

Defining maximum and minimum values with calculus

Many of the equations you produce in mechanics of materials are smooth and continuous polynomials. Fortunately, the power rule I discuss in "Basic differentiation and tangents to functions" works especially well on polynomials.

Remember that when you differentiate a function, you're actually computing the slope of the function. If the derivative is set equal to zero, you're looking at a point where the slope of the function is actually a horizontal line:

$$\frac{dy}{dx}=f'(x)=0$$

If the tangent line (the slope) is zero at a specific point, you've actually uncovered a maximum or minimum. These points are especially useful when you're dealing with generalized equations (such as the ones I demonstrate in Chapter 3) because they can predict the peak internal loads, which you need when you start using mechanics of materials in the design process (see Chapter 19).

In order to find the location of a maximum or minimum value, all you need is the first derivative of the original function, the ability to set that first derivative equal to zero, and the ability to find the value(s) of the independent variable x that satisfy that equation. After you determine the locations, simply plug those x values back into the original function $f(x)$ and compute the value of that function.

Working with Units in Mechanics of Materials

A major challenge for someone just becoming familiar with mechanics of materials involves the two competing systems of measurement used in different locations around the world: the SI system and U.S. customary units. I cover them both in the following sections.

SI units

The *International System of Units* (SI) is a system of standardized units that uses measurements exclusively from the metric system. The SI abbreviation is short for the French system *Système International d'Unités* and is used extensively in many parts of the world.

The SI system uses base units for all areas of measurement (mass, force, distance, and so on). Table 2-1 presents some common base units and abbreviations you may come across in the SI unit system.

Table 2-1 SI Base Units and Abbreviations

Measurement	*SI Units*	*SI Abbreviations*
Length	Meter	m
Force	Newton	N
Time	Second	s
Mass	Gram	g

When working with SI units, you have to be able to convert between base units with different prefixes. After choosing a proper base unit from Table 2-1, you attach a series of prefix values to that base unit to create a *scaled unit* (a larger or smaller unit than the base SI unit). In Table 2-2, you can see some common SI prefixes, including some for getting larger (mega- and kilo-) and some for getting smaller (milli- and micro-), that you encounter in mechanics of materials.

Table 2-2 SI Conversions

Prefix	*Symbol*	*Multiplier*	*Exponential Conversion*
Getting Bigger			
mega-	M-	1,000,000	10^6
kilo-	k-	1,000	10^3
Getting Smaller			
milli-	m-	0.001	10^{-3}
micro-	μ-	0.000001	10^{-6}

Within the SI system, you always need to be familiar with a subset of conversions. To increase from a smaller prefix to a larger prefix, you must multiply by the exponential conversion shown in Table 2-2. The first term in the conversion is always the starting unit. The second term is always the conversion to go from the starting units back to the base units. Here's the formula:

(starting units) · (conversion to base unit) · (conversion to final unit) = final units

U.S. customary units

The *U.S. customary system*, often referred to as *English units,* is the unit system widely used in the United States. Like the SI system, the U.S. customary system also has common base units, which you can see in Table 2-3.

Table 2-3 U.S. Customary Base Units and Abbreviations

Measurement	*U.S. Customary Units*	*U.S. Abbreviation*
Length	Foot	ft
Force	Pound	lb (or #)
Time	Second	s
Mass	Slug (1 lb s^2/ft)	Slug

Micro and kip: Noting two exceptions

Not all units fall cleanly into the SI or U.S. customary categories. For example, the *kip* is a hybrid unit for expressing very large forces. It's actually an abbreviation for the *kilo-pound,* a combination of the SI prefix kilo- and the

U.S. customary force unit pounds. *Kilo* means 1,000, so 1 kip equals 1,000 pounds. Most engineering books also abbreviate the kip with the unit *k*, so don't get it confused with the abbreviation for the SI prefix kilo-, which is also *k*. Just remember that the *k* for kip always comes after a numeric answer and doesn't appear with any other units, whereas the kilo- prefix always comes before a base unit.

Another exception is the *micro*. Although the micro is actually one of the SI prefixes in Table 2-2, it can also be a sort of unit for strain in its own right (represented by the Greek letter mu, μ), typically when calculations are dealing with very small values. Technically, strains actually have no reported units because they're measured as either m/m or in/in. Because these units are the same in the numerator and denominator, they cancel each other. The "unit" micro is just a signal to multiply the strain value by 10^{-6}, which is conveniently the conversion factor for the SI prefix micro-. So don't be alarmed when you see a unit represented as 200μ. In this case, you're actually saying that the strain is 200×10^{-6} (which is a very small number indeed).

All the derived mechanics units you'll ever need

Several common statics units are based on calculations involving the base units listed in Table 2-1. For example, the Newton is actually a *derived unit* created from a combination of other units and expressed as

$$1\text{ N} = (1\text{ kg})\cdot\left(1\ \frac{\text{m}}{\text{s}^2}\right)$$

As you may notice, this expression uses the mass unit of kilograms even though the SI base unit for mass is actually grams. The second term is a unit for acceleration. When you compute a force in Newton units, you must express the mass in kilograms.

A few more commonly used derived units are as follows:

- **Moments:** A *moment* is an action that causes rotation. In SI units, the standard base unit for a moment is the Newton-meter (N-m), and in the U.S. customary system, the base unit is the foot-pound (ft-lb or lb-ft — the order doesn't matter).
- **Distributed force effects:** You express these units as a force per linear distance. Their SI unit is Newton per meter (N/m), and their U.S. customary unit is pounds per foot (lbs/ft). Another common representation for lbs/ft is *plf*, which is an abbreviation for *pounds per linear foot*. Similarly, in the event of larger forces, you may also encounter a unit of *klf*, or *kip per linear foot*.

- **Pressure effects:** A pressure effect is expressed as a force per area. The SI unit for pressure effects (including stress, which is a measure of the intensity of a force acting over an area) is Newton per square meter (N/m^2). This unit is also known as the *pascal* and may be abbreviated as Pa. The U.S. customary representation is usually either pounds per square foot (lb/ft^2 or psf) or pounds per square inch (lb/in^2 or psi).
- **Volumetric effects:** A volumetric effect is expressed as a force per volume and includes quantities such as the density or specific weight of materials. The SI unit is Newton per cubic meter (N/m^3), and the U.S. customary unit is usually pounds per cubic foot (lb/ft^3 or pcf).

Converting angular units from degrees to radians (and back again)

A common pitfall for the mechanics and materials student is the distinction between different angular units. Units for angles can be expressed in either degrees or radians. Both of these units are actually related to each other, but if employed incorrectly at the wrong times, they can destroy your calculation results. A *radian* is the measure of the internal angle at the center of one-half of a circle. This same internal angle corresponds to a measurement of 180 degrees (because a whole circle contains 360 total internal degrees or 2π radians). Thus

$$1 \text{ radian} = \frac{360^\circ}{2\pi} = 57.296...^\circ$$

Most calculators are capable of performing calculations in both degrees and radians, and in some models, switching between the two is as easy as pushing a single button — which often happens accidentally and when you least expect it. So before you get wild with those trigonometric functions in this chapter, take a moment to verify your calculator setting. (You may need to consult your calculator's instruction manual.)

Chapter 3

Brushing Up on Statics Basics

In This Chapter

- Drawing free-body diagrams
- Using equilibrium to solve for reactions and internal forces
- Finding internal loads by using generalized equations and area calculations.

Simply put, without statics, you have no mechanics of materials. To perform even the most basic analysis with mechanics of materials, you must have a firm understanding of free-body diagrams, equilibrium, and internal forces. Although I have to assume that you already had a grasp of statics prior to reading this book, I use this chapter to help you dust the cobwebs off a few of the more-important skills you need to use on a regular basis.

In this chapter, I provide a basic review of statics fundamentals involving equilibrium while refreshing your memory on how to calculate support reactions and internal forces of objects. I then show you how to create generalized equations, which you use to work several types of mechanics problems, including deflections of beams in Chapter 16. I conclude the chapter with a quick method for determining internal force diagrams for simple statics problems.

Sketching the World around You with Free-Body Diagrams

Before you can begin applying the principles of mechanics of materials, you have to complete some sort of static analysis. The first steps of any static analysis are always to construct a free-body diagram (F.B.D.) and then solve for as many of the support reactions as you can.

As you're constructing any F.B.D., remember that you should include four categories of forces in addition to dimensions and angular information. Those forces include external loads, internal loads, support reactions, and self weight, and I cover them in the following sections.

External loads

External loads are applied loads that act directly on an object. The force of one beam pushing on another and the torsion applied to the end of a power-transmission shaft are both examples of external loads. You can classify external loads into two basic categories:

- **Applied forces:** An *applied force* is a behavior that wants to move an object in the direction of the force. A *concentrated force* is a force that acts at a single point (or on a very, very small area), and you always represent it as a single arrow acting on your free-body diagram (see Figure 3-1). Concentrated forces resulting from one object pushing on another are known as *contact forces*.

  A *distributed force* is a force that acts over a length as shown in Figure 3-1. Distributed forces can come in a wide variety of shapes, with the *uniform distribution* (or constant intensity) being the most common. The *linear distribution* is a distribution that varies linearly from a maximum at one end of the distribution to a minimum value at the other. Applied forces can also be spread over areas (known as pressure effects), and in some cases, such as self weight, they can act over a volume (known as volumetric effects).

  The total magnitude (known as the *resultant*) of a distributed force or moment is equal to the area of the load under its loading diagram. This resultant magnitude is then located at the centroid (see Chapter 4) of the distribution. The magnitude of the distributed load is called the *intensity* and is measured as a force per length or moment per length.

- **Applied moments:** An *applied moment* is a behavior that causes an object to curve (a *bending moment*) or to twist about a longitudinal axis (a *torsional moment* or *torque*). (Flip to Chapters 9 and 11 for more on bending moments and torsion, respectively.) Like forces, moments may be either concentrated or distributed. On a free-body diagram, you depict a *concentrated moment* as a circular arrow applied at a single point.

  You can use a similar method to represent a concentrated torsional moment as you do for bending moments, but they can be difficult to represent in two dimensions. As a result, using a double-headed arrow notation with straight arrows parallel to the longitudinal axis as I show in the torsion example of Figure 3-2 becomes more convenient.

  To help you determine the signs of the moments, you can employ a trick known as the right-hand rule for moments. Set up your right hand as you do for the similar right-hand rule in Chapter 5 (for establishing Cartesian coordinates). Then curl your fingers in the direction of the moment (your fingers point in the same direction as the arrow head on the circular moment arrow) around the axis of rotation about which your moment is acting. If the thumb on your right hand is pointing toward the positive end of one of the Cartesian axes, it's a positive moment about that axis. If the thumb points toward the negative end, you're dealing with a negative moment. Figure 3-3 illustrates the right-hand rule for moments.

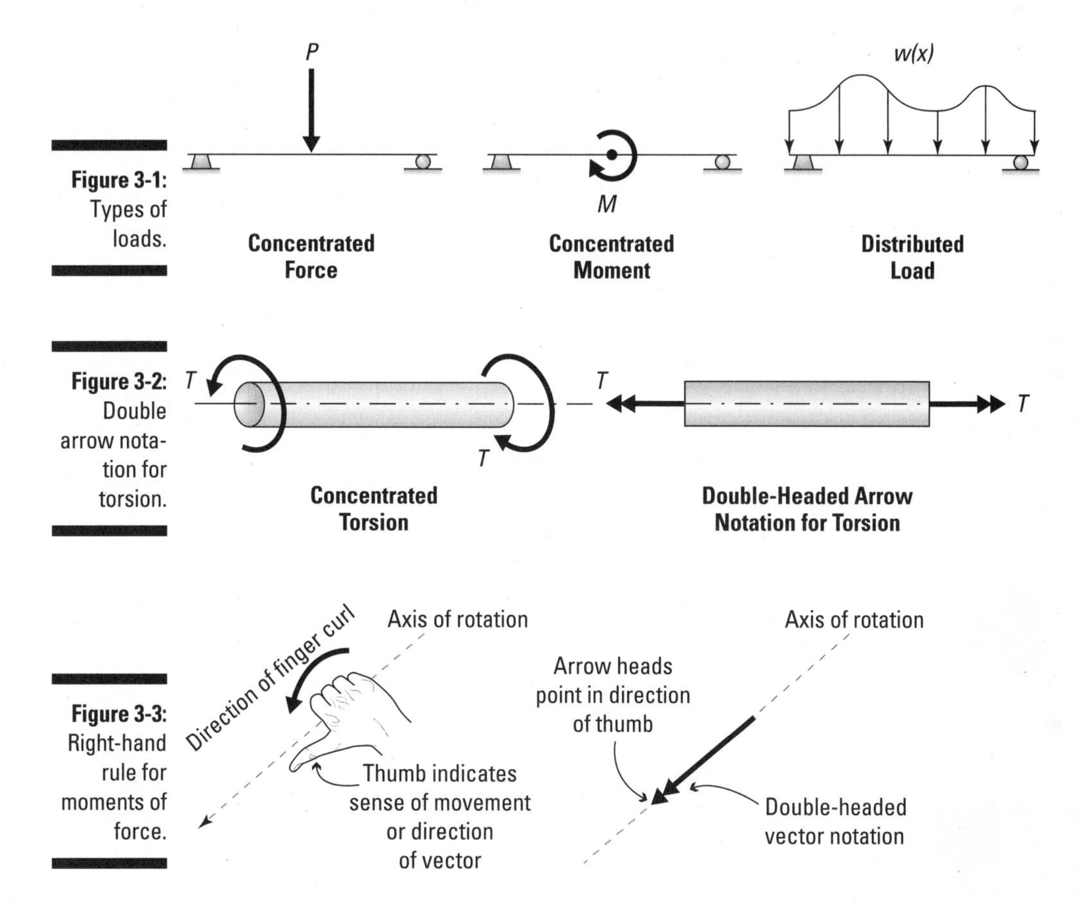

Figure 3-1: Types of loads.

Figure 3-2: Double arrow notation for torsion.

Figure 3-3: Right-hand rule for moments of force.

Internal loads on two-dimensional objects

One of the most important statics skills you need to work within the realm of mechanics of materials is the ability to determine and calculate internal loads. For a complete system such as the one in Figure 3-4a, internal forces are balanced and never visible. However, if you cut an object, internal forces appear to help maintain equilibrium.

You can separate the internal forces that develop into three categories: axial forces, shear forces, and internal moments:

- **Axial forces:** An *axial force* is an internal force that acts parallel to the longitudinal axis of the member and tends to make the object either increase or decrease in length. An object subjected to a positive axial force causing elongation is said to be in *tension*. Conversely, an object subjected to a negative axial force that causes shortening is said to be in *compression*. Figure 3-4b shows the axial force N_X that develops in response to a support reaction (A_x in this case). I show you more about working with axial forces in Chapter 8.
- **Shear forces:** A *shear force* is an internal force that acts parallel to the exposed surface of the cross-sectional area at the cut location. As you can see in Figure 3-4c, the shear force V_X is necessary to keep the object balanced in equilibrium in the vertical (or a second) direction. In a two-dimensional problem, you have only one shear force, while three-dimensional problems can have two shear forces on a cut plane. Check out Chapter 10 for more about working with shear forces.
- **Moments:** As I note in the preceding section, a *moment* is a behavior that causes a member twist or flex. The internal moment on the exposed face (shown in Figure 3-4d) is in response to the eccentric vertical support reaction and the couple it creates with the internal shear force V_X and other external loads. In a three-dimensional member, two moments can cause bending, and a third moment (a torsional moment) can cause twisting. I explain more about how to work with internal bending moments Chapter 9 and with twisting moments in Chapter 11.

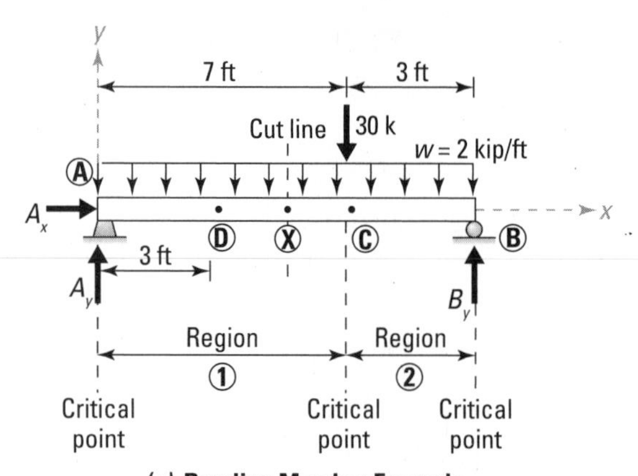

(a) **Bending Member Example**

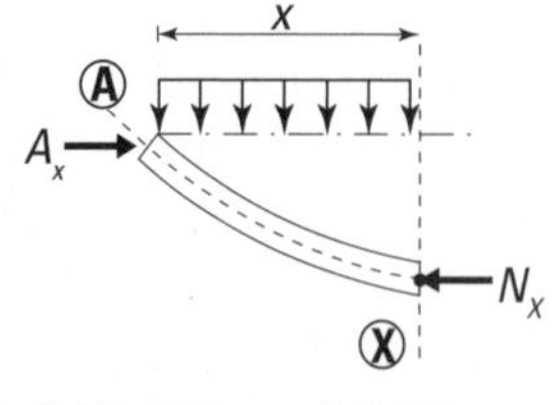

(b) **To Balance Axial Force**

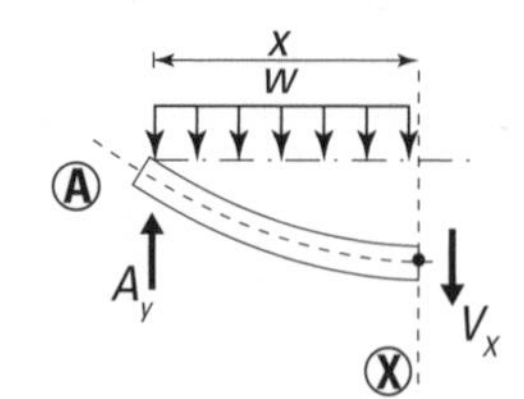

(c) **To Balance Shear Force**

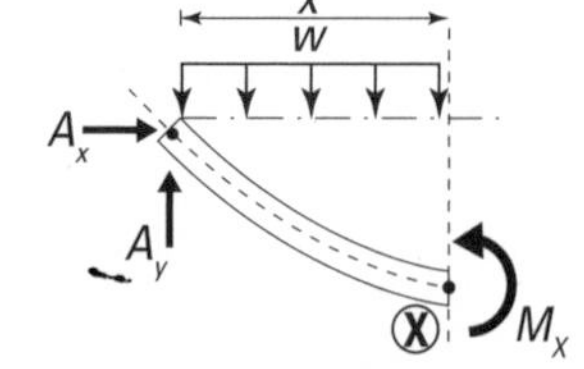

(d) **To Balance Moment**

Figure 3-4: Internal forces of a bending member.

Support reactions

A *support reaction* (also known as a *support condition*) is a force or moment that develops at a location in an attempt to restrain a body from moving or rotating when a load or moment is applied to it. In two dimensions, support reactions usually come in three flavors:

- **Roller support:** A *roller support* is free to translate in one direction while being restrained in another. The support is also free to rotate. A roller support has one unknown support reaction that acts perpendicular to the support surface, as shown in Figure 3-5a. Examples of roller supports are the wheels on your car or your favorite pair of inline skates.
- **Pinned support:** The *external pinned support* (sometimes referred to as a *simple support*) is a support reaction that restrains translation in two different directions while being free to rotate, such as the movement of the hinges on your door. An external pin support has two support reactions as shown in Figure 3-5b. The *internal pinned support* (or *internal hinge*) is a type of connection made between two or more members. Like the external pinned support, the internal hinge restrains translation while allowing rotation.
- **Fixed support:** The *fixed support* (or *cantilever support*) can't rotate and is restrained in two directions from translation. The fixed support has two nonparallel support reactions and a resisting moment, as shown in Figure 3-5c.

Note that the diagrams in Figure 3-5 indicate only the forces and moments associated with each support reaction. When you start to draw your free-body diagrams, you definitely need to make sure to include all internal forces as well.

You can always identify other support types by their resistance to motion. If a specific direction is restrained (or partially restrained) from translation, a resisting force must be present. If a direction is restrained from rotation, a resisting moment must exist.

Self weight

Self weight is a force that the effects of gravity cause on the mass of an object. Depending on your static analysis, you may treat self weight either as a single concentrated force located at the center of mass of the object or as a distributed force along the entire length.

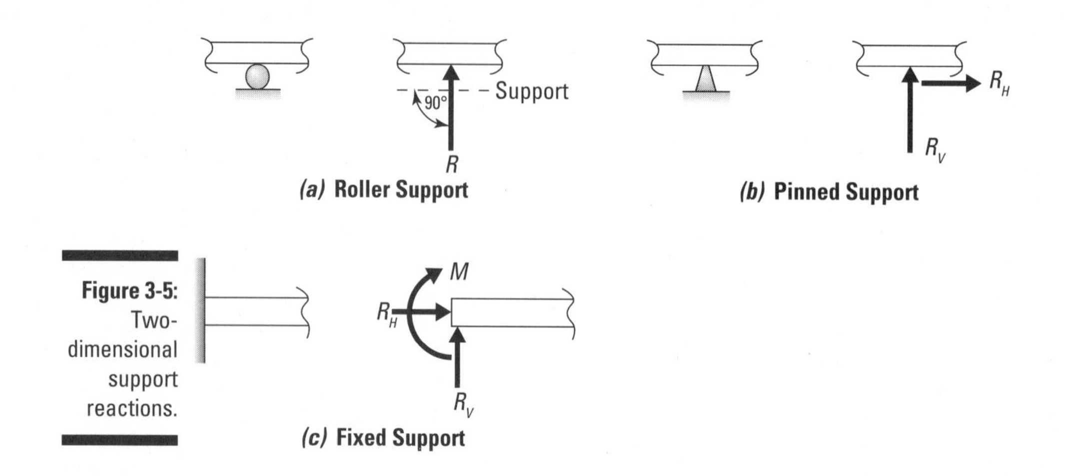

Figure 3-5: Two-dimensional support reactions.

Reviewing Equilibrium for Statics

After you have the free-body diagram drawn, the next step is usually to compute the values of the support reactions. (Check out the earlier section "Support reactions" for more on these conditions.) To calculate support reactions, you use Newton's laws of motion, particularly his second law, which says that force is directly related to acceleration (which in statics is equal to zero) and mass. This law is the basis for defining a state of balance for an object, known as equilibrium.

In statics, *equilibrium* means that an object or system experiences no net motion or acceleration. You can classify *motion* in statics into two major categories:

- **Translation:** *Translation* is a linear or straight-line movement of an object. Translational motion is a response to unbalanced forces acting on a system. So, to define translational equilibrium, you simply need to have a system that has balanced forces. More specifically, in a two-dimensional application, you need to ensure that translational equilibrium is maintained in at least two nonparallel directions; an object that is balanced in only one direction may not necessarily be balanced in a different direction. The most convenient directions are often parallel to the Cartesian axes of your assigned coordinate system (usually as x and y for two dimensions and x, y, and z for three dimensions).

  To ensure translational equilibrium, you use the classic statics force equations:

  $$\sum F_x = 0 \text{ and } \sum F_y = 0 \text{ and } \sum F_z = 0$$

  For two-dimensional applications, you need to sum forces in the two directions in the plane of the problem (usually x and y) only. For three dimensions, you need all three.

- **Rotation:** *Rotation* is a spinning or turning movement about a reference point or axis. To provide equilibrium in rotation, the net rotational effects (or moments) must also be balanced, a state you can verify with the moment equilibrium equations:

  $$\sum M_x = 0 \text{ and } \sum M_y = 0 \text{ and } \sum M_z = 0$$

  For two dimensions, you need to sum moments about the axis perpendicular to the plane of the problem (usually z) only. For three dimensions, you also need to use all three moment summation equations.

For an object to be in a state of equilibrium, it must meet both of the requirements for translational equilibrium while simultaneously satisfying rotational equilibrium. If your object isn't in equilibrium in all three equations, the object isn't actually balanced.

For the beam of Figure 3-4 earlier in the chapter, you can determine the support reactions by writing each of these three equilibrium equations for the free-body diagram. For the moment equation, you can choose any point you want, but I like to choose the pinned support at Point A because it eliminates both A_x and A_y from the moment equation (the moment of each of these forces about Point A is zero). Here are the calculations:

$$\sum F_x = 0 \rightarrow A_x = 0$$

$$\sum M_A = 0 \rightarrow B_y(10 \text{ ft}) - \left(2\frac{\text{kip}}{\text{ft}}\right)(10 \text{ ft})\left(\frac{10 \text{ ft}}{2}\right) - (30 \text{ kip})(7 \text{ ft}) = 0$$

$$\Rightarrow B_y = +31 \text{ kip}$$

$$\sum F_y = 0 \rightarrow A_y + B_y - (30 \text{ kip}) - \left(2\frac{\text{kip}}{\text{ft}}\right)(10 \text{ ft}) = 0 \Rightarrow A_y + 31 \text{ kip} =$$

$$50 \text{ kip} \Rightarrow A_y = +19 \text{ kip}$$

With these equations, you're ready to begin to solve for unknown support reactions on an object. If you have more equations of equilibrium than you have support reactions, the object is said to be *statically determinate.* If you have more unknown reactions than available equations, the structure is *statically indeterminate,* and you can't solve it completely without applying the basic principles of mechanics of materials that I outline in Chapter 17.

Locating Internal Forces at a Point

After you have the support reactions computed for the beam in Figure 3-4, you're ready to determine the internal forces at any desired location. As I note earlier in the chapter, internal forces appear whenever you cut the object. You then apply the equations of equilibrium on this new, cut member in a similar fashion to how you determine the support reactions (see the preceding section).

For example, to determine the internal forces at Point D in Figure 3-4, you cut the beam at that location and add the internal effects: an axial force (N_D), a shear force (V_D), and an internal moment (M_D) to your free-body diagram (as I show in Figure 3-6).

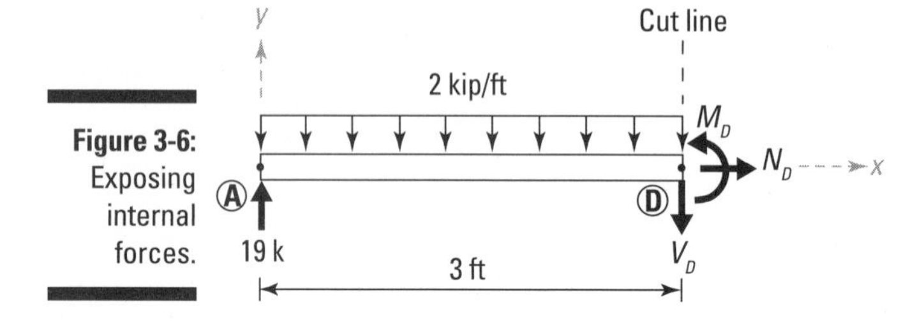

Figure 3-6: Exposing internal forces.

After you have this free-body diagram, you simply rewrite your equilibrium equations for the cut member AD. For example, if you wanted to calculate the internal moment M_D at Point D,

$$\circlearrowleft + \sum M_D^{AD} = 0 \Rightarrow M_D - (19\text{ kip})(3\text{ ft}) + V_D(0) + N_D(0) + (2\text{ kip/ft})(3\text{ ft})\left(\frac{3\text{ ft}}{2}\right) = 0$$

$$\Rightarrow M_D = +48\text{ kip}\cdot\text{ft}$$

Remember that positive signs in your answers indicate that the direction you assumed on the free-body diagram is correct. If you get a negative sign, you assumed the direction backward.

You can then write the translation equilibrium equations to determine the internal shear and axial forces at this location in a similar fashion.

Finding Internal Loads at Multiple Locations

The method for finding internal forces as I describe in the preceding section does have one serious limitation for design. When you start the design process, you want to make sure that you're designing for the most severe internal loads. Unfortunately, you often don't immediately know where this location is within an object. Rather than cutting the object at a thousand (or more) locations in the hope of accidentally selecting the correct spot, finding a more consistent and predictable methodology is useful. The following sections show you how to create equations that allow you to calculate internal forces at any location within a member.

Writing generalized equations

In this section, I show you how to generate equations (known as *generalized equations*) for internal forces. You can write a generalized equation for areas where the free-body diagram looks similar, except for the length of the member, which you replace with a variable dimension x. I usually define this variable with respect to one of the supports; for beams, I often use the leftmost support on the beam. The first challenge is knowing which and how many places you need to cut a section.

You can define these regions for the generalized equations by locating *critical points,* which are locations where the internal loads change. The following are all considered critical points:

- Ends of the structural object (usually the leftmost and the rightmost points for a horizontal beam)
- All support reactions and internal hinge locations
- All concentrated forces and moments
- Start and end points for distributed loads
- Changes in cross section geometry
- Changes in material properties
- All locations where internal shear force is equal to zero.

For the beam of Figure 3-4 earlier in the chapter, you can determine the generalized equation by cutting the beam and drawing a generalized free-body diagram for each of the regions. Figure 3-4 has two regions defined between critical points, as I show in Figure 3-7.

After you have the generalized F.B.D., you simply apply the equations of equilibrium again and solve for the internal forces at the general cut location at Point X (which is the general location at a position x from your reference location) on the member. When you apply the equilibrium equations, the variable distance x appears that represents the internal forces within the beam as a function of its position.

You must measure the distance x to the same reference point for each region. I like to use the leftmost support as my reference when working with beams.

The following equation is the generalized moment equation for Point X in Figure 3-7. You can then repeat the process and construct a generalized moment equation for each of the other regions as well.

$$\circlearrowleft + \sum M_z = 0 \Rightarrow M_X - (19\text{ kip})(x) + V_X(0) + N_X(0) + (2\text{ kip/ft})(x\text{ ft})\left(\frac{x\text{ ft}}{2}\right) = 0$$

$$\Rightarrow M_X = 19x - x^2\text{ kip}\cdot\text{ft}\ (0 \le x < 7\text{ ft})$$

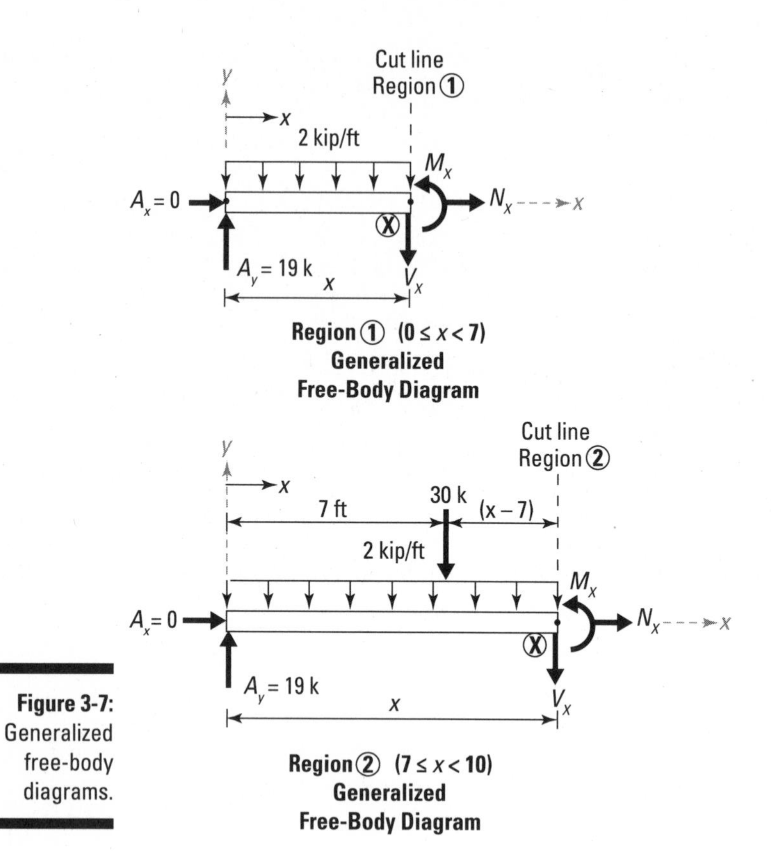

Figure 3-7: Generalized free-body diagrams.

Note that the subscripts in this equation refer to the internal forces at Point X and do not necessarily imply the Cartesian *x*-direction (although N_X does act in that direction).

In fact, the generalized moment equation is actually a necessary part of determining the deflection of members subjected to bending, which I discuss in Chapter 16.

You can also calculate generalized shear force equations, generalized axial force equations, or even generalized torsion equations in a similar fashion by writing the translational or rotational equilibrium equations for each generalized region. After you have all the generalized equations created for each region along the entire length of the object, you can then create a plot of these equations with respect to length or apply the principles of calculus to find the equations' maximum or minimum values.

Drawing simple shear and moment diagrams by using area calculations

One of the most repeated steps in the design of beams and other flexural members is constructing shear force and bending moment diagrams. For beams that have a very simple loading — such as point loads and uniform or linearly distributed loads, you can actually determine the shear and moment diagrams without ever writing the generalized equations that I discuss in the preceding section.

When you're working with area calculations, keep in mind the following:

- Arrange the diagrams vertically with the load diagram on top, the shear force diagram in the middle, and the bending moment diagram on the bottom.
- Locate the critical points on each of the three diagrams.
- Concentrated loads cause a jump in shear force diagrams in the direction of the point load for an amount equal to the magnitude of the point load.
- Finish the shear force diagram first.
- Start from the left and work to the right.
- The shear force and bending moment diagrams you draw and the calculations you perform must compute to zero at the right end (or last point) of the diagram.

You can determine internal load values at every location within an object by using a few simple geometric area calculations in conjunction with a bit of statics and following a few simple steps that I outline in the following sections. The beam in Figure 3-8 illustrates this procedure.

Constructing a simple shear force diagram

The first diagram that you must construct is the shear force diagram, which you establish from the loading diagram directly above it. To construct the shear force diagram for the beam of Figure 3-8, you follow these basic steps:

1. **Starting at Point 1, place a point at $V_1 = 0$ and then examine the first critical region (between Point A and Point C).**

   A concentrated point (the vertical support reaction A_y) load occurs at Point A, so the value in the shear force diagram experiences an instantaneous jump at this location in the amount of +19 kip.

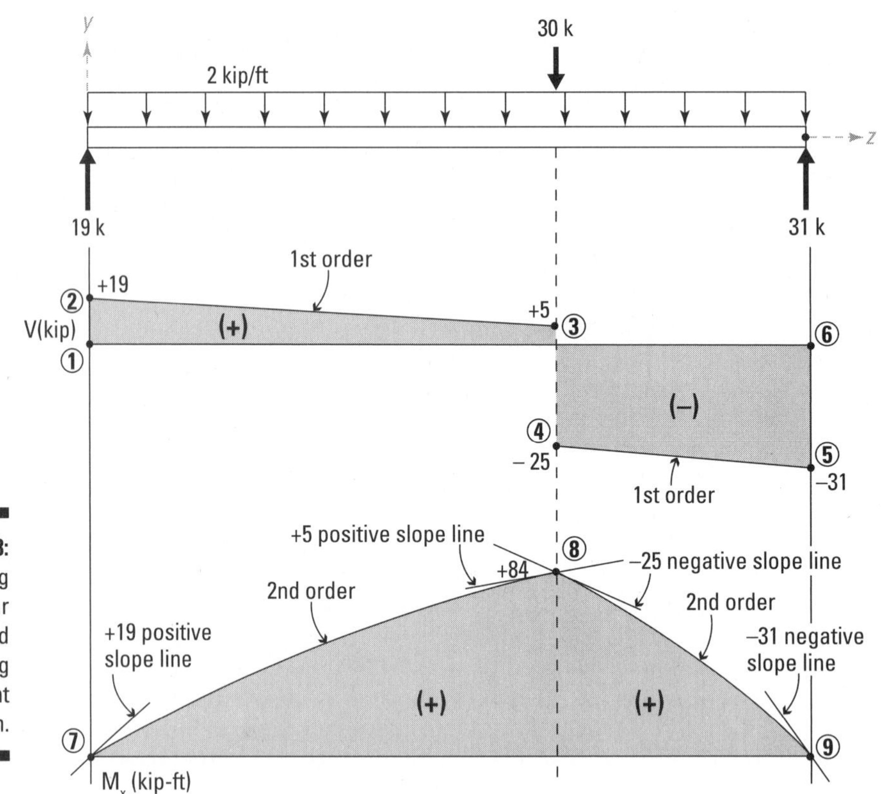

Figure 3-8: Constructing a shear force and bending moment diagram.

2. **At Point 2 place a point at a shear force value of V_2 = 0 + 19 kip = +19 kip; draw a line to connect Points 1 and 2.**

   At Point A, a reaction of 19 kip acts upward on the beam. This reaction is the same as a concentrated load and causes the value in the shear force diagram to jump instantly upward.

   Beginning at Point 2, which has a value of V_2 = +19 kip, you can see that the region between Point A and Point C is subjected to a uniformly distributed load. The area under this load (or the resultant) is equal to the change in shear force value and helps you calculate the shear at Point 3. For this example, the resultant of the distributed load on this region is (–2 kip/ft)(7 ft) = –14 kip. Because this uniform load is acting downward, the resultant must also be acting downward. This fact means that the total change from Point 2 to Point 3 must be –14 kip. The value of shear at Point 3 is then V_3 = (+19 kip – 14 kip) = +5 kip.

The shape of the shear force diagram between Points 2 and 3 depends on the shape of the load between those points. The order of the shear function is always one order higher than the load function for the same interval. Thus, a uniform distributed load (with an order of zero) results in a linear (or first-order) shear function. So a straight line connecting Point 2 and Point 3 is correct.

3. **At Point 3 (located at Point C), place a point at V_3 = +5 kip.**

   At this point, the 30-kip concentrated load is acting downward, so the shear force diagram must instantly jump downward an amount of –30 kip.

4. **At Point 4, place a point at a shear force value of V_4 = +5 kip –30 kip = –25 kip.**

   Beginning at Point 4, which has a value of V_4 = –25 kip, the load in this critical region is uniformly distributed as well. Just as with Step 2, you can calculate the total change in shear as the area of the load between Point C and Point B. The area under this load is (–2 kip/ft)(3 ft) = –6 kip. Thus the change in shear between Point 4 and 5 is equal to –6 kip.

5. **At Point 5, place a point at a shear of V_5 = –25 kip – 6 kip = –31 kip.**

   At this point (Point B), you're at the end of the beam at Point 5, which has a value of –31 kip. Even though you've reached the end of the beam, remember that a vertical reaction B_y = +31 kip is also acting upward. So, the value of Point 6 is equal to the value of Point 5 plus the effect of the concentrated point load due to the reaction. Thus, the shear at Point 6 is V_6 = –31 kip + 31 kip = 0, which means the shear force diagram ends on a zero value.

6. **Denote any areas of positive shear with a plus sign (+) inside the region and areas of negative shear with a negative sign (–).**

   This notation helps with the moment calculations I discuss in the next section. I also like to shade the areas to make them a bit more visible.

7. **Look for a secondary critical point that occurs at locations of zero shear.**

   At any locations where the shear force diagram has a value of zero, you may need to add a new critical point if one isn't there already. In this diagram, Point 1, Point 6, and the line between Points 3 and 4 are already critical points. However in some shear and moment diagrams, you may have a zero shear force value occur at noncritical points. The critical points at shear forces equal to zero are actually locations of minimum or maximum moment on the moment diagram, which I cover in the following section.

Creating a simple moment diagram

After you have the shear force diagram (see the preceding section), you can create the moment diagram from that shear force diagram. You base the moment diagram directly off the shear force diagram above it, and all your calculations come from the shear force diagram. However, if your loading diagram has any concentrated or distributed moments, you still need to look back up to the original loading diagram for those values as well.

Concentrated moments cause an instantaneous jump on the moment diagram because they act the same way that concentrated loads behave on the shear force diagram. (This behavior is why you draw the critical points across all three diagrams: to help remind you of critical locations on your diagrams.)

Follow these steps to create a moment diagram from the shear force diagram in the preceding section:

1. **Starting at Point 7 (the left end of the moment diagram), place a point at $M_7 = 0$ and then examine the first critical region (between Point A and Point C).**

   By looking at the area under the shear force diagram within the critical region between Point A and Point C, you can see that the shear force is positive. That positive designation means that the change in area between Point 7 and the next point at Point 8 must be positive, which means Point 8 must be more positive in value than Point 7.

   The change in moment is equal to the area under the shear force diagram or

   $$\Delta M = (0.5)(19 \text{ kip} + 5 \text{ kip})(7 \text{ ft}) = +84 \text{ kip-ft}.$$

2. **At Point 8, place a point at a moment of $M_8 = 0 + 84$ kip-ft = +84 kip-ft.**

   At this point, you need to do a bit of detective work to find the shape of the moment diagram. The first clue is in the shape of the shear force diagram. Remember that moment diagrams are always one order higher than the shear force diagram in the same region, so if the shear force diagram is linear (first-order), as in this case, the moment diagram must be parabolic (second-order).

   To deduce which second-order curve actually fits, you need to look at the slope of the moment diagram at each point. The slope of the moment diagram is equal to the value of the shear force at that point. Thus, the slope of the moment diagram at Point 7 is +19 and the slope at Point 8 is +5; therefore, the slope at Point 7 is more steeply positive than at Point 8, which creates an inverted parabola as shown on the moment diagram in Figure 3-8. It has the highest positive slope on the left end of the region, where the shear force was the highest positive value.

Be careful, though — this second-order curve doesn't necessarily work for every linear shear force function. You have to look at the slopes to make the decision!

3. **Starting at a moment value (M_8 = +84 kip-ft), compute the change in moment from Point 8 to Point 9 as the area under the shear force diagram in this region (between Point C and Point B).**

   Because the area under the shear force curve is negative, you can expect the change in moment to also be negative:

   $$\Delta M = (0.5)(-25\text{kip} - 31\text{ kip})(3\text{ ft}) = -84\text{ kip-ft.}$$

4. **At Point 9, place a point at a moment of M_9 = +84 kip-ft – 84 kip-ft = 0 kip-ft.**

   Hence, the moment diagram ends at a value of zero. You can deduce the shape of the second-order curve in the same manner as in the preceding step. The final value of zero indicates that the work you did is most likely correct. Be sure to label and shade your positive and negative moment regions as a useful reminder when you're done.

By looking at the shear force diagram, you can see that the maximum positive shear force V_{MAX+} is +19 kip at Point A (or Point 2), and the maximum negative shear force V_{MAX-} equals –31 kip at Point B (or Point 5). Similarly, the maximum positive moment M_{MAX+} is +84 kip-ft at Point C (or Point 8), and the maximum negative moment M_{MAX-} is 0 kip-ft at Point A and Point B (or Point 7 and Point 9). With these values determined, you're ready to begin applying the principles of mechanics of materials to this beam.

Chapter 4

Calculating Properties of Geometric Areas

In This Chapter

- Establishing a Cartesian coordinate system
- Determining cross sections with cutting planes
- Computing the cross-sectional areas of discrete and continuous regions

In mechanics of deformable bodies, you're always interested in how an object is behaving on the inside, so you need to become very adept at determining specific cross sections and computing specific cross section properties such as the cross-sectional area.

In this chapter and the next, I show you the methods and techniques for calculating the section properties that many mechanics of materials formulas use; I discuss their actual use on a case-by-case basis in later chapters.

Determining Cross-Sectional Area

Imagine you're in the kitchen preparing a snack and decide you're craving sliced carrots. The question now is, do you prefer circular disks or long strips? Simply cutting across the carrot produces a circular shape or cross section (assuming your carrot isn't one of those funky, crooked ones), while slicing down the length produces a generally more rectangular cut surface.

The actual shape of your sliced carrot is determined by the orientation of geometric planes. A *geometric plane* is a two-dimensional flat surface that extends endlessly in each direction (both positively and negatively). A plane can have any orientation in space, but mechanics uses several planes, known as *Cartesian planes,* more commonly than others. In the Cartesian coordinate system, the XY Cartesian plane is a geometric plane that is parallel to the plane containing both the *x*-axis and *y*-axis. Similarly, the YZ Cartesian plane is parallel to the plane containing the *y*-axis and the *z*-axis, and the XZ Cartesian plane is parallel to a plane containing (you guessed it) the *x*-axis and the *z*-axis.

One especially useful geometric plane is one that I refer to as a *cutting plane.* A cutting plane is any plane used to cut an object into separate pieces (which proves very handy when you are trying to expose internal forces). Another use of this cutting plane is to develop a region known as the *cross section,* which is created when a solid three-dimensional object intersects this cutting plane. The computed area of this region is known as the *cross-sectional area.* Being able to recognize and compute different cross-sectional areas is an instrumental part of solving mechanics of materials problems, as the following sections explain.

Depending on which cutting plane you use, you can create distinctly different cross sections. In Figure 4-1, I show you the different cross sections that can be created when you can cut a rectangular object with dimensions of b (width) x h (height) x L (length) with different cutting planes.

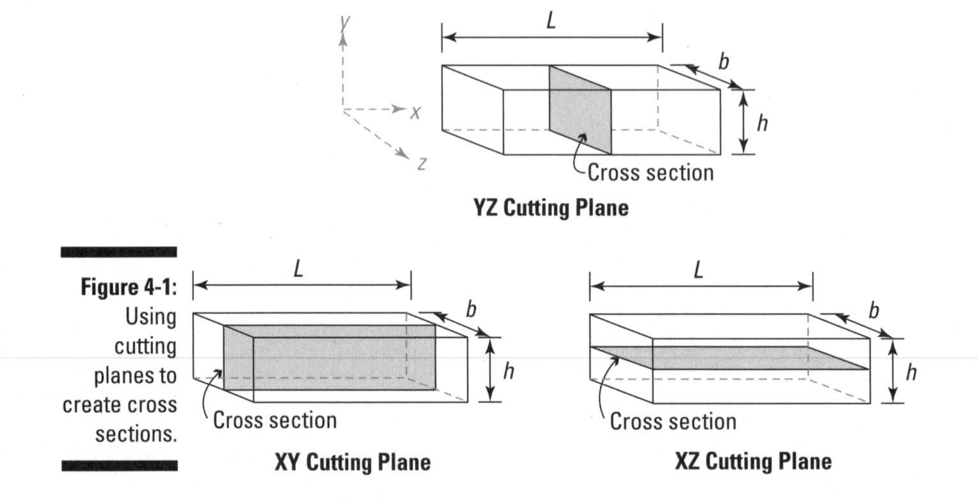

Figure 4-1: Using cutting planes to create cross sections.

Classifying cross-sectional areas

The cross-sectional area is perhaps the most commonly computed geometric property in all of mechanics of materials. Most cross-sectional areas can be classified into one of two categories: discrete and continuous. Figure 4-2 shows examples of discrete and continuous regions (or cross sections).

- **Continuous regions:** A *continuous cross section* (also known as a *general cross section*) is any cross section that you can't completely divide into simple geometric subregions such as rectangles, circles, and triangles (shapes with easily computed areas and centroids). You must express the boundaries of these regions as mathematical functions. To compute the area (A_{TOT}) of a continuous region requires integration:

  $$A_{TOT} = \int_A dA$$

- **Discrete regions:** A *discrete region* or *discrete cross section* (which is sometimes referred to as a *composite* or *compound area*) is a cross section that you can break down into multiple simple-shaped subregions (called *primitives*) such as rectangles, circles, or triangles while still accounting for all the original combined area. Discrete regions can also include holes or openings as long as these subtracted subregions are also discrete. To compute the area of a discrete section, you simply add up the areas of the subregions by using the following formula:

  $$A_{TOT} = \sum_{i=1}^{n} A_i = A_1 + A_2 + \ldots + A_n$$

  where n is the total number of subregions in the discrete cross section.

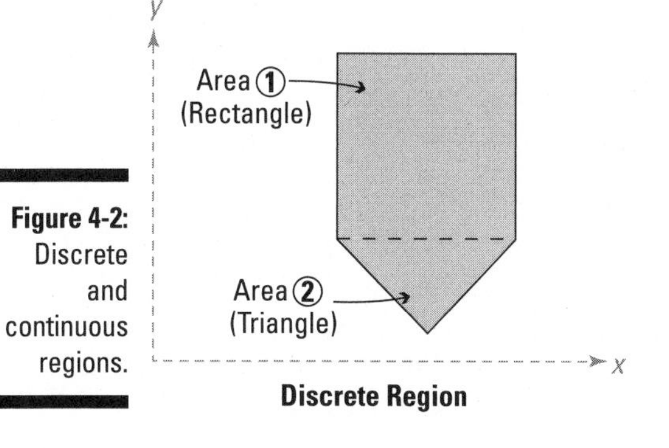

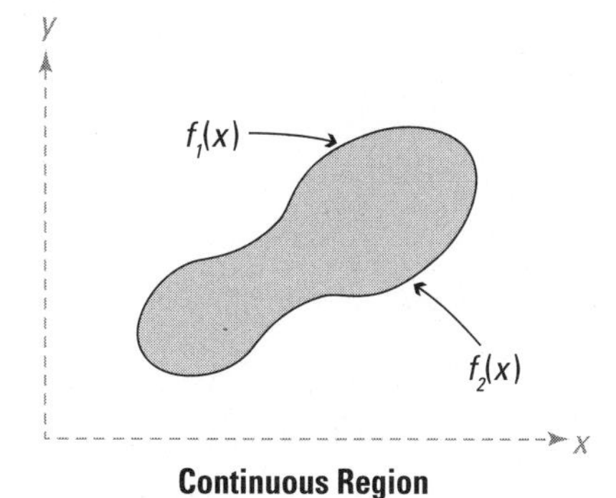

Figure 4-2: Discrete and continuous regions.

Computing cross-sectional areas

Computing cross-sectional areas is fairly straightforward when the cutting planes are aligned perpendicular (or *normal*) to a longitudinal axis of a member. However, sometimes you also need to be able to calculate cross-sectional areas at *oblique* (or non-perpendicular) orientations. Oblique cross sections prove useful in the development of transformation equations for stress, which I discuss in Chapter 7.

After you establish the cross section with a cutting plane, as done in Figure 4-1, you can then compute the cross-sectional area.

Figure 4-3 shows how you can cut a rectangular object with dimensions of *b* (width) x *h* (height) x *L* (length) in multiple directions, creating distinctively different cross sections.

- Figure 4-3a shows a block cut by a cutting plane in the Cartesian YZ plane, resulting in a cross-sectional area of $A = (b)(h)$.
- If the same object is cut by a cutting plane in the Cartesian XY plane (as shown in Figure 4-3b), the cross-sectional area is $A = (L)(h)$.
- If the same block is cut by a Cartesian XZ plane as in Figure 4-3c, the cross-sectional area is $A = (L)(b)$.

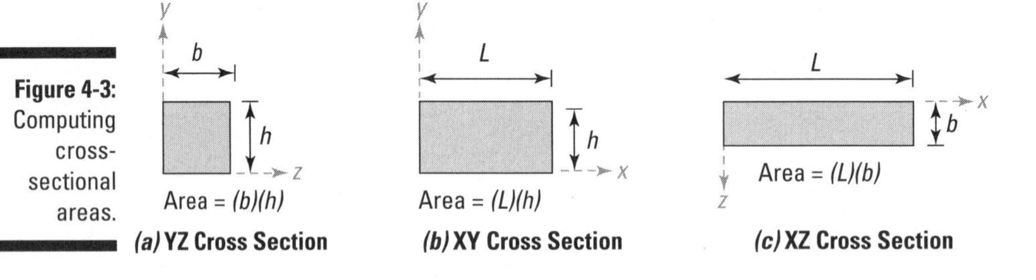

Figure 4-3: Computing cross-sectional areas.

Many objects in engineering can have drastically different cross sections when sliced by cutting planes that are parallel to each other. Consider the hollow tube of Figure 4-4a, which has a length of *L*, an outer radius of r_o, and a wall thickness of *t*.

- When you cut the tube with any Cartesian YZ plane, the same cross section (shown in Figure 4-4b) is revealed at every location along the longitudinal axis.
- However, if you cut through the center of the tube with XZ Plane 1 as shown in Figure 4-4c, the cross-sectional area is $A = 2(L)(t)$.
- Cutting with XZ Plane 2 as shown in Figure 4-4d, you reveal that the cross-sectional area is $A = 2(L)(t_1)$ where $t_1 > t$.

Although the two formulas for the hollow tubes cut by Planes 1 and 2 may look similar, the difference is in the t_1 and *t* terms. Remember that for tubes and pipes, you measure the wall thickness along the diameter of the circle. For Plane 1, the cutting plane is along the diameter. For Plane 2, the cutting plane isn't along the diameter, so the cut thickness t_1 becomes bigger than the wall thickness *t*.

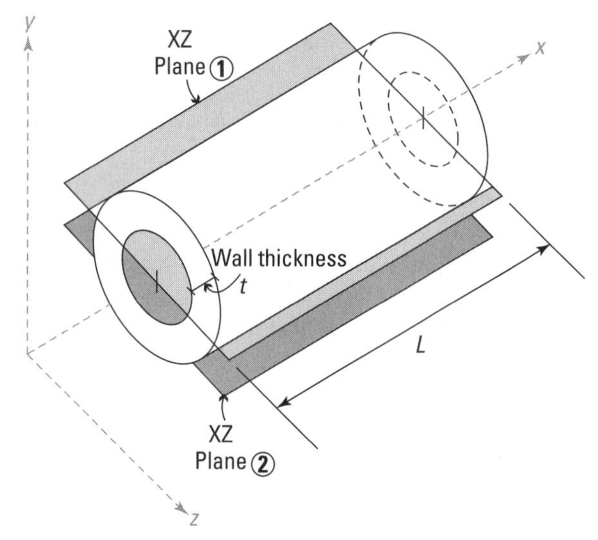

(a) **Establishing Planar Cross Sections**

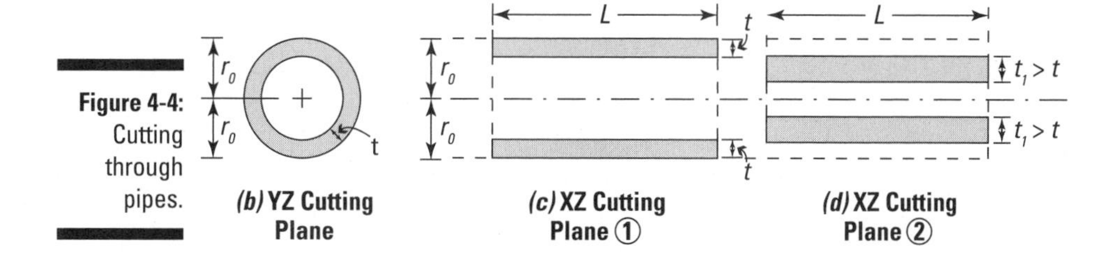

Figure 4-4: Cutting through pipes.

(b) **YZ Cutting Plane** *(c)* **XZ Cutting Plane ①** *(d)* **XZ Cutting Plane ②**

Considering prismatic members

Many of the formulas I discuss throughout this book are derived from statics and are based on the assumption that the members are prismatic. A *prismatic member* (or *prismatic section*) is a member that has the same geometric cross section along at least one axis, and the axis(es) can be oriented in any direction in space.

In the case of the members of Chapter 9, the axis along the member's length (or longest dimension) is known as a *longitudinal axis,* and the beam must be prismatic along this axis.

Defining symmetry of cross sections

A geometric region is said to have *symmetry* (or *be symmetrical*) when an imaginary line, sometimes referred to as the *axis of symmetry*, can divide that region into two identical mirror images, or reflections. An object can have multiple axes of symmetry that can be oriented in any number of directions.

Common structural shapes such as a T-section or an L-section with equal legs have one axis of symmetry. Other structural shapes such as I-shaped sections may have two axes of symmetry. Some cross sections can actually have many different axes of symmetry (see Figure 4-5).

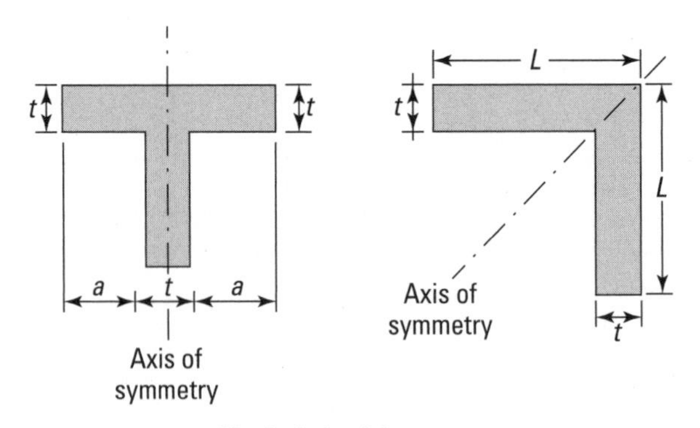

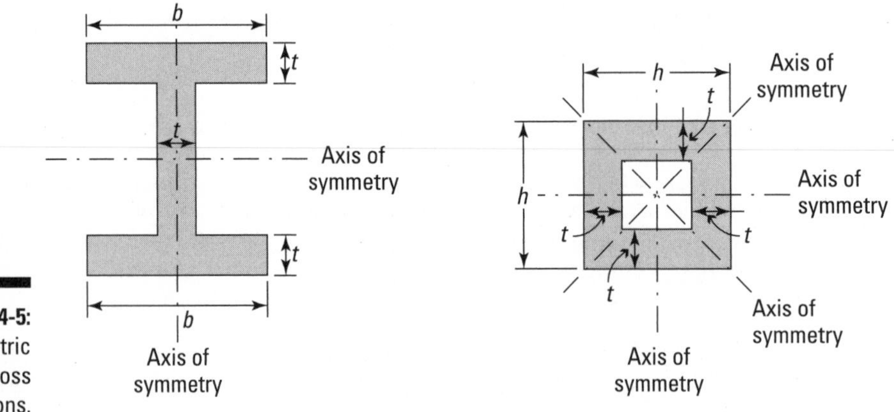

Figure 4-5: Symmetric cross sections.

Finding the Centroid of an Area

After you determine the cross-sectional area (see "Determining Cross-Sectional Area" earlier in the chapter), the next step is to find the *centroid* (or *center of area*), which is the geometric center of a region. You express it as a Cartesian coordinate. You start the majority of your section property calculations (which I discuss later in this chapter as well as in Chapter 5) by first finding the centroid of a cross-sectional area.

In fact, many of the formulas for stress in strain that I discuss in Parts II and III of this book are in their simplest form when you place the origin of your reference Cartesian coordinate system at the centroid of the cross section. So that's what I do in this book.

Making discrete region calculations

The simplest type of region you encounter is the discrete regions that I describe in "Classifying cross-sectional areas" earlier in the chapter. You can divide discrete regions into smaller subregions consisting of simple shapes such as the ones in Figure 4-6. This figure also includes information about the subregions' cross-sectional areas, as well as the *x*- and *y*-coordinates for their centroids as measured from their indicated origins in the figure.

You want to be very careful when using tables of geometric areas that you find in textbooks. Be sure to make special note of where the author of a specific figure has established the origin for the basic shapes in the table. The formulas a table gives are derived exclusively for the indicated Cartesian coordinate system (and especially the location of the origin). If the reference origin location changes, you must recompute the formulas in the tables because they change as well. With the exception of the circular region, I take the origin as the lower-left corner of the object as my reference point. For circles, I use the center point of the circle because that location is usually referenced on a drawing.

To actually compute the centroid of a discrete, two-dimensional region, you utilize the following basic formulas:

$$\bar{x} = \frac{\sum_{i=1}^{n} x_i A_i}{\sum_{i=1}^{n} A_i} \quad \text{and} \quad \bar{y} = \frac{\sum_{i=1}^{n} y_i A_i}{\sum_{i=1}^{n} A_i}$$

where x_i and y_i are the centroidal distances of the subregion and A_i is the area of a particular subregion. Each of these summations must include all of the subregions of the discrete cross section.

Shape	Area	Centroid Location
Rectangle/Square	$A = b(h)$	$x_i = \frac{b}{2}$ $\quad y_i = \frac{h}{2}$
Triangle	$A = \frac{1}{2}(b)(h)$	$x_i = \frac{2}{3}b$ $\quad y_i = \frac{1}{3}h$
Circle	$A = \pi r^2 = \frac{\pi d^2}{4}$	$x_i = 0$ $\quad y_i = 0$
Half Parabolic Area	$A = \frac{2}{3}bh$	$x_i = \frac{3}{8}b$ $\quad y_i = \frac{3}{5}h$
Half Parabolic Complement	$A = \frac{1}{3}bh$	$x_i = \frac{3b}{4}$ $\quad y_i = \frac{3h}{10}$
Quarter Circle	$A = \frac{\pi r^2}{4}$	$x_i = \frac{4r}{3\pi}$ $\quad y_i = \frac{4r}{3\pi}$

Rectangle/Square

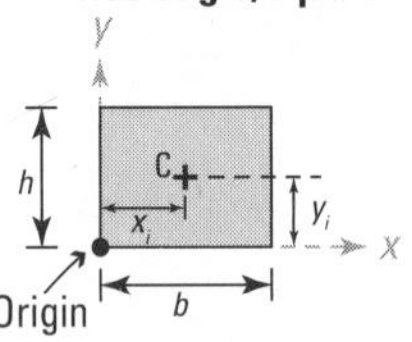

Triangle

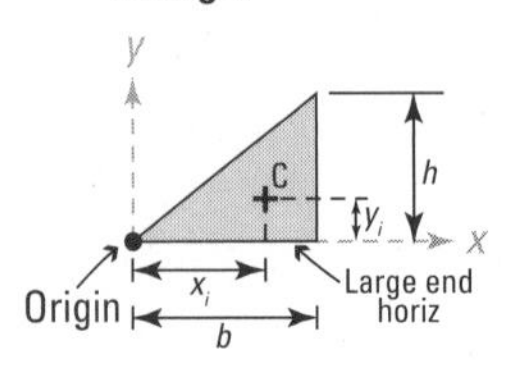

Circle

Origin at center

d, r, C, x, y

Half Parabolic Area

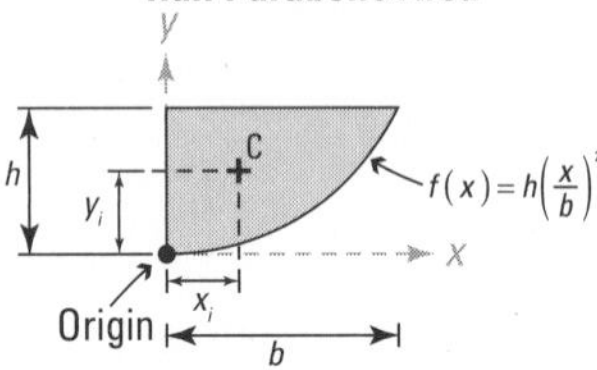

Half Parabolic Complement

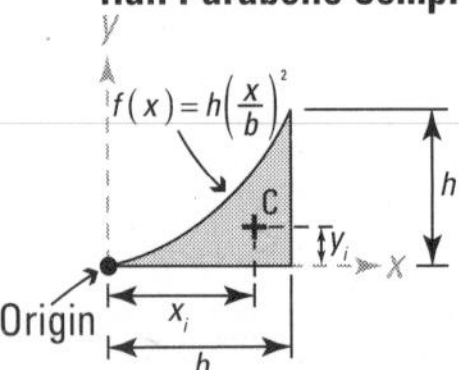

Quarter Circle

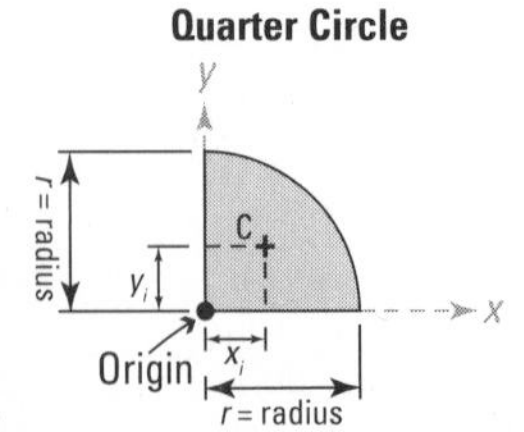

Figure 4-6: Area and centroid equations for common shapes.

Consider the discrete region shown in Figure 4-7a that consists of a circular quadrant and trapezoidal area and contains a hole.

To simplify the centroidal calculation process for discrete regions, I recommend that you establish a basic coordinate table such as the one shown in Table 4-1.

After you select an origin for the discrete region (you can use any point you want), you can complete the basic table by following several simple steps:

1. **Divide the combined region into discrete subregions and add one line to the table for each subregion, making sure to include any holes or open regions in your calculations.**

   Select a reference coordinate system and divide the composite region into a subregion with simple shapes. You can represent a hole or open region as a region with a negative area on top of any region with a positive area. Make sure you account for the total area of the composite region. Figure 4-6 earlier in the chapter can give you an idea of what subregion shapes to look for.

2. **In the second column, record measurements of the *x*-distance from your established origin to the centroid of each subregion.**

   For Region 1 of Figure 4-7b, you can use the information in Figure 4-6 to determine the centroid of the circular quadrant of radius $r = 40$ millimeters as follows:

   $$x_i = \frac{4r}{3\pi} = \frac{4(40 \text{ mm})}{3\pi} = 16.99 \text{ mm}$$

   which is measured from the vertical edge of the quadrant.

   WARNING! At this point, you must be very cautious. In Figure 4-7b, the orientation of the circular quadrant isn't the same as the circular quadrant of Figure 4-6. In this example, the straight edge is on the right side, so you actually measure the distance of 16.99 millimeters from the right-side edge of the circular quadrant, while the origin is actually on the lower-left corner. For this reason, you must perform an additional calculation to determine the correct centroid distance from your established origin by subtracting the x_i value of your previous calculation from the radius of the circle:

   $$x_1 = \text{distance between origins} - x_i = 40.00 \text{ mm} - 16.99 \text{ mm} = 23.01 \text{ mm}$$

   which is measured from the origin. This value is shown in the centroid table in the second column.

   The distance to the centroid of a subregion can actually be a negative value. This situation happens when the centroid of the subregion lies to the left or below the origin.

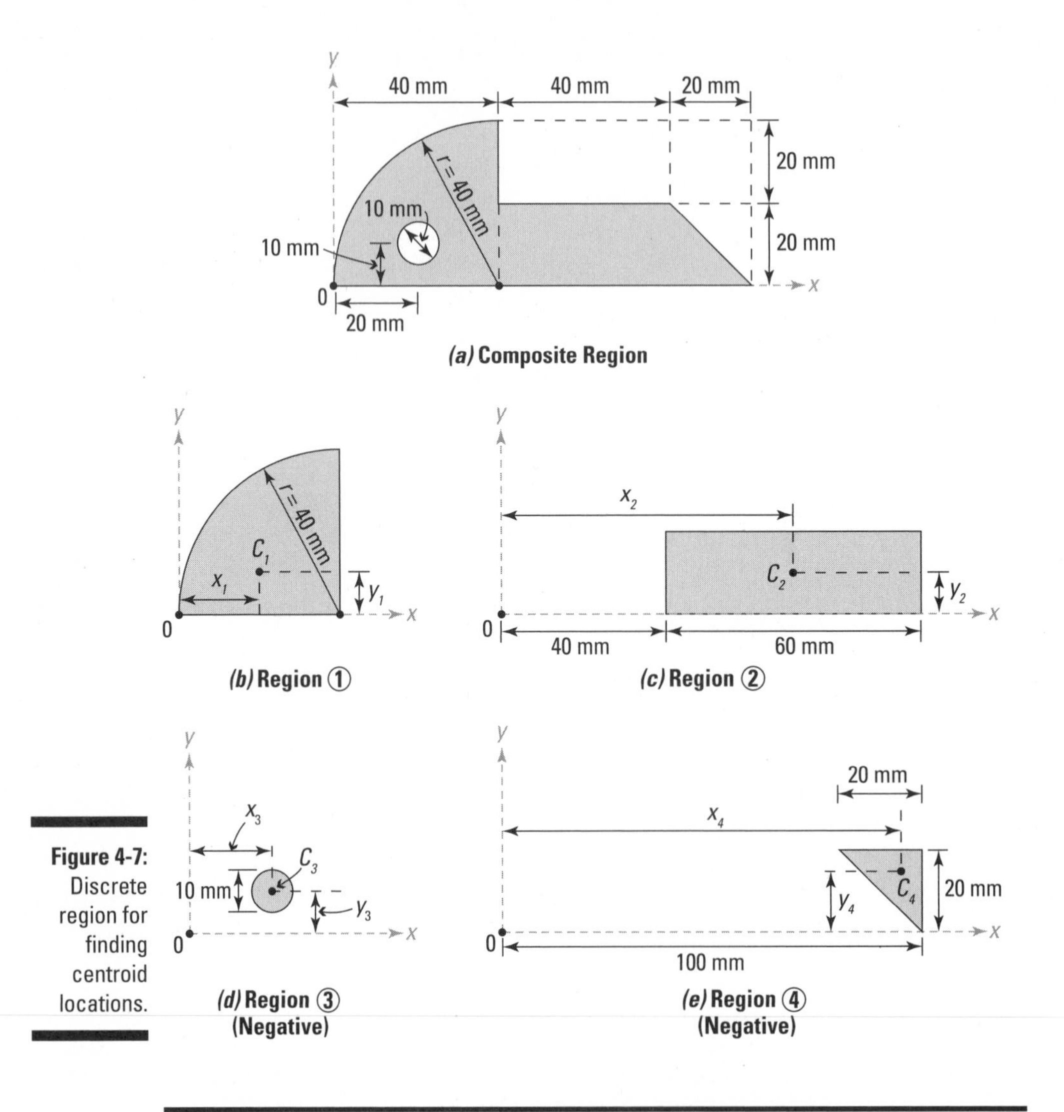

Figure 4-7: Discrete region for finding centroid locations.

Table 4-1 Centroid Coordinate Table

Region	*x_i (mm)*	*A_i (mm2)*	*x_iA_i (mm3)*
1 (circle quadrant)	23.01	1,256.00	28,900.56
2 (rectangle)	70.00	1,200.00	84,000.00
3 (circular hole)	20.00	–78.54	–1,570.80
4 (triangle overestimate)	93.33	–200.00	–18,666.00
TOTAL	-------------------	ΣA_i=2,177.46	Σx_iA_i=92,633.76

3. **Compute the area for each subregion and place this value in the third column.**

   For Region 1, the area of the circular quadrant is

   $$A_1 = \frac{1}{4}\pi r^2 = \frac{1}{4}\pi(40\text{ mm})^2 = 1{,}256\text{ mm}^2$$

   For areas that are computed for holes or other subtracted regions (such as the overestimated triangle of Region 4), you input the values as negative values. In Table 4-1, notice that both Region 3 and Region 4 are negative values.

4. **Multiply the values of the second and third columns (as computed in Step 2 and Step 3) and record this value in the fourth column.**

5. **Sum the values of the third column for each subregion and record this value as the TOTAL for that column.**

6. **Repeat Step 5 for the values of the fourth column.**

7. **Compute the location of the centroid (or the $\bar{x}$ coordinate) to determine the centroid distance from the established origin by dividing the TOTAL from column 4 (computed in Step 6) by the TOTAL from column 3 (calculated in Step 5).**

   $$\bar{x} = \frac{92{,}633.76\text{ mm}^3}{2{,}177.46\text{ mm}^2} = +42.54\text{ mm}$$

From the result of Step 7, you now know that the *x*-coordinate of the centroid of the discrete region is located a distance of 42.54 millimeters from the origin. To find the *y*-coordinate of the centroid, you must create another table, similar to Table 4-1, with the corresponding *y* data.

The fact that this value is a positive dimension indicates that the horizontal centroid is located to the right of the origin for this problem. This conclusion should make sense because I established the origin at the lower-leftmost point on the object. If you had calculated a negative value, that result would indicate that the centroid is located to the left of the origin.

Working with continuous (general) regions

When you can't divide an object's cross section into the simple regions described in Figure 4-6 earlier in the chapter, calculating the centroid can become a bit more mathematically complex. To define a continuous region, you need to express the boundaries of the object as mathematical functions. To compute the centroid, you then need to make use of the integral forms of the centroidal calculations:

$$\bar{x} = \frac{\int_A x \cdot dA}{\int_A dA} \quad \text{and} \quad \bar{y} = \frac{\int_A y \cdot dA}{\int_A dA}$$

These equations actually appear very similar to the summation equations of the discrete centroid calculations (see the preceding sections) because you still need an area, expressed by the *dA* in the equation, and a distance to the center of that area from the origin of your established Cartesian coordinate system — *x* for the *x*-centroidal coordinate and *y* for the *y*-centroidal coordinate.

Consider the object indicated by the shaded region in Figure 4-8a, which is bounded by the curves $f_1(x) = x(mm)$ on the lower bound edge and by $f_2(x) = x^2+2\ (mm)$ on the upper bound edge.

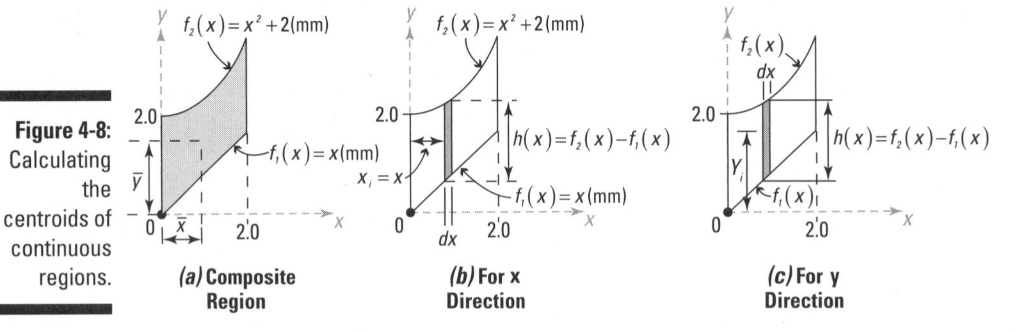

Figure 4-8: Calculating the centroids of continuous regions.

The biggest challenge in working with continuous regions is developing the expressions for the incremental area *dA,* indicated by the shaded rectangular region in Figures 4-8b and 4-8c. You can compute the incremental area as the rectangular area from

$$dA = h(x) \cdot dx = \left(f_2(x) - f_1(x)\right) \cdot dx = \left(x^2 + 2 - x\right) \cdot dx$$

The distance *x* in this equation is the horizontal distance from the origin to the centroid of the rectangular incremental area *dA*.

With this equation developed, you've now successfully transformed the area integral (using *dA*) into an integration with a single variable, *x*. As a result, you now need to change the limits of integration. For this example, the upper limit of the linear integration is 2.0 and the lower limit is 0. You're now ready to compute the *x*-direction centroidal coordinate of Figure 4-8b:

$$\bar{x} = \frac{\int_0^2 x\left(x^2+2-x\right)\cdot dx}{\int_0^2 \left(x^2+2-x\right)\cdot dx} = \frac{\left.\frac{1}{4}x^4+x^2-\frac{1}{3}x^3\right|_0^2}{\left.\frac{1}{3}x^3+2x-\frac{1}{2}x^2\right|_0^2} = \frac{5.333\text{ mm}^3}{4.667\text{ mm}^2} = 1.142\text{ mm}$$

To determine the *y*-direction centroidal coordinate of Figure 4-8c, you basically follow the same procedure. However, this time you need to incorporate the *y*-distance for the incremental area in the equation:

$$\bar{y} = \frac{\int_0^2 y\left(x^2+2-x\right)\cdot dx}{\int_0^2 \left(x^2+2-x\right)\cdot dx}$$

Because the centroid distance *y* isn't a constant value but rather a function of the variable *x,* you need to make an additional transformation by using the following expression:

$$y = \frac{1}{2}\left(f_1(x)+f_2(x)\right) = \frac{1}{2}\left(x^2+2+x\right)$$

Substituting into the centroid integral equation

$$\bar{y} = \frac{\int_0^2 \left(\frac{1}{2}\left(x^2+2+x\right)\right)\left(x^2+2-x\right)\cdot dx}{\int_0^2 \left(x^2+2-x\right)\cdot dx} = \frac{\left.\frac{1}{2}\left(x^4+3x^2+4\right)\right|_0^2}{\left.\frac{1}{3}x^3+2x-\frac{1}{2}x^2\right|_0^2} = \frac{14.000\text{ mm}^3}{4.667\text{ mm}^2} = 3.000\text{ mm}$$

In most cases, the math required for the solution of these integral equations remains fairly simple. However, if the math seems like it's getting ugly in terms of the algebra and polynomials you're working with (such as in the earlier *y*-centroidal integral calculation), you may find that slicing the area in another direction makes your work a bit easier. Remember from calculus that you aren't necessarily required to use a vertical slice for the estimate of *dA*. In fact, in some situations, it may be more mathematically convenient to slice the area horizontally. This setup changes your equations for *dA* a bit, and you need to modify your expressions for *x* and *y* for the center of the incremental area accordingly. You must adjust your limits of integration as well.

Regardless of which slice you use to compute *dA,* your final computation should yield the same numerical answer.

Using symmetry to avoid centroid calculations

If you're able to identify that an object has an axis of symmetry (flip to the earlier section "Defining symmetry of cross sections"), you can assume that the centroid location in the perpendicular direction must be located somewhere on that axis of symmetry.

This assumption proves to be very handy for common structural shapes such as T-sections and I-sections because an axis of symmetry for these shapes is often right down the middle. So if you know that the object is 10 inches wide and is symmetric about a vertical axis down the middle (or 5 inches from either edge), you automatically know that the horizontal centroidal coordinate occurs at 5 inches from either side.

In the event that your object has two or more axes of symmetry (as is the case with tubes, pipes, and certain hollow sections), you know that the centroid occurs where these axes intersect.

Chapter 5

Computing Moments of Area and Other Inertia Calculations

In This Chapter

- Relating the centroidal axis to inertia calculations
- Handling moment of inertia equations
- Working with different types of inertia calculations
- Explaining the radius of gyration

One of the difficulties that most students have when first starting out in mechanics of materials is in the computation of section properties. That's why I devote two whole chapters to getting you up to speed on this important topic.

You can actually physically see and even measure certain geometric properties, such as the cross-sectional area that I describe in Chapter 4. (If you haven't already read Chapter 4, I recommend you do so before proceeding; it can help provide a foundation for the topics in this chapter.) The cross-sectional properties covered in this chapter aren't as easily measured, but they are equally important in helping you predict the effects of a variety of different loading situations.

When you're working with mechanics of materials equations, you must make sure you use the correct cross-sectional property for the calculation you're performing. For example, you need the first moment of area for studying flexural shear (which I cover in Chapter 10), while you use the second moment of area to compute stresses caused by moments. If you accidentally use the wrong section property in your calculations — an easy mistake to make — you may end up computing completely wrong values in your analysis. In Parts II and III of this book, I show you how to determine which property you actually need for a given loading.

Dog-ear this chapter. As you read through the book, I point out exactly where a specific property is required so that you can refer to this chapter to allow you to see more clearly how you actually compute it.

Referencing with the Centroidal Axis

In addition to the cross-sectional area property that I discuss in Chapter 4, you repeatedly use three additional section properties in mechanics of materials calculations: first moment of area, second moment of area, and radius of gyration. (I describe each of these cross-sectional properties in more detail later in this chapter.) Although you may calculate each of these section properties differently, each calculation is dependent on the location of the *centroidal axes* (the axes that pass through the *centroid,* or geometric center of the cross section) introduced in Chapter 4.

The cross-sectional area always contains at least two of the centroidal axes. So a cross section contained in a Cartesian XZ plane contains centroidal axes that are measured parallel to the Cartesian *x*-axis and the Cartesian *z*-axis. A third centroidal axis relates one cross section in a plane to another cross section in the same plane at a different position along the member. For this example, the *y*-axis (which is perpendicular to the XZ plane) is sometimes referred to as the *longitudinal axis* of the member.

A cross section can be oriented in any plane, and the longitudinal axis must be oriented perpendicularly to this cross section. You can use the plane of the cross section to help you determine the direction of the longitudinal axis.

Figure 5-1 shows a basic rectangular bar (or prism) with a cross-sectional area contained in the XY plane. For the cross section in this figure, the *x*- and *y*-centroidal axes are the important centroidal axes for your section property calculations in this chapter; they're the ones contained in the cross-sectional area. For this figure, the *z*-centroidal axis is the longitudinal axis.

The centroidal axes are the axes of a Cartesian coordinate system with its origin placed at the centroid of the cross section. Two of the axes are placed in the plane of the cross section, and the third establishes the *longitudinal axis* for the cross section. By establishing these axes in a particular orientation (which I explain later in Parts II and III), the equations I introduce are in their simplest form.

To help you establish the orientation of these axes, which are usually aligned with respect to an *x*-, *y*-, and *z*-Cartesian direction, you must make sure that you choose them in the proper orientation to each other. To help remember the proper orientation, I like to use a relationship known as the *right-hand rule.* This analogy basically says that if you take your right hand and form an L-shape with your thumb and forefinger, you can assign the positive *x*-direction to the tip of your thumb and the positive *y*-direction to the tip of your forefinger. Then, if you bend your middle finger so that it's perpendicular to your hand, the tip of your middle finger points in a positive *z*-direction. As long as you maintain this relationship between your fingers, your axes will always be properly aligned with respect to each other.

For the calculations in this chapter, I've gone ahead and oriented the Cartesian axes such that the cross section I'm working with is always contained in the XY plane, and the longitudinal axis is along the Cartesian *z*-axis is the longitudinal axis of the member. However, computer programs and other textbooks may use a completely different orientation. If a resource doesn't establish Cartesian axes for you, you're free to use any orientation you please. Just realize that the subscripts I establish in this chapter are for a cross-section following the conventions listed here.

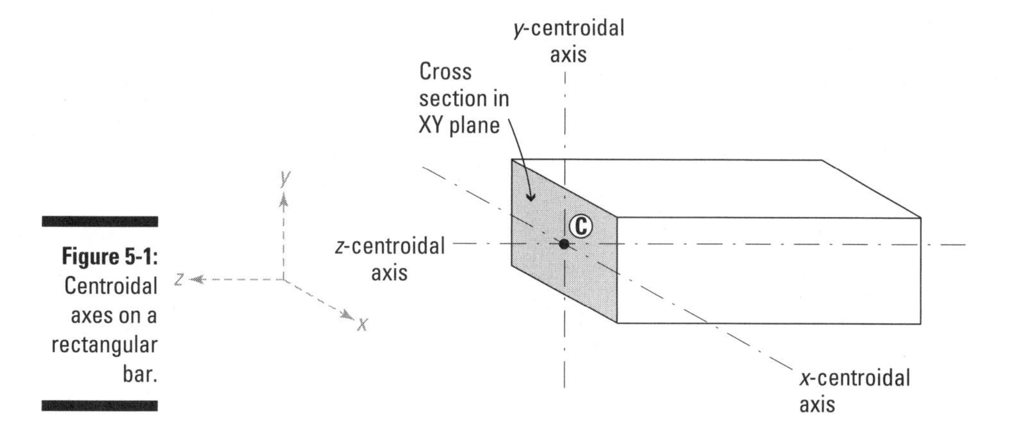

Figure 5-1: Centroidal axes on a rectangular bar.

Computing Q, the First Moment of Area

The *first moment of area* (or *static moment of area*) is a property that measures the moment of an area with respect to a reference axis of a cross section. You use it in the computation of centroids (or centers of area) as well as for determining the internal effects of shear forces on flexural members (see Chapter 10). Most classic textbooks (as well as this text) assign the variable Q to the first moment of area.

You can perform multiple Q calculations for a given cross section. The most common of these Q values are usually calculated with respect to the centroidal axes (depending on the direction of the applied internal shear force) within the plane of the cross section. For example, you use the *x*-centroidal axis or the *y*-centroidal axis for a cross section contained in the XY plane.

The units associated with the first moment of area are typically in in^3 for U.S. customary units and in m^3 for SI units, although in many engineering objects, you may also come across Q expressed as mm^3 as well.

Establishing the equations for Q

The basic formula for the first moment of area for a region in the XY plane about a reference *x*-axis is given by

$$Q_{xx} = \int_A y \cdot dA$$

where *dA* is a differential area within the cross section and *y* is the perpendicular distance from the centroid of the area *dA* to the reference *x*-centroidal axis. For discrete (or composite) regions, you can modify the formula and actually replace integration with simple summation for the *n* subregions of the cross section:

$$Q_{xx} = \sum_{i=1}^{n} y_i A_i$$

where A_i is the area of each of the subregions and y_i is again the perpendicular distance from the centroid of the subregion *i* to the *x*-centroidal axis of the cross section. (Flip to Chapter 4 for more on continuous and discrete regions.)

You can also calculate an additional first moment of area about a *y*-reference axis in a similar fashion. For continuous (or general) regions

$$Q_{yy} = \int_A x \cdot dA$$

where you plug in *x* (the perpendicular distance from the centroid of the area *dA* to the reference *y*-axis). Similarly, for discrete (or composite) areas

$$Q_{yy} = \sum_{i=1}^{n} x_i A_i$$

Revisiting centroid calculations with first moment of area

You may actually recognize the basic equations for computing the first moment of area that I describe in the preceding section from Chapter 4. And for good reason: The centroidal calculations actually use the first moment of area.

$$\bar{x} = \frac{Q_{yy}}{\sum_{i=1}^{n} A_i} \quad \text{and} \quad \bar{y} = \frac{Q_{xx}}{\sum_{i=1}^{n} A_i}$$

In the centroidal calculations of Chapter 4, you calculate the first moment of area with respect to an arbitrary reference axis (such as the bottom of the cross section or the left edge, depending on which centroid coordinate you're computing) because at the time of those calculations you usually don't know the centroid's location yet. After all, that's why you're doing those calculations, right? Conversely, the Q that you compute in this chapter is always computed with respect to a centroidal axis, which requires that you first know the centroid's location.

Determining Q within a cross section

To calculate the value of Q for a discrete subregion with respect to a specific centroidal axis, you must first establish a reference axis at the location of interest (such as a specific point within a cross section) and align that reference axis so that it's parallel to the centroidal axis of the cross section that you're interested in. This reference axis divides the cross section into two distinct subregions, one on either side of the reference axis. To compute the value of Q, you compute the area A_i of either of these two subregions (which one you choose doesn't matter) and determine the perpendicular distance from the centroid of your selected subregion to the centroidal axis of the entire cross section.

After I locate the centroid of the cross section (using the principles in Chapter 4), I often find that relocating the Cartesian reference axes to the centroid of the cross section is convenient. This move allows me to measure all my x and y distances directly to the centroid. However, you can always leave the reference at the same location; you just need to adjust the coordinates accordingly.

Consider the hollow cross section contained in the XY plane as shown in Figure 5-2. For this example, you first establish the Cartesian axes with the origin at the lower-left corner of the cross section and then compute the centroid of this cross-sectional area to be at Point C, which has coordinates of (+40,+40) with respect to the lower-left corner.

To determine Q_{xx} (or the first moment of area about the x-centroidal axis) for a subregion established by a reference axis through Point A in the cross section, just follow these steps:

1. **Locate a new Cartesian coordinate system at the centroid of the cross section.**
2. **Establish a reference axis through the desired location and parallel to the centroidal axis you want to use.**

   In this case, because you want the first moment of area with respect to the x-centroidal axis, your reference axis should be parallel to the x-centroidal axis (or horizontal for this example) and acting through Point A.

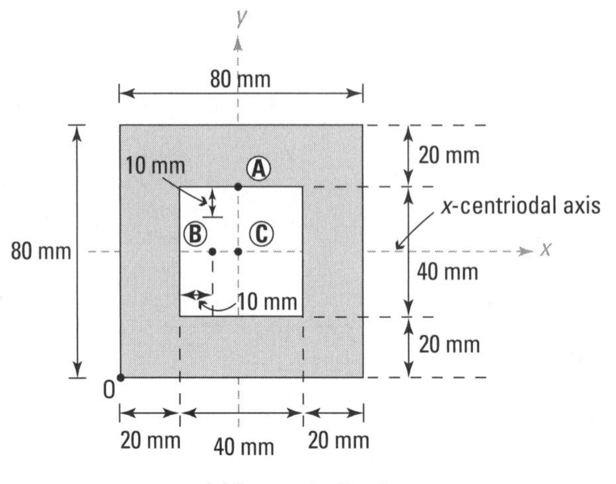

(a) **Composite Region**

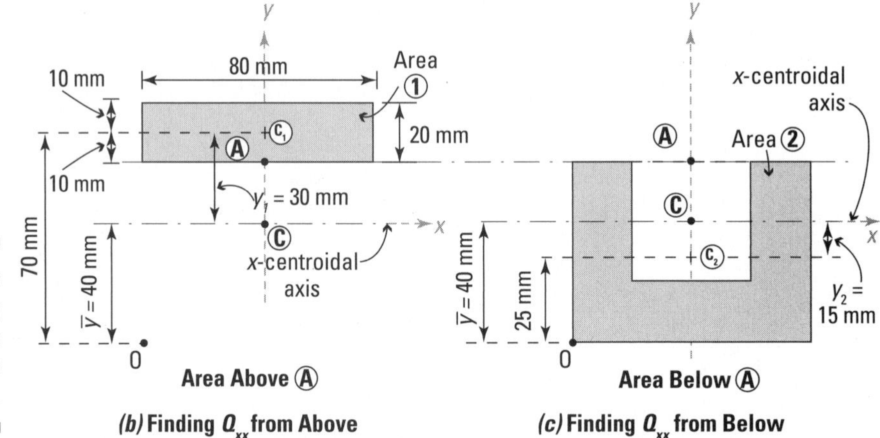

(b) **Finding Q_{xx} from Above** *(c)* **Finding Q_{xx} from Below**

Figure 5-2: Computing $Q_{xx,A}$ within a cross section.

3. **Compute the area and locate the centroid of one of the subregions on either side of the reference axis from Step 2.**

   You can choose whichever subregion you want; the calculation works for either. Just make sure you divide the cross section at the proper location and with a reference axis in the correct direction (parallel to the desired centroidal axis). The math involved produces the same result.

   For this example, I choose the area above Point A because I'm interested in calculating the first moment of area about the *x*-axis and because (as shown in Figure 5-2b) it appears to be the simplest of the two areas to work with (it has fewer subregions, and I can easily determine the centroid of the shaded area).

If I had chosen the area below (shown in Figure 5-2c), my centroid calculations for the subregion would be more involved because I'd have to deal with multiple subregions.

4. **Compute the perpendicular distance y_i from the centroid of the area of Step 3 to the centroidal axis that you're working with.**

   In this example, you're working with the *x*-centroidal axis, so you need to compute the y_1 distance:

   $$y_1 = +30 \text{ mm}$$

   which indicates that the centroid area above Point A is located +30mm above the *x*-centroidal axis of the cross section.

5. **Compute the area A_i.**

   For this example, the area A_i is the area A_1 of the shaded region, which has an area of

   $$A_1 = (80 \text{ mm})(20 \text{ mm}) = 1{,}600 \text{ mm}^2$$

6. **Repeat Steps 4 and 5 for any additional subregions that make up the shaded area for the selected subregion from Step 3.**

7. **Compute the first moment of area by multiplying the results of Steps 4 and 5 for each subregion and summing the total.**

   For this example

   $$Q_{xx,A} = \sum_{i=1}^{1} y_i A_i = (30 \text{ mm})\left(1{,}600 \text{ mm}^2\right) = 48{,}000 \text{ mm}^3$$

Creating a table for calculating Q about a centroidal axis

As with the centroidal calculations of Chapter 4, establishing a table can be especially useful when you're calculating the first moment of area about a centroidal axis.

For example, consider the section in Figure 5-2 in the preceding section. Suppose you're interested in calculating the first moment of area of a subregion created by a vertical reference axis through Point B with respect to the vertical centroidal axis of the cross section (or a *y*-centroidal axis in this case). In this example, Point B is located a distance of 30 millimeters to the right of the left edge of the cross section. You start by choosing one of the two subregions created by the reference axis — that is, either to the left or to the right of the point of interest (Point B). For this example, I arbitrarily choose the subregion to the left (as shown in Figure 5-3).

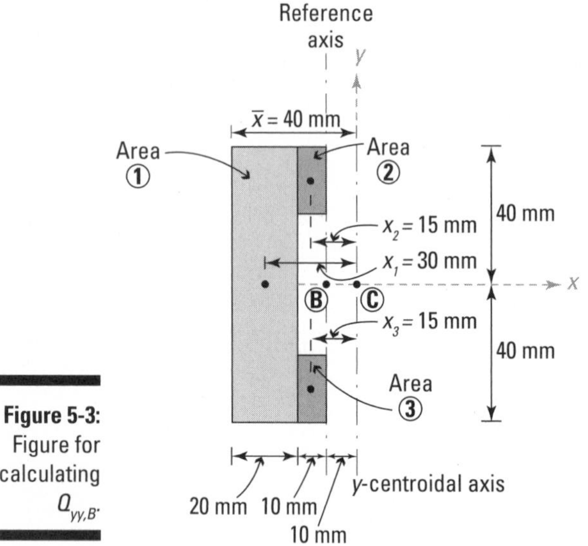

Figure 5-3: Figure for calculating $Q_{yy,B}$.

Table 5-1 looks very similar to the table for centroid calculations that I introduce in Chapter 4.

Table 5-1 Table for Computing $Q_{yy,B}$ about Centroidal Axis

Subregion	x_i *(mm)*	A_i *(mm²)*	x_iA_i *(mm³)*
1	30	(80)(20) = 1,600	48,000
2	15	(20)(10) = 200	3,000
3	15	(20)(10) = 200	3,000
TOTAL	----------------	-------------------	Σ = 54,000 mm³

The second column is the perpendicular distance from the centroid of each subregion created by the vertical reference axis to the *y*-centroidal axis in this example. If you want to calculate the first moment of area about the *y*-centroidal axis, Q_{yy}, you need to use a horizontal distance (or *x*-distance).

If you relocate the origin of your Cartesian coordinate system to the centroid of the cross section, you can use the absolute value of the *x*-coordinate of the subregion's centroid for the values in the second column. The third column contains the area calculations for each subregion. The fourth column is the product of the second and third columns. The final numerical value for the first moment of area of either of the subregions created by a vertical reference axis located at Point B is given by the sum total of the fourth column. (Head to Chapter 4 for more details on creating this kind of table.) For this example, the computed first moment of area at Point B about the *y*-centroidal axis is

$$Q_{yy,B} = \sum_{i=1}^{3} x_i A_i = 48{,}000 \text{ mm}^3 + 3{,}000 \text{ mm}^3 + 3{,}000 \text{ mm}^3 = 54{,}000 \text{ mm}^3$$

When calculating the value of the first moment of area, remember that if your point of interest within the cross section lies at an extreme edge of the cross section, the value of the first moment of area of a subregion created by a reference axis at that point, which is parallel to that edge, is always zero. This situation occurs because at an extreme edge, the area of one of the subregions created by the reference axis is equal to zero, and the perpendicular distance from the centroid of the other subregion to the corresponding centroidal axis is equal to zero, which makes Q equal zero for both cases. Just make sure you're using a centroidal axis that is parallel to the edge that you're working with.

Encore! Encore! I, a Second Moment of Area

One of the most important cross-sectional properties in all of engineering is the *second moment of area* (or the *area moment of inertia*) with respect to the centroidal axis of a cross section. The second moment of area is a calculation that you use frequently for computing *deflections* (displacements from the original positions) caused by bending moments (see Chapter 16) and for determining the internal effects of bending moments (see Chapter 9) on flexural members. It also happens to be one of the more complex section property calculations you can make. If you combine that with the fact that mechanics frequently uses several different types of area moments of inertia, including basic moments of inertia, product moments of inertia, polar moments of inertia, and even the radius of gyration, understanding why so many students are frustrated by inertia calculations is easy.

You may remember from your physics class that *inertia* is a measure of an object's resistance to change. In this book, as with most textbooks, I assign the variable I to the moment of inertia.

Many areas of physics and engineering make use of the terms *inertia* and *moments of inertia,* so you need to be very aware of which particular inertia calculation you need for a given application. In mechanics of materials, you're more interested in the area moment of inertia, which is a measure of an object's resistance to bending and deflection and utilizes the basic geometric dimensions of a cross section as well as the location of the centroid. In dynamics, moments of inertia refer to the mass moment of inertia, which relates rotational characteristics, mass, and an axis of rotation to the resistance of an object. See the nearby sidebar "Separating from Euler's mass moment of inertia" for more on this difference. In this book, if you see the words *moment of inertia,* I'm actually referring to the second moment of area.

Conceptualizing on area moments of inertia

The *area moment of inertia* is a geometrical measure of an object's resistance to bending and deflection. An object may behave differently when subjected to bending in one direction than it does when subjected to bending in another, simply due to the fact that the relevant area moments of inertia of the cross section may be drastically different.

In special cases, you may be calculating the area moment of inertia about one reference axis, such as at the base of a cross section, before ultimately modifying the calculation so that the moment of inertia is referenced about the object's centroidal axis. You do this modification on only the very special occasion that I discuss in the following section.

The beams of Figure 5-4 illustrate a simple explanation of the moment of inertia's effect on an object. Each of the beams has the same support reactions (a pinned support at one end and a roller support at the other) and span and is loaded identically at midspan. The cross section of each beam has the same dimensions of $b \times h$. For one of the beams, b is the width and h is the height. For the other, b is the height and h is the width. The only physical difference between these beams is in the orientation of each beam's cross section. In Figure 5-4a, the beam is oriented such that h is vertical, while in Figure 5-4b, the beam is oriented such that the h dimension is now horizontal.

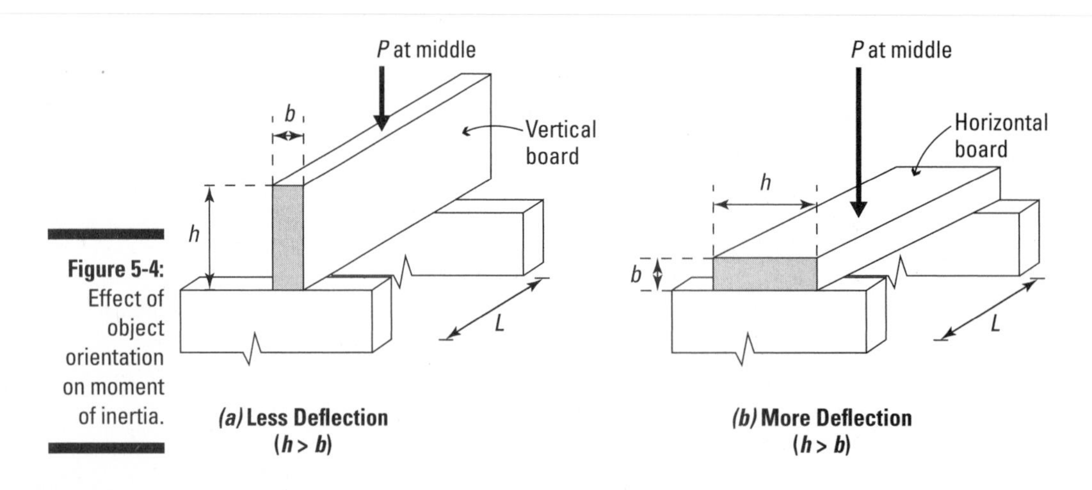

Figure 5-4: Effect of object orientation on moment of inertia.

Separating from Euler's mass moment of inertia

Swiss mathematician and physicist Leonhard Euler first proposed the moment of inertia in his 1730 book *Theoria Motus Corporum Solidorum Seu Rigidorum* (or *Theory of the Motion of Solid or Rigid Bodies*). He introduced the mass moment of inertia, which related the rotational acceleration characteristics of an object to its mass. The basic form of his equation is

$$I = m \cdot k \cdot r^2$$

where m represents the mass of the object, r is the distance between the axis and the rotational mass, and k is a numeric constant that defines the shape that is being rotated.

In engineering mechanics, the moment of inertia has a slightly different meaning (although the variable I is still widely used for both — including in this text — and many engineers reference this calculation as a moment of inertia). The difference is that in most structural objects, the mass density of the object is usually assumed to be constant throughout the member, and the geometric dimensions of the cross section are what affect the changes in this section property. The equation for the area moment of inertia that I discuss nearby actually arises because of the involvement of these dimensions in the derivation of the Euler-Bernoulli beam theory.

Under the shown load, when the height h is greater than the width b as shown in Figure 5-4a, the deflections are smaller than the same beam in Figure 5-4b simply because the beam of Figure 5-4a has a larger moment of inertia value. I show you how to actually calculate these values in the later section "Calculating Basic Area Moments of Inertia." The difference in these deflections is a result of the effect of the area moment of inertia. Beams with larger moments of inertia deflect less than beams with smaller moments of inertia, assuming that all other parameters remain the same, of course. I explain more about how to actually calculate the deflections of beams in Chapter 16.

Categorizing area moments of inertia

Depending on the type of application you're studying, using the proper moment of inertia is of crucial importance. In mechanics, you encounter several types of area moments of inertia that I explain throughout this book:

- **Basic moment of inertia:** The *basic moment of inertia* (which is often simply referred to as the *moment of inertia*) involves calculating the second moment of area about a centroidal axis. If a cross section lies in the XY plane, the resulting basic moments of inertia are usually about an x- and y-centroidal axes.

- **Product moment of inertia:** The product moment of inertia (along with the basic moments of inertia) helps determine the maximum and minimum values of second moments of area (known as the *principal moments of inertia*) for a given shape. The product moment of inertia becomes especially important in the study of nonsymmetric bending (or *asymmetric bending*).
- **Polar moment of inertia:** The *polar moment of inertia* is used in determining the resistance of an object to twisting moments (or *torsion,* the rotations about a member's longitudinal axis), which I discuss in Chapter 11.

In the remainder of this chapter, I show you how to mathematically calculate the different moments of areas, including tips and techniques for simplifying the process. After you know how to actually compute them, I show you how to put them to work in specific applications starting in Part II.

Calculating Basic Area Moments of Inertia

In the U.S. customary units of measure, the units for the second moment of area are in^4; in the SI system, the units are usually taken as m^4. Just as with the first moment of area (covered earlier in the chapter), you may encounter cross-sectional dimensions measured in millimeters, so the appropriate units for the moment of inertia in this case become mm^4.

Keeping inertia simple with basic shapes and centroidal axes

You calculate the second moments of area for a cross section with respect to the centroidal axes. For this section's calculations and corresponding figures, I assume that the cross section lies in the Cartesian XY plane. The subscripts assigned to the variable I refer to the axis about which the calculation is being performed (as shown in Figure 5-5).

The basic equation defining the moments of inertia with respect to their x- and y-centroidal axes are given by the following:

$$I_{xx} = \int_A y^2 \cdot dA \text{ and } I_{yy} = \int_A x^2 \cdot dA$$

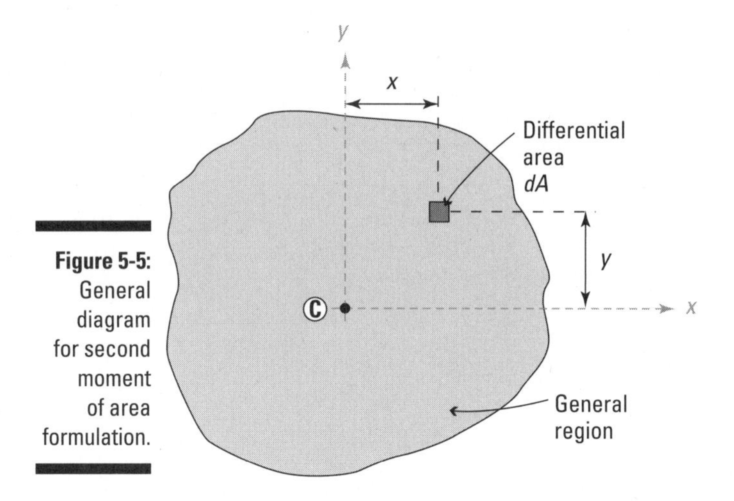

Figure 5-5: General diagram for second moment of area formulation.

where y is the perpendicular distance from the centroid of the differential element dA to the x-centroidal axis of the cross-section. Similarly, x is the perpendicular distance from the centroid of the differential element dA to the y-centroidal axis of the cross section.

I like to use a double subscript to indicate the centroidal axis about which I'm performing the calculation. In some textbooks and resources, you may see this value represented as an I_x with a single subscript, but I find this setup can be confusing because on some occasions, you may actually need to compute the moment of inertia about some other x-axis (such as the base of a rectangular cross section) that is different from the x-centroidal axis. This double subscript also helps distinguish these calculations from the product moment calculations (which have a subscript of "xy" for a cross section in the XY plane) but involves perpendicular distances in both the x and y directions from the centroid — hence the double letters. Realistically, how you label your values doesn't matter as long as you compute and apply the values correctly.

In all cases, these second moment of area calculations about their x- and y- reference axes always produce a positive value. However, the product moment I_{xy} (which I discuss later) may sometimes be negative.

Although the integral form always works, it can be a bit mathematically intimidating at times. For several simple shapes, you can find these evaluated integrals in Figure 5-6 with respect to the centroidal axes at Point C. If you need help locating this point, flip to Chapter 4.

Figure 5-6 also shows the results of the product moment of inertia I_{xy}, which I discuss in more detail in the later section "Having It Both Ways with Product Moments of Area."

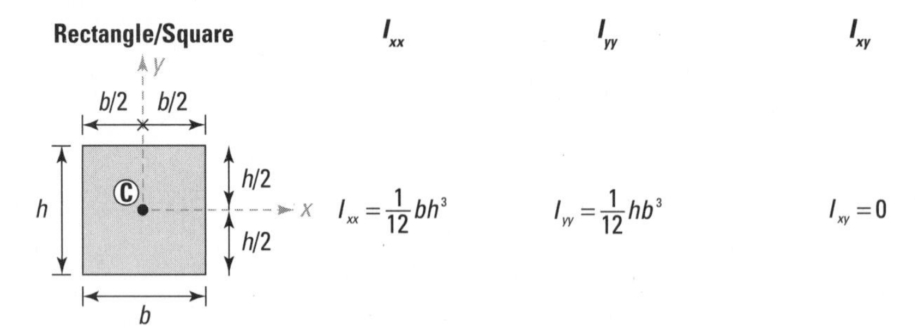

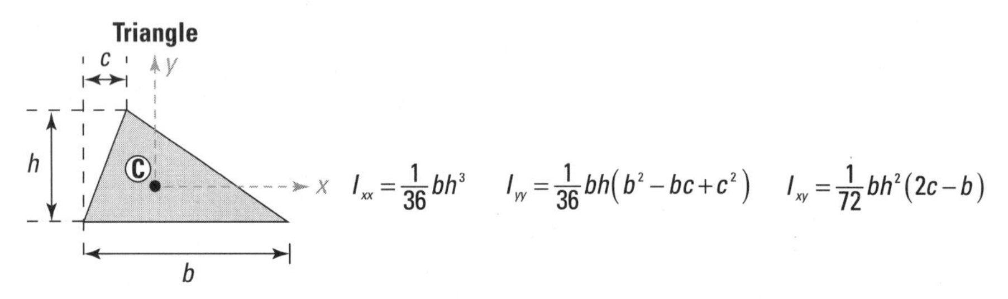

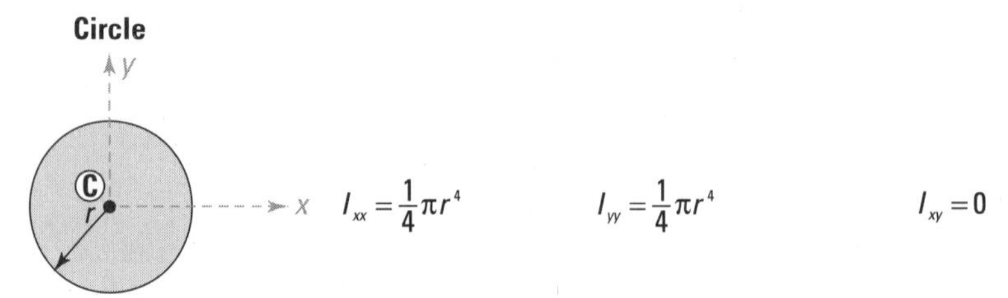

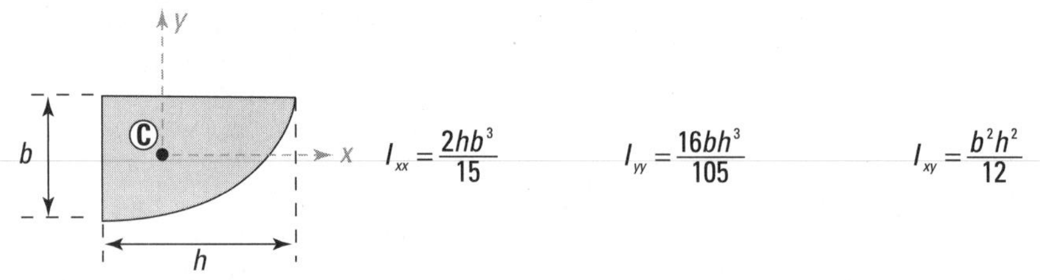

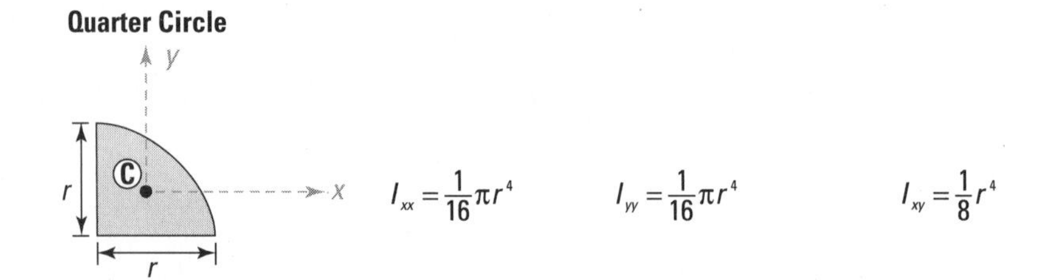

Figure 5-6: Moment of inertia values for common shapes.

For simple shapes, the calculation of the second moment of area about a centroidal axis is actually as simple as plugging in the necessary dimensions into the basic formula. Just be sure that the orientation of your region matches the orientation of the region in the table.

One particular set of axes that are especially important when it comes to calculating second moments of area are the centroidal axes. At these locations, the smallest values for the second moment of area occur.

If you have multiple shapes that share a common centroidal axis, you can compute the moment of inertia with respect to that centroidal axis by simply adding the inertia values of each of the regions together. When you encounter a shape that has a hole or opening that shares the centroid of the solid object (such as in Figure 5-7), you subtract the inertia value of the hole from the region around it.

For cases where a hole or opening does not share the same centroidal axes, such as the subregion that contains the hole or opening, you need to perform an additional calculation as I show in "Transferring reference locations with the parallel axis theorem" later in this chapter.

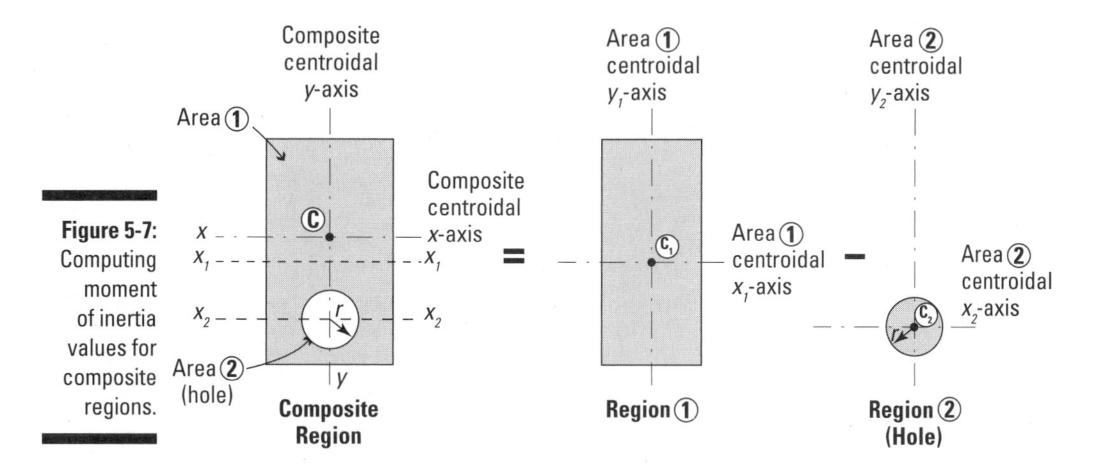

Figure 5-7: Computing moment of inertia values for composite regions.

Figure 5-7 shows a composite region consisting of a rectangle with a centroid located at Point C_1 and a hole with radius *r* with a centroid located at Point C_2. The composite centroid is located at Point C (which is different from the centroids of the subregions). Although the *x*-centroidal axis for the each of the regions is different, they all share a common *y*-centroidal axis (which passes through Point C, Point C_1, and Point C_2). Thus, you can compute the moment of inertia about the *y*-centroidal axis (I_{yy}) by simply summing the individual moments of inertia with respect to the *y*-centroidal axis as follows:

$$I_{yy} = I_{yy,1} + I_{yy,2} = \frac{1}{12}hb^3 + \left(-\frac{\pi}{4}r^4\right)$$

For the moment of inertia about the *y*-axis, you can simply add the two regions because they share the same centroidal *y*-axis. In this equation, the moment of inertia of the hole counts as a negative value because it's being subtracted from the solid region.

However, for the moment of inertia I_{xx} about the *x*-axis (for the composite region), you can't do this addition because the x_1-centroidal axis and the x_2-centroidal axis aren't the same as the centroidal *x*-axis for the combined region.

You can only add or subtract inertia calculations about a specific centroidal axis if they both share that same centroidal axis. If the centroidal axis is different for the shapes, you have to transfer the reference locations as I describe in the following section.

Transferring reference locations with the parallel axis theorem

The basic computations of simple shapes such as rectangles, circles, and triangles are fairly easy to calculate with respect to the shapes' own centroidal axes. Simply look up the value in a table (or perform the basic integration), and you're all set.

But what happens when one of these simple shapes is part of a bigger discrete (or composite) shape that has a different centroidal axis than the basic element? In this situation, you can use a basic transfer formula known as the parallel axis theorem (or *Steiner's rule*).

The *parallel axis theorem* takes the area moment of inertia about an object's own centroidal axis and allows you to determine the equivalent area moment of inertia about another axis (such as a centroidal axis of a combined region). The basic equation for moment of area from the parallel axis theorem is given as follows:

$$I = I_o + A(d)^2$$

where I_O is the area moment of inertia of a region about its own centroidal axis; A is the area of the region for which I_O was computed; and d is the perpendicular distance from the centroidal axis of A to the new parallel axis location.

Suppose you want to find the area moment of inertia about the *x*-centroidal axis I_{xx} for the cross section of Figure 5-8.

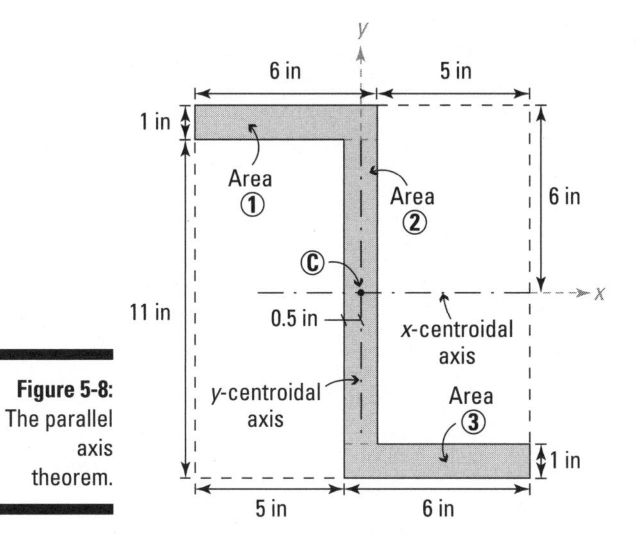

Figure 5-8: The parallel axis theorem.

I have also divided the region into three subregions, Area 1, Area 2, and Area 3 as shown in Figure 5-9.

As with centroid calculations (see Chapter 4) and first moment of area calculations (which I discuss earlier in this chapter), you can greatly simplify your work when you perform calculations with the parallel axis theorem by making a table such as the one in Table 5-2 and following a few simple steps.

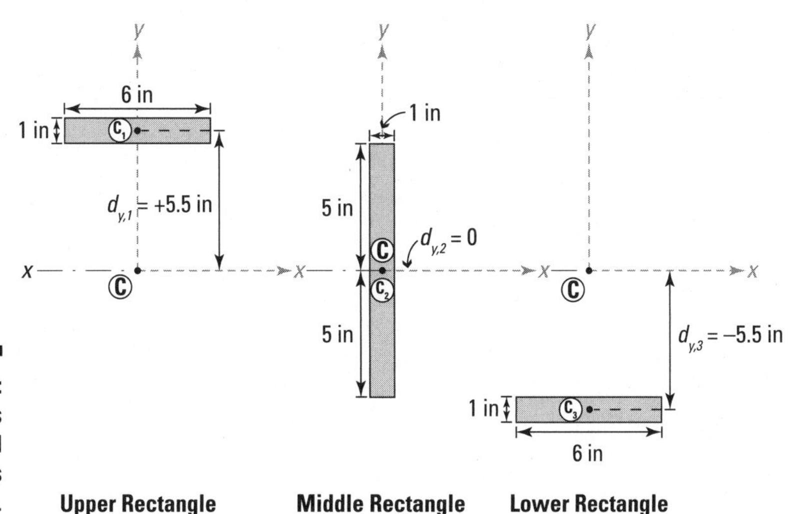

Figure 5-9: Subregions for parallel axis example.

Table 5-2 Table for computing I_{xx} about Centroidal Axis

Region	$I_{0,xx}$ (in^4)	A_i (in^2)	$d_{y,i}$ (in)	$I_{0,xx}+A_i(d_{y,i})^2$ (in^4)
1 (upper rectangle)	$\frac{1}{12}(6)(1)^3 = 0.50$	6.00	+5.50	$0.50 + 6.00(5.50)^2 =$ 182.00
2 (middle rectangle)	$\frac{1}{12}(1)(10)^3 = 83.33$	10.00	0	$83.33 + 10.00(0)^2 =$ 83.33
3 (bottom rectangle)	$\frac{1}{12}(6)(1)^3 = 0.50$	6.00	–5.50	$0.50 + 6.00(-5.50)^2 =$ 182.00
TOTAL	------	-----	------	$\Sigma = 447.33$

1. **Locate a new Cartesian coordinate system at the centroid (Point C) of the cross section.**
2. **For each subregion, determine the area moment of inertia $I_{O,xx}$ about its own *x*-centroidal axis and record this value in the second column.**

   For Region 1, the area moment of inertia I_{O1} as shown in Figure 5-9 for a rectangular region about its own *x*-centroidal axis is given as

   $$I_{O,xx,1} = \frac{1}{12}(b)(h)^3 = \frac{1}{12}(6 \text{ in})(1 \text{ in})^3 = 0.50 \text{ in}^4$$

   Even though all the subregions of this example are discrete regions that are shown in Figure 5-6, you can just as easily include a more general region as one of the subregions in this table.

3. **Compute the area A_i for each subregion and record this value in the third column of the table.**

   For Region 1, the area of the region is given as

   $$A_1 = (6 \text{ in})(1 \text{ in}) = 6.00 \text{ in}^2$$

4. **Determine the perpendicular distance d_i from the centroid of the area to the *x*-centroidal axis of the composite cross section and record this value in the fourth column.**

   Because you're working with the *x*-centroidal axis in this example, you must use the *y*-distance in your calculation.

   $$d_{y,1} = +5.50 \text{ in}$$

   Note that for Region 2, this distance is actually 0.0 because the centroid of Region 2 is the same as the centroid of the combined region.

5. **Complete the calculation by using the parallel axis theorem and record this value in the fifth column.**

   For Region 1, you can compute the moment of inertia with respect to the centroid of the combined cross-section $I_{xx,1}$ as follows:

$$I_{xx,1} = I_{O,xx,1} + A_1(d_{y,1})^2 = (0.50\text{ in}^4) + (6.00\text{ in}^2)(5.50\text{ in})^2 = 182.00\text{ in}^4$$

6. **Repeat Steps 2 through 5 for each subregion.**
7. **Sum the total of all values in the fifth column to compute the total area moment of inertia I_{xx} for the composite cross section with respect to its own centroidal axis.**

$$I_{xx} = \sum_{i=1}^{3} I_{xx,i} = (182.00 + 83.33 + 182.00)\text{ in}^4 = 447.33\text{ in}^4$$

As the subtotals in the fifth column of Table 5-2 indicate, despite the fact that Region A_2 had the largest area, it didn't contribute as much to the total inertia of the cross section as the other two subregions did because this region's centroid location was the same as the centroid of the combined cross section (making the $d_{y,2} = 0$). This result actually illustrates why many common structure shapes utilize flanged members. By putting the same area at a larger distance from the centroid, you can get a significant increase in the moment of inertia, even for a very small area.

You can add as many subregions to your table as you need. In fact, you can also add additional general subregions for each of the discrete regions included in the same parallel-axis table. Just remember that for every subregion, you need three important pieces of information: the subregion's area, the subregion's area moment of inertia about its own centroidal axis, and the perpendicular distance to the desired centroidal axis of the composite section.

Having It Both Ways with Product Moments of Area

The *product moment of inertia* (sometimes referred to as the *mixed second moment of area*) is another area moment of inertia calculation that is necessary for calculating the internal effects of bending moments for nonsymmetrical sections and is also used to compute the maximum or minimum moments of inertia as well.

The product moment of inertia has the same units as the basic area moment of inertia calculations (see "Calculating Basic Area Moments of Inertia" earlier in the chapter): in^4 for U.S. customary units and m^4 or mm^4 for SI units.

The sign convention for calculating the product moment is fairly normal as well. In this text, I assume positive x is to the right and positive y is upward. Where this section property differs from the others is that the product moment of inertia can actually have a negative value; the other area moments of inertia are always positive.

Including x- and y-axes for product moment calculations

The basic formula for the product moment appears very similar to the area moment of inertia calculations as shown in the following equation:

$$I_{xy} = \int_A (xy) \cdot dA$$

where dA is a differential area within a region, x is the perpendicular distance from the centroid of dA to the y-centroidal axis, and y is the perpendicular distance from the centroid of dA to the x-centroidal axis. Figure 5-6 earlier in this chapter shows the results of evaluating this integral for several simple cross sections.

The basic computations for most area moments of inertia utilize either an x-centroid distance or a y-centroid dimension but not both at the same time. The product moment of inertia, on the other hand, does deal with both simultaneously.

As with the basic moment of area computations, product moments of inertia can also be calculated with a parallel axis theorem:

$$I_{xy} = I_{O,xy} + A(d_x)(d_y)$$

where $I_{O,xy}$ is the product moment of the subregion about its own centroidal axis; A is the area of the region; d_x is the x-distance from the centroid of the subregion to the centroid of the composite region; and d_y is the y-distance from the centroid of the subregion to the centroid of the composite region.

This calculation is almost identical to the parallel axis theorem for the basic moment of inertia I discuss earlier in this chapter. However, you must be careful about the signs associated with d_x and d_y in this formula because I_{xy} can actually be a negative value.

Computing the product moment of area

For general regions, the integrals can become rather complex because of the presence of both centroid variables (x and y). As with many calculations, creating a simple table makes the computations much easier. Table 5-3 helps you figure the product moment of inertia for Figure 5-8 earlier in the chapter by using the parallel axis theorem for product moments.

Table 5-3 Table for Computing I_{xy} about Centroid

Region	$I_{0,xy}$ (in^4)	A_i (in^2)	$d_{x,i}$ (in)	$d_{y,i}$ (in)	$I_{0,xy}+A_i(d_{x,i})(d_{y,i})$ (in^4)
1 (upper rectangle)	0	6.00	–2.50	+5.50	0 + (6)(–2.50)(+5.50) = –82.50
2 (middle rectangle)	0	10.00	0	0	0 + (10)(0)(0) = 0.00
3 (bottom rectangle)	0	6.00	+2.50	–5.50	0 + (6)(+2.50)(–5.50) = –82.50
TOTAL	------	-----	------		Σ = –165.00

The following steps help you do the math:

1. **Locate a new Cartesian coordinate system at the centroid of the cross section.**
2. **For each subregion, determine the product moment of inertia, $I_{0,xy}$ about its own centroid and record this value in the second column.**

   For the three subregions of this example, all three shapes are symmetrical, which means that the product moment of inertia with respect to their individual centroids is automatically zero for all three shapes. If the subregions weren't symmetrical, you'd need to include the appropriate calculation based on the formulas from Figure 5-6 earlier in the chapter.

3. **For each subregion, compute the area A_i and record this value in the third column.**
4. **In the fourth column, record the *x*-distance, $d_{x,i}$ from the centroid of each subregion to the centroid of the composite shape.**

   If the centroid of the region is to the left of the centroid of the composite shape, this distance is negative. Conversely, if the centroid of the region is to the right of the centroid of the composite shape, this distance is positive.

   For Region 1, $d_{x,1} = -2.5$ in

   Be very careful that you get the sign of the *x*-distance correct because it has a big influence on your final computation value of the product moment of area.

5. **In the fifth column, record the *y*-distance $d_{y,i}$ from the centroid of each subregion to the centroid of the composite shape.**

If the centroid of the region is above the centroid of the composite shape, this distance is positive. But if the centroid of the region is below the centroid of the composite shape, the distance is negative. For Region 1, $d_{y,1} = +5.5$ in. Like in Step 4, make sure you use the correct sign.

6. **Compute the product moment of the subregion about the centroid of the composite region by using the parallel-axis theorem for product moments and record this in the sixth column.**

   For Region 1,

   $$I_{xy,1} = I_{O,xy,1} + A_1(d_{x,1})(d_{y,1}) = 0 + (6.00\ \text{in}^2)(-2.50\ \text{in})(+5.50\ \text{in}) = -82.50\ \text{in}^4$$

7. **Compute the total product moment by summing the values of the sixth column.**

   The final computed value gives you the product moment of area for the combined cross section.

   $$I_{xy} = \sum_{i=1}^{3} I_{xy,i} = (-82.50 + 0.00 - 82.50)\ \text{in}^4 = -165.00\ \text{in}^4$$

Putting a Twist on Polar Moments of Inertia

The polar moment of inertia is another of the second moments of area that you use in mechanics of materials. The *polar moment of inertia* is a measure of an object's resistance to twisting phenomenon and is important in calculations involving *torsional moments* (moments acting about a longitudinal axis; see Chapter 11 for more).

For most structural shapes, you typically compute the polar moment of inertia about a specific (and easily defined) longitudinal axis. That is, for a cross section in the XY plane, the polar moment is acting about the longitudinal or *z*-axis. In these cases, I refer to the polar moment as a second moment of area with the subscripts describing the axis about which I've calculated it — in this case, I_{zz}.

The basic formula for the polar moment about a longitudinal *z*-axis is given by the following equations:

$$I_{POLAR} = I_{zz} = \int_A r^2 \cdot dA$$

where dA is the incremental area and $r^2 = x^2 + y^2$ as shown in Figure 5-10.

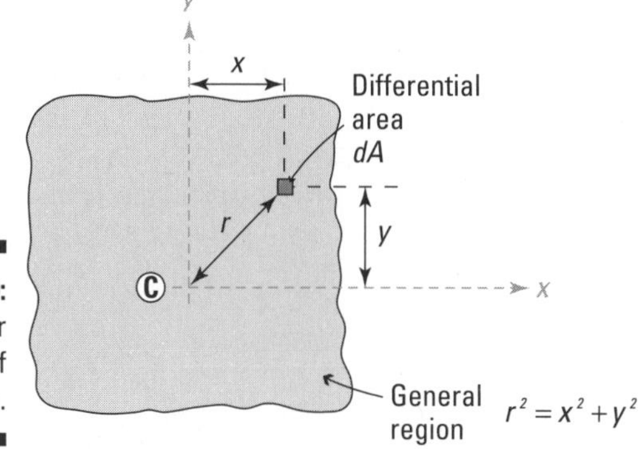

Figure 5-10: The polar moment of inertia.

As with all other second moment of area section properties, the units for the polar moment are in^4 for U.S. customary units and m^4 or mm^4 for SI units.

For a solid circular region having a radius *c*, you can compute the polar moment of inertia as follows:

$$I_{POLAR} = I_{zz} = \frac{\pi}{2}(c)^4$$

And for a concentrically hollow circular shape (such as a pipe) with outer radius c_{OUT} and an inner radius c_{IN}, you can compute the polar moment of inertia by computing the moment of inertia of a solid shaft with radius c_{OUT} and subtracting the polar moment of inertia of the hollow portion having a radius c_{OUT}:

$$I_{POLAR} = I_{zz} = \frac{\pi}{2}(c_{OUT})^4 - \frac{\pi}{2}(c_{IN})^4 = \frac{\pi}{2}\left((c_{OUT})^4 - (c_{IN})^4\right)$$

Many textbooks express the polar moment of inertia with the variable *J*, which can be confused with another section property known as the torsion constant. In circular cross sections, $J = I_{POLAR} = I_{zz}$. However, *J* is not the same as the polar moment of inertia for non-circular cross sections; in fact, that property can be vastly different, so be mindful of this common, albeit troublesome, substitution that you may come across. I explain more about the torsion constant in Chapter 11.

You can add and subtract regions as long as the polar axis (which is in the *z*-direction for this example) is the same for all regions.

Fortunately, computing the polar moment of inertia for more-complex shapes is fairly simple because of a simplification known as the *perpendicular axis theorem*. For a region contained in the Cartesian XY plane, you can rewrite the polar moment of inertia about the *z*-axis as

$$I_{POLAR} = I_{zz} = I_{xx} + I_{yy}$$

where I_{xx} is the second moment of area with respect to the *x*-centroidal axis and I_{yy} is the second moment of area with respect to the *y*-centroidal axis. With this simplification, the polar moment is now a function of two other second moment of area calculations that are often either already known or easily computed.

Just as with the basic moment of inertia (which I cover earlier in the chapter), you can subtract polar moments of negative subregions and holes if they share a common axis, which for polar moments must be the longitudinal axis. As with other area moment of inertia calculations, you can transfer reference locations of polar moments of inertia with a modified form of the parallel axis theorem:

$$I_{POLAR} = I_{zz} = I_{zz,O} + Ar^2$$

where $I_{zz,O}$ is the polar moment of inertia of the region about its own centroidal axis *A;* is the area of the region; and *r* is the linear distance (not necessarily parallel to the *x*- or *y*-axes) between the centroid of the region and the centroid of the composite section.

You may notice that this expression looks very familiar to the parallel axis theorem in its basic form (see "Transferring reference locations with the parallel axis theorem" earlier in the chapter). You start with a moment of area about a centroid of a subregion, and you can transfer its location by incorporating the area of the subregion and its distance from the new location. It's just a matter now of figuring out which distances you need for your calculations.

Computing Principal Moments of Inertia

For unsymmetrical shapes, another tremendously important moment of area exists. In your mechanics calculations, you need to be able to determine the maximum and minimum values (known as the *principal moments of inertia*) and the orientation angle (known as the *principal angles*) at which they occur within the cross section. You compute these principal moments of inertia from both the basic moment of inertia and the product moment of inertia for a cross section. You use the principal moments of area to study the effects of a cross section under combined bending.

For a symmetrical cross section, the principal moments of inertia about the centroidal axes are often the same as the basic moments of inertia that I show you how to compute earlier in the chapter. They also occur at the same orientation. However, for unsymmetrical cross sections, the principal moments don't occur at the same orientation.

Calculating principal moments of inertia

To calculate the principal moments of inertia, I_{p1} and I_{p2}, you use the following equation:

$$I_{p1,p2} = \frac{I_{xx} + I_{yy}}{2} \pm \sqrt{\left(\frac{I_{xx} - I_{yy}}{2}\right)^2 + \left(I_{xy}\right)^2}$$

One of these values is the maximum principal moment of inertia, and the other is the minimum principal moment of inertia. The larger value of I_{p1} and I_{p2} determines an orientation known as the *strong axis,* which indicates that an object is more resistant to loads in one direction than another. Likewise, the smaller of I_{p1} and I_{p2} refers to an orientation known as the *weak axis.* The weak and strong axes are always *orthogonal* (or perpendicular). I explain how you determine which is which in the following section.

Consider the example shape shown in Figure 5-8 earlier in the chapter. Earlier sections show you how to determine two of the necessary values that you need to compute the principal values: I_{xx} = 447.33 in^4 and I_{xy} = –165.00 in^4. (Tables 5-2 and 5-3 show you these values, respectively.) You can calculate the second moment of area about the *y*-centroidal axis similarly, but for now, I can tell you that I_{yy} = 111.83 in^4.

Plug those numbers into the corresponding parts of the principal moments equation to get the following:

$$\frac{I_{xx} + I_{yy}}{2} = \frac{447.33 \text{ in}^4 + 111.83 \text{ in}^4}{2} = 279.58 \text{ in}^4$$

$$\frac{I_{xx} - I_{yy}}{2} = \frac{447.33 \text{ in}^4 - 111.83 \text{ in}^4}{2} = 167.75 \text{ in}^4$$

Thus, to determine the principal moments of area,

$$I_{p1,p2} = 279.58 \text{ in}^4 \pm \sqrt{\left(167.75 \text{ in}^4\right)^2 + \left(-165.00 \text{ in}^4\right)^2} = 279.58 \text{ in}^4 \pm 235.30 \text{ in}^4$$

$$\Rightarrow I_{p1} = 514.88 \text{ in}^4 \text{ (maximum)}$$

$$\Rightarrow I_{p2} = 44.28 \text{ in}^4 \text{ (minimum)}$$

Finding the principal orientation angles

The final step is to determine the principal angles associated with these principal values. You can calculate one of the principal angles, θ_p, from the following relationship:

$$\theta_p = \frac{1}{2}\tan^{-1}\left(-\frac{2I_{xy}}{\left(I_{xx} - I_{yy}\right)}\right)$$

To compute the principal orientation angles for Figure 5-8 earlier in the chapter, you use the following math:

$$2\theta_p = \tan^{-1}\left(-\frac{2\left(-165.00\text{ in}^4\right)}{447.33\text{ in}^4 - 111.83\text{ in}^4}\right) = \tan^{-1}(0.984)$$

$$\Rightarrow 2\theta_p = 44.54°$$

$$\Rightarrow \theta_p = 22.27°$$

Note that within the range of $0 < 2\theta_p < 360°$, two values of $2\theta_p$ actually satisfy the requirement of $\tan^{-1}(0.984)$. The first value (the one that I calculate in the preceding equation) is 44.54°. The second value occurs exactly 180° from that value or at $2\theta_p = 224.54°$.

Remember that the two expressions are based on $2\theta_p$, so you must divide your angles from these formulas by 2!

$$\theta_p = 22.27° \quad \text{and} \quad \theta_p = 112.27°$$

You typically don't report the angle as a number larger than 360°. When this situation happens, recognize that 360° occurs at the same orientation as 0°, so you can start recounting after you pass 360° when the numbers get too large.

Determining moments of area at specific orientation angles

After you find the angles of orientation (see the preceding section), you need to know which angle goes with which principal moment of inertia. To determine which angle goes with which principal value, you use the following transformation equation:

$$I_{x1x1} = \frac{I_{xx} + I_{yy}}{2} + \frac{I_{xx} - I_{yy}}{2}\cos 2\theta - I_{xy}\sin 2\theta$$

where I_{xx}, I_{yy}, and I_{xy} are the moments of area of the objects based on their centroidal Cartesian axes (in the *x*- and *y*-directions). The angle θ is the orientation of a new Cartesian axis (labeled as the *x1* axis) with respect to the original *x*-Cartesian axes. A new *y1* axis is oriented 90° from the new *x1* axis, or the same angle θ from the original *y*-axis.

For the previous example, you simply plug in your known values of I_{xx}, I_{yy}, and I_{xy}. Finally, choose one of the two principal angles (it doesn't matter which), and compute the corresponding moment of area at that orientation.

$$I_{p1} = 279.58 \text{ in}^4 + \left(167.75 \text{ in}^4\right)\cos\left(2(22.27°)\right) - \left(-165.00 \text{ in}^4\right)\sin\left(2(22.27°)\right) = 514.88 \text{ in}^4$$

This result indicates that I_{p1} at an angle of 22.27° from the original orientation corresponds to the maximum principal value (or the strong axis value). You can also verify that at an angle of 112.27° is where the minimum principal moment value I_{p2} = 44.28 in^4 (or the weak axis) occurs.

Figure 5-11 shows the orientation of the strong and weak axes with respect to their original *x*- and *y*- axes based on these principal angle calculations.

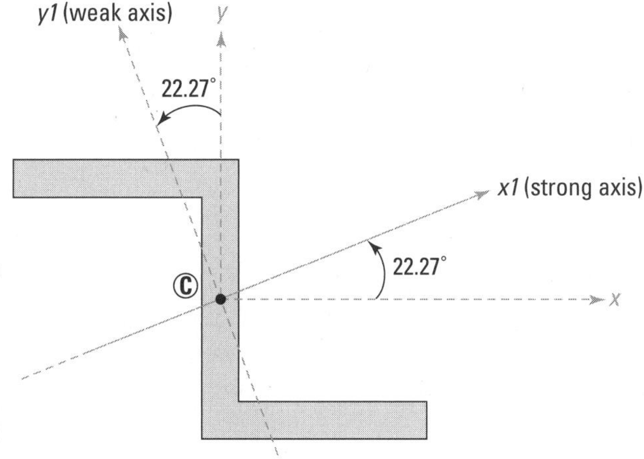

Figure 5-11: Strong and weak axes orientation.

You can also compute the product moment at any given axis orientation I_{x1y1} with its own transformation equation.

$$I_{x1y1} = \frac{I_{xx} - I_{yy}}{2}\sin 2\theta + I_{xy}\cos 2\theta$$

You can make an interesting observation if you plug a principal angle into the previous transformed product moment equation:

$$I_{x1y1} = \left(167.75 \text{ in}^4\right)\sin\left(2(22.27°)\right) + \left(-165 \text{ in}^4\right)\cos\left(2(22.27°)\right) = 0.00 \text{ in}^4$$

As you can see, the corresponding product moment at a principal angle is exactly zero. That is, if you're working with principal moments of inertia, the corresponding product moment is always zero. This tidbit proves especially useful when you recognize that the moments of inertia that you calculate about a centroidal axis of symmetry are automatically principal moments of inertia.

Rounding Up the Radius of Gyration

The *radius of gyration* (sometimes called the *gyradius*) is a derived section property based on the second moment of area (or area moment of inertia) and the cross-sectional area. The radius of gyration is a section property that describes the distribution of a cross-sectional area about its own centroidal axis. You use it frequently in the analysis of columns (or members subjected to compression), which I describe in Chapter 18. Its units are units of length: inches for U.S. customary units and meters for SI units. As with other SI units in this chapter, you may encounter the radius of gyration in millimeters.

You can compute r_x, the radius of gyration with respect to the *x*-centroidal axis and r_y, the radius of gyration with respect to the *y*-centroidal axis, from the following relationships:

$$r_x = \sqrt{\frac{I_{xx}}{A}} \quad \text{and} \quad r_y = \sqrt{\frac{I_{yy}}{A}}$$

where I_{xx} is the second moment of area about the *x*-centroidal axis; I_{yy} is the second moment of area about the *y*-centroidal axis; and A is the cross-sectional area. Because this chapter and Chapter 4 already show you how to determine I and A, I haven't included the math here. A smaller area located at a large distance can have the same I_{yy} value as a larger area located at a small distance. Despite having the same moment of inertia, these two regions behave extremely differently under certain load conditions such as compression and buckling in columns (which I discuss in Chapter 18). The radius of gyration is a section property that takes this consideration into effect.

The base units of I_{xx} and I_{yy} must match the units associated with the cross-sectional area A. For SI units, if the area is expressed in mm^2, you must make sure that you express the second moment of area in mm^4 in order for the units to cancel out. This fact is especially important with U.S. customary units when one value may be expressed in feet and others are expressed in inches.

Part II

Analyzing Stress

"I'll finish this stress transformation equation and then we'll step back to see the bigger picture."

In this part . . .

Stresses are one of the fundamentals of mechanics of materials because they relate internal forces to specific section properties. I start this section by describing the basic categories of stress and then show you how to determine the maximum and minimum values (known as *principal stresses*) and their orientation angles through several basic transformation techniques. I conclude the part by demonstrating how to calculate stresses for different load types, including axial forces, bending moments, shear forces, and torsional moments.

Chapter 6

Remain Calm, It's Only Stress!

In This Chapter

- Defining the basics of stress
- Computing average stresses
- Grasping stress's units
- Working with stresses at a point
- Understanding assumptions about plane stress problems

In the "old days," engineers used a trial-and-error approach or relied on previous experience. But you don't really want to build a bridge and load it until it fails just to find out how much it could carry before you broke it. That proposition sounds expensive (not to mention potentially dangerous)! So how do you actually determine whether a particular object made from a particular material can carry a particular load? A more scientific approach involves calculating the actual intensity of the force (known as stress) on an object and then comparing this stress to the intensity of a force that a material is capable of withstanding before it fails. With that information, you can begin predicting the most-critically stressed location in a particular object, which is a fundamental design skill for engineers.

In this chapter, I introduce the basic concept of stress because the intensity of a force affects different types of objects in different ways. I start the explanation with the simplest of stress calculations: average stresses. You can then use these average stresses to develop a general relationship about the state of stress at a particular point. Finally, I introduce the concept of plane stress, which is a significant assumption in the formulation of many of the equations and design relationships in Part III.

Dealing with a Stressful Relationship

In psychological terms, stress is defined as a mental response to external stimuli. To most people, stress is what they feel when they get stuck in traffic or their bosses start piling on the work. But to engineers, stress has a significantly different meaning. In mechanics of deformable bodies, *stress* is a measure of the effect of loads on an object; more specifically, it's a measure of the intensity of an internal force or moment.

Unfortunately you can't actually see a stress, and experimentally measuring it is no real picnic either — in fact, doing so is downright impossible. However, you can see the resulting effect of stress. When an object becomes stressed, it may change shape or break, depending on the magnitude of stress on the object and the material the object is made from. Most commonly, deformable objects experience *deformation* (which is why they're called *deformable*). Fortunately, these deformations are often measureable, and you can use them to compute corresponding stresses (as I show in Chapter 14).

Calculating stress

Simply put, the basic relationship for stress can be expressed as

$$\text{stress} = \frac{\text{internal load}}{\text{section property}}$$

To calculate a stress, you need two pieces of information: internal loads and section properties.

- **Internal loads:** As you learned in statics (and I discuss in Chapter 3), *internal loads* come in the form of axial forces, shear forces, and moments, and they are created in response to the external applied loads on an object. Internal loads can vary in type, magnitude, and direction within an object, and they can cause two different types of stress, both of which I define in the following section.

  The kind of internal load you need in order to compute this calculation actually depends on the type of stress you want to calculate (as I explain later in this chapter). You also encounter many situations where more than one of these internal loads can occur at the same time. (Don't sweat it here; I show you how to handle these combined problems in Chapter 15). Just remember that in order to calculate the magnitude of a stress in an object, you must first determine these internal loads.

- **Section properties:** As with the internal loads, the *section property* (such as the area, first moment of area, or moment of inertia) that you need to use depends entirely on which type of stress you're working with. For example, an average normal stress (which I discuss in "Remaining Steady with Average Stress" later in the chapter) requires the cross-sectional area, while bending and torsion require one of the moments of inertia as well as the position within a given cross-section. Flip to Chapters 4 and 5 for more on calculating these section properties. Some other stresses require additional section properties from those already listed here, which I explain on a case-by-case basis as they appear.

The major issue you encounter when performing stress calculations, however, is "Which internal load do I need? And which section property do I use?" If you can answer these two basic questions, you're well on your way to dealing with stresses.

When calculating stresses, you must make sure that you use the appropriate internal load with the correct section property. If you accidentally use the wrong internal load or section property, your calculations are doomed!

Defining the types of stress

In mechanics, stresses can be classified into one of two major types: normal stress and shear stress. The simplest of these stresses are the *average normal stress* and the *average shear stress,* which represent a constant and uniform (or average) stress intensity over a given region.

Average normal stress

An *average normal stress* is an average stress that results from a force component F_{INT}, which is acting *normal* (or perpendicular) to a cross section as shown in Figure 6-1. The force must be a *component force* (or a part of a bigger force that acts in a particular direction) that is measured parallel to a longitudinal axis. In most texts, the lowercase Greek symbol sigma, or σ, indicates the normal stress. Normal stresses can be either *tensile* (causing elongation) or *compressive* (causing shortening).

Although bending moments can also cause normal stresses, normal stresses from bending are never constant and uniform along a cross section and thus aren't average normal stresses. Average normal stresses are the result of axial load effects, which I discuss in Chapter 8.

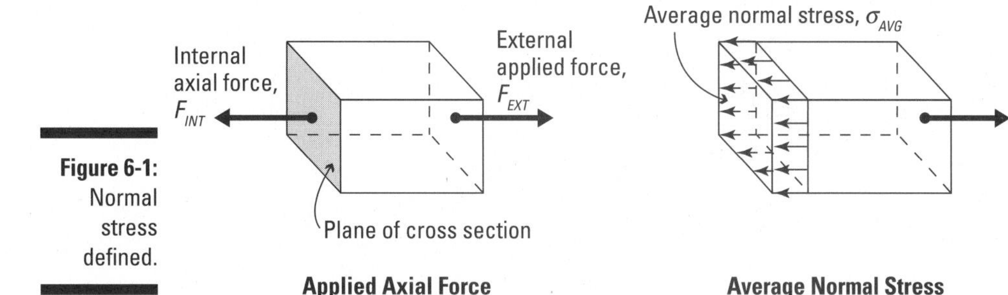

Figure 6-1: Normal stress defined.

Average shear stress

An *average shear stress* is an average stress that results from a force component V that's acting parallel to (or *shearing*) a cross section. The force must be a component force lying in the plane of the cross-sectional area. In most texts, the lowercase Greek symbol tau, or τ, indicates the shear stress. (Refer to Figure 6-2.)

Torsional moments and flexural loads can also cause shear stresses, but shear stresses from these sources are never constant and uniform along a cross section. For that reason, they aren't considered average shear stresses. Average shear stresses result from direct shear load effects (flip to Chapter 10).

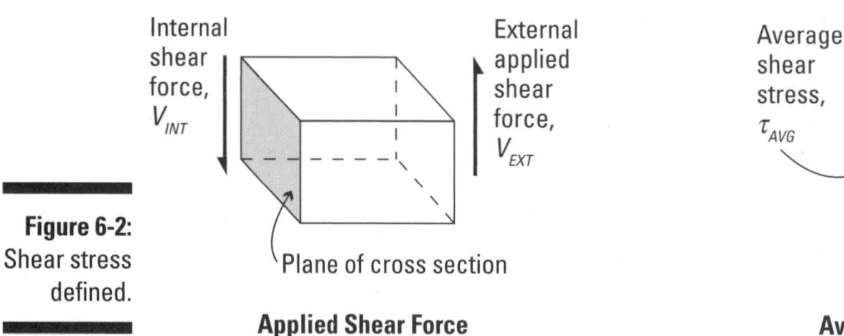

Figure 6-2: Shear stress defined.

Understanding the units of stress

As with any calculation, you always need to be mindful of what units you're working with when you're dealing with stress. The units of stress are force per area measurements. In U.S. customary units, the units of stress are pounds per square inch (lbs/in^2), which is often abbreviated *psi.* You also commonly see stress expressed as *ksi,* which is short for *kip per square inch.* You may recall that 1 kip equals 1,000 pounds, so 1 kip per square inch equals 1,000 pounds per square inch. You can use either ksi or psi as long as you're consistent with your units. In SI units, stress is often measured in meganewton (or 1.0 x 10^6 Newton) per square meter (MN/m^2), which is actually a megapascal (or MPa for short).

SI units can sometimes actually present a bit of a dilemma. In engineering, most objects that you encounter aren't measured on cross sections that are accurately measured in square meters. In fact, most practical engineering examples that use SI units are usually measured in centimeters or millimeters, so you often have to make a few unit conversions along the way when you're working with SI units.

Remaining Steady with Average Stress

The first type of stress that you encounter in the world of mechanics of materials is the average stress. As I note earlier in the chapter, the *average stress* takes the entire force on a cross section and distributes it evenly across the entire cross-sectional area. Basically, you're taking an average value of the force over the entire cross section. Average stresses are pretty straightforward to calculate. The only real trick is determining the cross-sectional area and choosing the appropriate internal force to use in the equation. The following sections dive into the details of working with average stress.

Computing average normal stress for axial loads

The average normal stress is most commonly computed for members subjected to axial tension or compression (both of which are *axial forces*). The average normal stress for a bar subjected to simple axial tension or axial compression (such as the one shown in Figure 6-1) is computed as

$$\sigma_{AVG} = \frac{F_{INT}}{A}$$

where F_{INT} is the magnitude of the internal force acting normal to the cross-sectional area as determined from statics, and A is the calculated cross-sectional area on which the internal force F_{INT} is acting. In these calculations, a tensile force is a positive F_{INT}, and a compressive force is a negative F_{INT}.

This calculation results in an even (or *uniform*) distribution of stress over the entire region. However, this equation works only as long as the internal force is both acting at the centroid of the cross section and oriented in a direction that's normal to the cross-sectional area.

Consider a round bar with a diameter of 100 millimeters and subjected to an axial tension of 30 kilo-Newton as shown in Figure 6-3a.

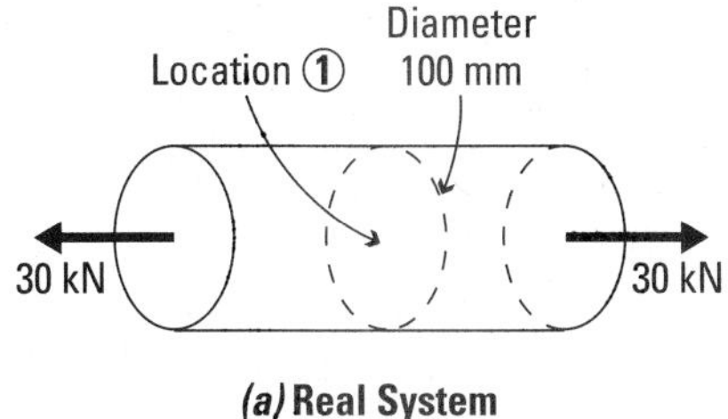

(a) Real System

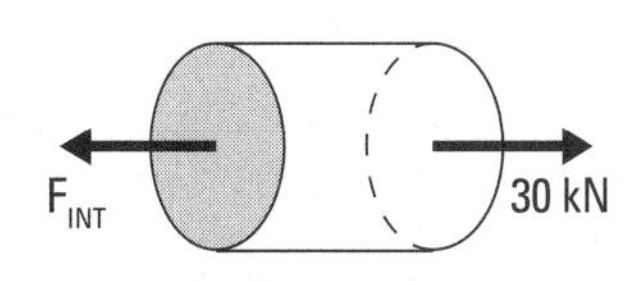

(b) F.B.D. at Location ①

Figure 6-3: Average normal stress example.

1. **Determine the internal axial force F_{INT} acting on the cross section of interest.**

   You find F_{INT} by slicing the bar at the cross section of interest and applying the equations of equilibrium as I show in Chapter 3. From statics, you can find that the internal axial force on the bar of this example is 30 kilo-Newton acting on the shaded cross section.

2. **Compute the cross-sectional area on which the force F_{INT} is acting.**

   The cross-sectional area for this example is the shaded region shown in Figure 6-3b. You compute the area of the round bar as follows:

$$A = \frac{\pi}{4}(d)^2 = \frac{\pi}{4}(100\text{ mm})^2 = 7{,}854\text{ mm}^2$$

3. **Compute the average normal stress on the cross section.**

   To calculate the average stress, you simply use the basic formula earlier in this section. Remember that the cross-sectional area that you computed in the previous step was measured in millimeters, so you need to convert those units to meters:

$$\sigma_{AVG} = \frac{F_{INT}}{A} = \frac{(+30\text{ kN})}{(7{,}854\text{ mm}^2)}\left(\frac{1{,}000\text{ mm}}{1\text{ m}}\right)^2 = +3{,}820\text{ kPa} = 3.82\text{ MPa (T)}$$

Determining average shear stress

The average shear stress, τ_{AVG}, is another stress you encounter regularly, usually when a shear force is acting on a relatively small cross-sectional area, such as a bolt, pin, or other thin object. You compute the average shear stress acting on cross-sectional area A due to an internal shear force V_{INT} as

$$\tau_{AVG} = \frac{V_{INT}}{A}$$

Just as with the average normal stress calculations earlier in the chapter, you compute the cross-sectional area at the location of interest and determine the internal shear force V_{INT} acting on that cross section. Beyond that, the average shear stress calculations should look very familiar to the average normal stress calculations. The only real difference is that for average shear stress, the internal force is now a force acting within the plane of the cross section instead of perpendicularly.

Consider the 20-millimeter (or 0.02 meter) diameter shaft subjected to the 200-Newton applied external loads shown in Figure 6-4a. Because these two applied loads are acting in opposite directions, they create a shear effect across the shaft (such as at Location 4).

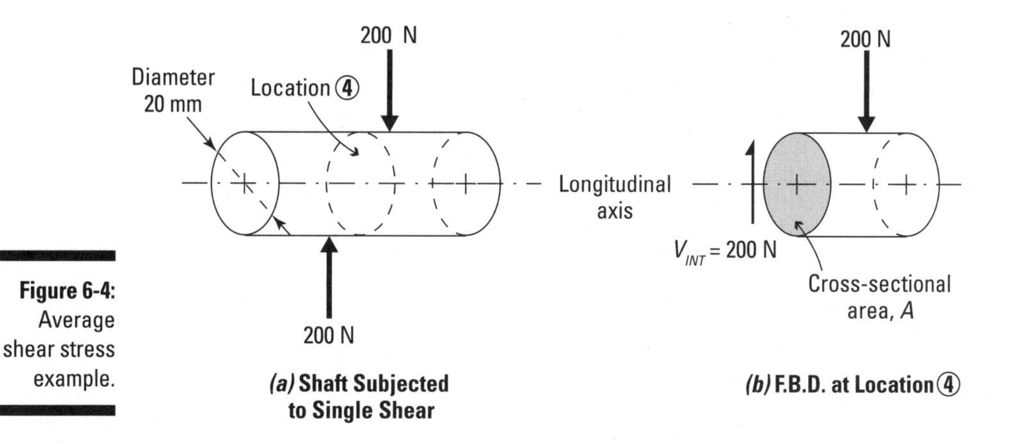

Figure 6-4: Average shear stress example.

Don't (normal) stress bending stress calculations

Unlike the average normal stresses at all points on the cross section produced by axial load, the normal stresses caused by a bending moment (see Chapter 9) vary from one position in the cross section to another in a direction that is normal (or perpendicular) to the cross section.That's why they're classified as normal stresses. However, normal stresses due to bending often actually change signs as well, meaning that at one edge of the member (such as the top edge in this figure), the normal stresses can be compressive (or negative) and at the opposite edge (the figure's bottom), they can be tensile (or positive).

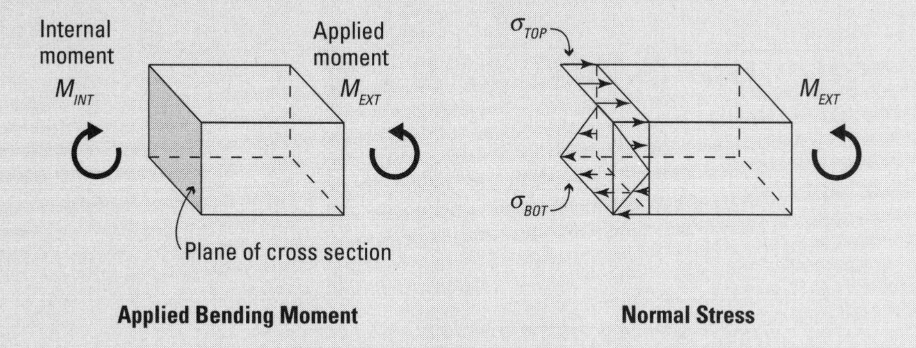

Follow these steps to calculate average shear stress:

1. **Determine the magnitude of the internal shear force V_{INT} acting in the plane of the cross section.**

   By slicing the shaft at Location 4 (see Figure 6-4b) and examining the free-body diagram, you can compute the internal shear force from equilibrium as

   $$+\uparrow\sum F_y = 0 \Rightarrow V_{INT} - 200\text{ N} = 0 \Rightarrow V_{INT} = 200\text{ N}$$

2. **Calculate the gross cross-sectional area on which the internal shear force is acting.**

   Cylindrical shafts have a circular cross section when a plane is oriented perpendicularly to the longitudinal axis (as in this example). Use the following formula to calculate that cross section:

   $$A = \frac{\pi}{4}(d)^2 = \frac{\pi}{4}(0.02\text{ m})^2 = 0.000314\text{ m}^2$$

3. **Compute the average shear stress at the point of interest.**

   In this case, you want the average shear stress across the shaft at Location 4. Here's the calculation:

   $$\tau_{AVG} = \frac{200\text{ N}}{0.000314\text{ m}^2} = 636{,}942\text{ Pa} = 0.637\text{ MPa}$$

Developing Stress at a Point

Although using the average stress is acceptable in certain applications (such as the ones I describe earlier in this chapter), many times you must be able to determine stresses acting on a single point. The following sections show you how.

Deriving stresses at a single point by using force components

To determine the internal forces acting at a single point (such as Point 1 in Figure 6-5), you need to cut the object with the three Cartesian cutting planes (XY, YZ, and XZ) to identify all the force components acting in that direction. For the x-direction, simply slice the three-dimensional object with a YZ cutting plane.

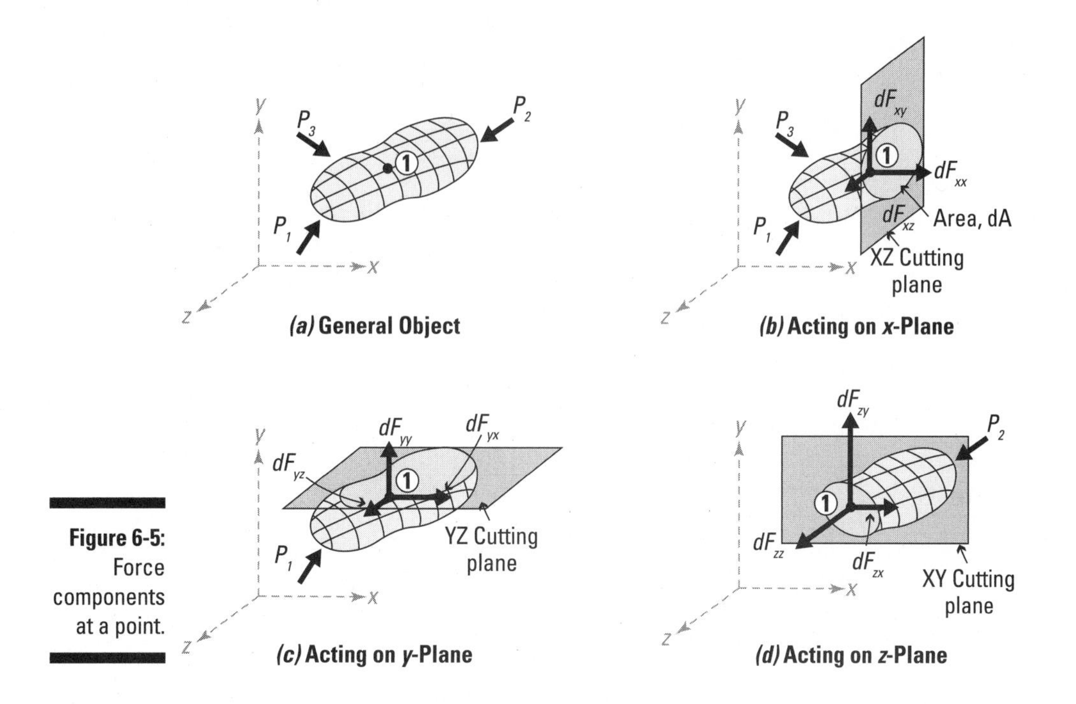

Figure 6-5: Force components at a point.

Next, you can determine the internal forces acting in each of the Cartesian *x*-, *y*-, and *z*-directions from the equilibrium equations of statics (turn to Chapter 3 if you need a quick reminder).

Labeling force components and stresses at single points

Try this simple sign convention to help you keep the labeling straight during your derivation. For an arbitrary incremental force dF, I like to add a set of subscripts to help remind me of the directions and planes that I'm dealing with: dF_{ij}, where i represents the orientation of the cutting plane that the internal force component is acting upon (such as the *x*-, *y*-, and *z*-cutting plane), and j represents the direction of the internal force component (in the *x*-, *y*-, and *z*-directions).

Exposing force components and stresses with Cartesian planes

For the *x*-cutting plane, you express the incremental forces acting on a differential area dA_x at Point 1 as dF_{xx}, which is acting in the +*x*-direction. Notice that the force dF_{xx} is acting perpendicular to the differential area, so this force component creates a normal stress. Conversely, dF_{xy}, which is acting in the +*y*-direction, and dF_{xz}, which is acting in the +*z*-direction, are acting in the *x*-plane of the differential area, which means that these two forces create an average shear stress on this area.

Rearranging the basic formula for average normal and shear stress, you can now compute the state of stress acting on the incremental area dA_x for the *x*-cutting plane (shown in Figure 6-5b):

$$dF_{xx} = \sigma_{xx} \cdot dA_x \qquad dF_{xy} = \tau_{xy} \cdot dA_x \qquad dF_{xz} = \tau_{xz} \cdot dA_x$$

Repeating this process for a *y*-cutting plane (which is the XZ plane) on the same object as shown in Figure 6-5c, you can determine the stress components that are acting on that plane as well:

$$dF_{yx} = \tau_{yx} \cdot dA_y \qquad dF_{yy} = \sigma_{yy} \cdot dA_y \qquad dF_{yz} = \tau_{yz} \cdot dA_y$$

And finally, for a *z*-cutting plane (which is the XY plane) as shown in Figure 6-5d, you can relate the incremental forces to the average stresses as

$$dF_{zx} = \tau_{zx} \cdot dA_z \qquad dF_{zy} = \tau_{zy} \cdot dA_z \qquad dF_{zz} = \sigma_{zz} \cdot dA_z$$

From this derivation, you can see that to fully define the state of stress at a single point in three dimensions, you must define a total of nine different stress components, which I show in Figure 6-6. You need information about three normal stresses and six shear stresses.

Figure 6-6 shows only the stresses acting on the faces in the positive edges of the element. However, for this element to be in equilibrium, another set of stresses is actually acting on the negative face of the element. These additional stresses have the same magnitude but opposite sense and direction in order to balance the stresses of the positive face shown.

With a few basic assumptions and equilibrium requirements from statics, you can simplify this process even further, as I show in the following section.

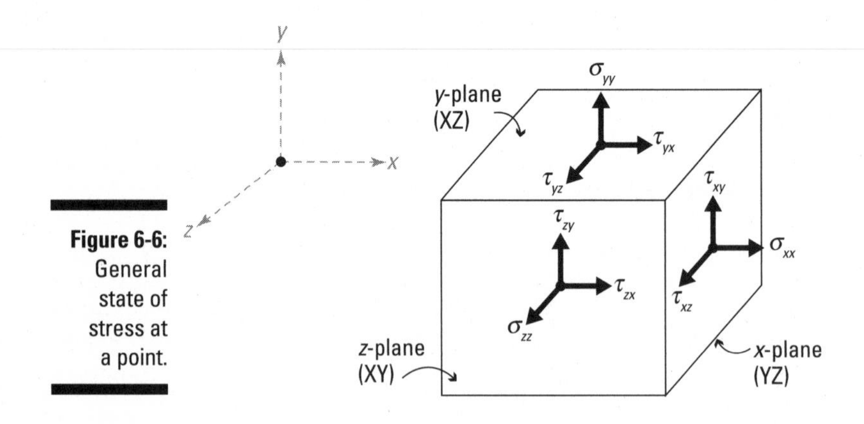

Figure 6-6: General state of stress at a point.

Looking at useful shear stress identities for stress at a point

To help simplify your gathering of the nine pieces of info you need to define stresses at a single point (see the preceding section), consider a small cube element within the object, having dimensions of Δx, Δy, and Δz as shown in Figure 6-7. For the stress element, I define the positive faces as follows:

- XY plane in the front face of the cube
- XZ plane on the top face of the cube
- YZ plane on the right side of the cube

I've indicated these faces on the cube of Figure 6-7a with the + sign.

If you look at just the XY plane (or the front face of the element) and superimpose the stresses of Figure 6-6, you notice that the element is actually unbalanced as shown in Figure 6-7b. (***Note:*** I've neglected the *z*-direction stresses because they're out of the plane of the element.)

To balance this element (shown in Figure 6-7c), you actually need to include additional forces on the left and bottom edges. If you pull the element on the right edge with a force component of $F_{xx} = \sigma_{xx}A_x$, you need an equivalent force component pulling on the left edge to balance the element. If the area A_x has the same dimensions, the stress σ_{xx} must be the same magnitude on the left side. Repeating this process for τ_{xy} acting upward on the right edge, you soon see that you need a second τ_{xy} acting on the left edge to balance the translational equilibrium component. Similarly, you need a σ_{yy} on the bottom (acting downward) to balance the σ_{yy} on the top and a τ_{yx} on the bottom acting to the left to balance the τ_{yx} on the top edge.

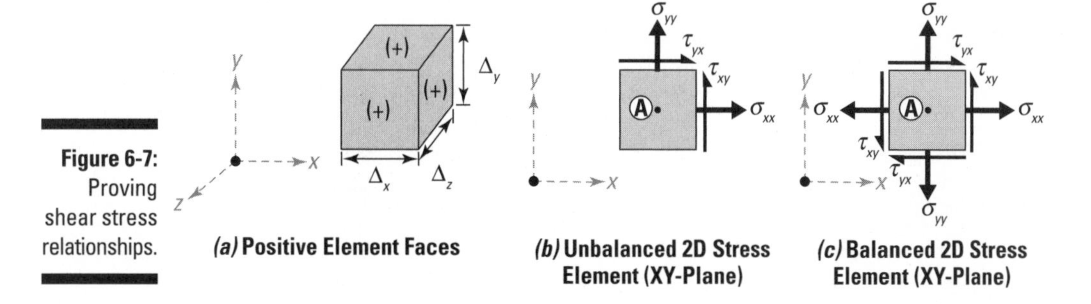

Figure 6-7: Proving shear stress relationships.

You can extend this logic to three dimensions to include all those stresses with a z in the subscript, but for the sake of this discussion, I am dealing with only two dimensions — the x- and y-directions.

You've established all these normal stresses based on providing translational equilibrium of the force components. But that's only one part of the battle — you also need to satisfy rotational equilibrium.

Equilibrium equations only work with forces, so you need to convert those stresses to forces by multiplying them by the areas they're acting on.

The normal stresses are all *concentric* (meaning they pass through the same point) through the middle of the element (Point A). However, the shear stresses are acting *eccentrically* at a given distance, which means they cause a rotational effect around that point. Writing the equation for rotational equilibrium at Point A,

$$\begin{aligned}\sum M_A &= \left(\tau_{xy}\cdot\Delta z\cdot\Delta y\right)\cdot\left(\frac{\Delta x}{2}\right)+\left(\tau_{xy}\cdot\Delta z\cdot\Delta y\right)\cdot\left(\frac{\Delta x}{2}\right)\\ &\quad-\left(\tau_{yx}\cdot\Delta z\cdot\Delta x\right)\cdot\left(\frac{\Delta y}{2}\right)-\left(\tau_{yx}\cdot\Delta z\cdot\Delta x\right)\cdot\left(\frac{\Delta y}{2}\right)=0\\ &\Rightarrow \tau_{xy}=\tau_{yx}\end{aligned}$$

This equation means that the force component from the vertical shear stress on the right edge is balanced by the force component from the horizontal shear stress on the top edge. Because you've assumed that the faces of the element all have the same area, these stresses must have the same magnitude. These shear stresses are also known as *complementary shear stresses* because they have equal values but act in different directions. The force components from these complementary shear stresses are what provide rotational equilibrium for the stress element.

This calculation also indicates a very important principle: Loads that cause shear stresses in one direction also create shear stresses in perpendicular directions at the same time.

If you repeat this process for the other shear stresses by using two-dimensional elements in the XZ plane and the YZ plane to establish your rotational equilibrium equations, you can also show that

$$\tau_{xz}=\tau_{zx} \qquad \tau_{yz}=\tau_{zy}$$

This setup actually means that for a three-dimensional state of stress at a point, you only have to compute a total of six stresses, three of which are the normal stresses (σ_{xx}, σ_{yy}, and σ_{zz}), and the other three are the shear stresses (τ_{xy}, τ_{yz}, and τ_{xz}).

Containing Plane Stress

The concept of plane stress is a very basic idea, but it provides the foundation for many of the equations and basic ideas of early mechanics and materials. *Plane stress* is a state of stress at a point in which the normal stress and shear stresses in one particular direction are assumed to be zero. By eliminating the stresses in one direction, the remaining stresses are all contained in a common plane. Figure 6-8 shows a plane stress condition for the XY plane.

You can have a plane stress situation on any Cartesian plane of the general stress diagram shown in Figure 6-8. In that figure, I show an example for the XY plane only. However, the YZ and the XZ planes can also be in a state of plane stress, depending on how the object is loaded.

In effect, you can use the concept of plane stress to recognize that a three-dimensional stress problem can be treated as a two-dimensional stress problem under the right circumstances. For example, if you're dealing with a plane stress problem with an incremental area contained in the XY plane, any stress term in the equations in this section that contains a z in either subscript is automatically equal to zero.

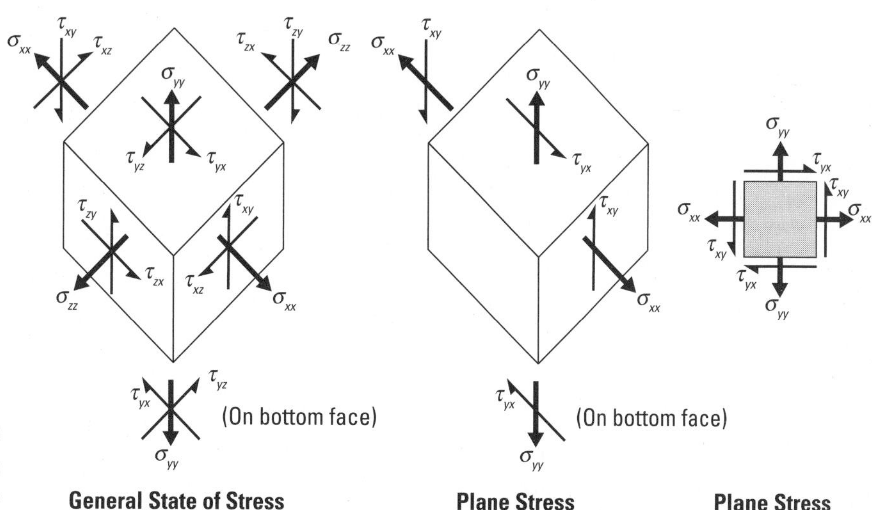

Figure 6-8: Plane stress diagrams.

You often assume a plane stress condition when you're dealing with problems involving thin or flat objects. Following are a couple of problem types where plane stress assumptions may be applicable:

- **Thin-walled pressure vessels**: *Thin-walled pressure vessels* are hollow containers or shell structures designed to hold fluids or gases, typically under pressure. An example of a pressure vessel is a propane gas tank or a scuba diver's oxygen tank. The wall thickness of a pressure vessel is typically significantly smaller than the other dimensions. By having a thin wall, shear stresses across the thickness are often neglected (assumed to be zero), which results in a state of plane stress being established in the wall of the vessel. I talk more about thin-walled pressure vessels in Chapter 8.
- **Beam problems:** Often, many problems involving the analysis of stress in beams can be simplified by using plane stress assumptions. I explain these problems in more detail in Chapter 9.

Chapter 7

More than Meets the Eye: Transforming Stresses

In This Chapter

- Understanding stress elements and sign conventions
- Transforming stresses
- Depicting transformed stresses graphically
- Dealing with principal stresses
- Using Mohr's circle for plane stress analysis in two and three dimensions

After you determine the stresses that loads cause on an object, you're one step closer to being able to perform engineering design. One of the main concerns in any design is that a member you select be able to support the desired loads while performing as it's intended. To verify this condition, you need to know the maximum stress the loads cause and to ensure that these maximum values are less than what the material of the object is able to support. In practice, the direct stresses from loads aren't necessarily the maximum stresses that an object feels.

In this chapter, I show you how to transform a set of stresses to determine these maximum stresses (as well as their orientations) by using stress transformation equations. You also discover how to use a graphical representation known as Mohr's circle for stress, which is one of the most fundamental (and most widely used) methods for performing basic stress transformations. Finally, I show how you can efficiently report the stresses' values through basic sketches.

Preparing to Work with Stresses

Before you can start to transform stresses, you need to understand a bit of common terminology and have a grasp of a basic sign convention for normal and shear stresses. In this section, I give you the rundown of these basics.

Building a stress block diagram

Statics analysis uses basic sketches called *free-body diagrams* (F.B.D.s) to help illustrate the forces (and their locations) that act on an object. With a properly constructed F.B.D., you can then apply the equations of equilibrium from Chapter 3 to determine the internal loads that act on the object.

Although you do work with free-body diagrams in mechanics of materials to determine internal loads, you're actually more interested at this point in the stresses in the object that result from those loads. That's where a stress block diagram (or *stress element*) comes in. The *stress block diagram* is a free-body diagram of sorts, except instead of showing forces as you do in a statics F.B.D., you actually indicate the magnitude and direction of normal and shear stresses on the Cartesian *x*-, *y*-, and *z*-planes (see Chapter 6 for more info on these basic types of stress). This diagram depicts all the stresses acting at a single point.

To draw a stress block diagram, you first sketch the basic element shape, which is a usually a cube with a width of dx, a height of dy, and a depth of dz.

For plane stress situations (where all stresses with respect to one of the Cartesian directions must be zero), you may find drawing a stress element as a simple square shape more convenient. This element is still actually a three-dimensional cube, but for in-plane stress problems, you're only looking at one of the cube's faces.

Identifying basic states of stress

Another term that you encounter quite regularly in mechanics of materials is *state of stress*. This term refers to the combination of normal and shear stresses that define all the stresses that act at a point within the object. You need to define three normal stresses and three shear stresses to fully describe the state of stress at a point. Figure 7-1 shows several common states of stress.

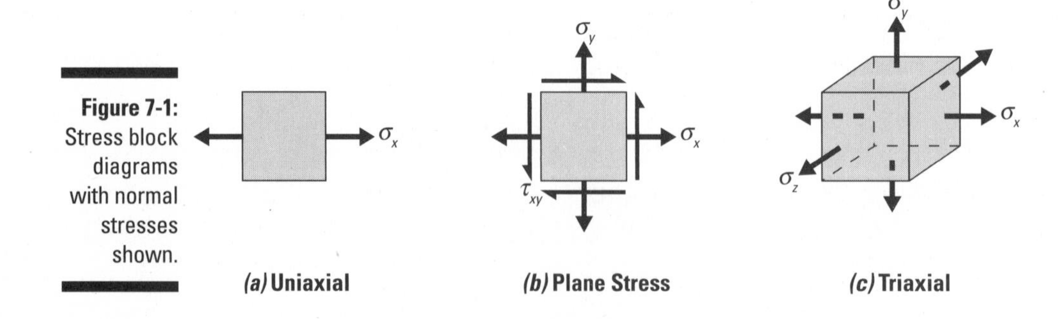

Figure 7-1: Stress block diagrams with normal stresses shown.

Before you can begin to work with stresses (see Chapter 6), you need to be able to identify the state of stress situation that you have on your hands. Determining the state of stress puts you one step closer to being able to choose the proper analysis equations and begin your work with mechanics of materials.

An infinite number of stress states are possible on a stress element. However, three states of stress appear frequently in mechanics of materials:

- **Uniaxial:** A *uniaxial state of stress* (see Figure 7-1a) illustrates an object that is subjected to a stress in a single direction. You can encounter uniaxial stress conditions in objects such as ropes or simple columns.
- **Plane stress:** A *plane stress state* (see Figure 7-1b) is a stress element that is subjected to stresses on no more than two Cartesian planes of the stress element. For an element of a plane stress state, you have at most two normal stresses and one shear stress (remember that horizontal and vertical shear stresses on an element must be the same magnitude).

  For a plane stress element (such as the one shown in Figure 7-1b) you indicate the stresses by referring to the horizontal surfaces (at the top and bottom of the element) and the vertical surfaces (on the left and right edge of the element). On each of these surfaces you can have a normal stress (denoted by the σ) and a shear stress (denoted by the τ).
- **Triaxial:** In a *triaxial state of stress* (see Figure 7-1c), the stress element is subjected to three normal stresses — one acting on each of the faces of a stress element — and at the same time has no shear stress. The triaxial state of stress is very important in the formulation of failure theories in advanced mechanics of materials.

Establishing a sign convention for stresses

Perhaps one of the most confusing aspects of mechanics of materials, particularly when you're working with stresses, is dealing with the signs of the stresses you're working with. Developing a basic, consistent sign convention can help.

Decoding sign conventions can be complicated because many reference books use custom conventions that are inconsistent. One book may consider a stress in tension to be positive, while other books base the signs on the direction of the normal stresses on a particular face (for example, upward on top of the element is positive). And shear stress conventions are a completely different matter altogether.

For plane stress elements in this book, I use the sign conventions for normal and shear stresses shown in Figure 7-2. The following sections delve further into the sign conventions for normal and shear stresses.

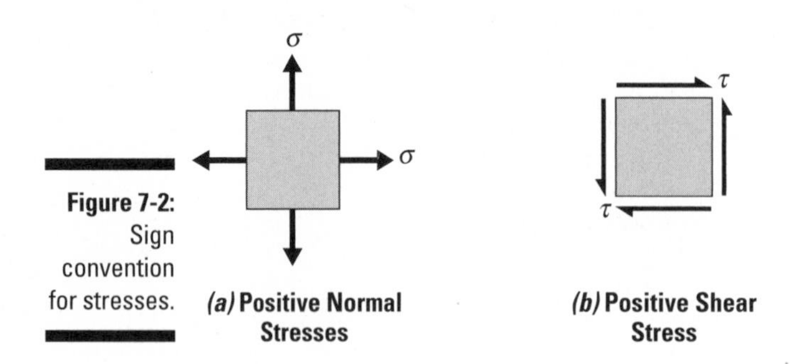

Figure 7-2: Sign convention for stresses.

Normal stresses

Normal stresses are considered positive if their equivalent force (or the stress multiplied by the area on which they're acting) results in a pulling action (or tension) on its respective face of the element. Normal stresses always act perpendicular to the face of an element.

Regardless of which face the normal stress is acting on, I consider a normal stress positive if its arrow acts perpendicular to a face (or edge) in a direction that is away (or outward) from the face of the element as shown in Figure 7-2a.

Shear stresses

Shear stresses are a little unusual and often cause students trouble, because on any given plane stress element, the shear stresses on opposite edges of a stress element are acting in opposite directions simultaneously. Shear stresses always act within the plane of the face of a stress element.

For a typical three-dimensional stress element, two shear stresses act in each face and in different Cartesian directions. For example, for a face in the XY plane (or the *z*-face), the first shear stress τ_{zx} is oriented in the *x*-direction, and the second shear stress τ_{zy} is oriented in the *y*-direction. These stress values may be either positive or negative depending on the sign convention you choose to work with.

For a stress element in the XY plane, I show a positive convention in Figure 7-2b. The second shear stress on each face is zero.

Because of the confusion of the signs that surround shear stresses, I use a logical and easy-to-remember method for determining the signs of the shear stresses acting on a plane stress element. To explain this sign convention for positive shear stresses, consider the stress element shown in Figure 7-2b. This simple element has shear stresses acting in all directions (one per side, of course). To help me remember which way should be considered positive, I break the shear stress diagram into two smaller diagrams, as shown in Figure 7-3.

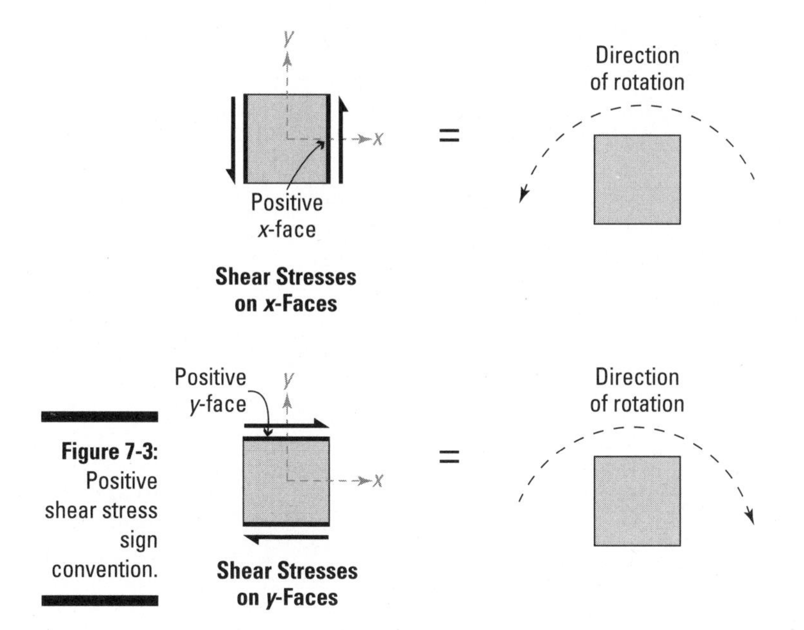

Figure 7-3: Positive shear stress sign convention.

The first diagram shows the shear stresses acting on the *x*-faces of an element that is situated in the XY plane. In this example, the *x*-faces are the vertical edges on the left and right side of the element. Notice that the right side is acting in an upward direction and the left side is acting in a downward direction. If you convert these stresses to forces, together they create a couple (or moment) that causes the element to rotate in a counterclockwise direction. This setup represents a positive shear stress orientation.

When the faces of an element are oriented horizontally and vertically, this procedure works very well because the positive *x*-face is easy to find (it's the one on the right side of the element). The vertical face that you should use as a reference is the face of the element that intersects the *x*-axis and is located toward the most-positive end of the axis.

However, for elements that are rotated at some other orientation, the positive *x*-face isn't quite so obvious. If you draw the reference *x*-axis and rotate it about its origin by the same amount and in the same direction as the angle you are rotating the object, the "vertical" face becomes the face that intersects the rotated *x*-axis and is located toward the most-positive end of the rotated axis. The faces that don't intersect the rotated *x*-axis become the "horizontal" faces.

As Chapter 6 indicates, to balance the shear stresses on vertical faces — and more specifically, their tendency to rotate the element in a counterclockwise direction — the shear stresses on horizontal faces must act together to rotate the element in the opposite (or clockwise) direction. For this example, the shear stress on the positive *y*-face (or the top horizontal face) of the element is acting to the right, and the corresponding shear stress on the bottom horizontal face is acting to the left. Together, the forces from these horizontal stresses create a couple that balances the forces from the stresses on the vertical faces.

The reason I consider the "vertical" faces to help establish the positive sign convention for shear stresses is because the couple they create wants to produce a counterclockwise rotation. If you remember the right-hand rule for rotation that I describe in Chapter 3, a rotation about the *z*-axis is positive if it acts counterclockwise when you're looking toward the origin from the end of the positive *z*-axis. In this example, the *z*-axis is perpendicular to the figure (or out of the page).

Stress Transformation: Finding Stresses at a Specified Angle for One Dimension

Manufacturers often craft objects such that multiple pieces of material are spliced together with glue or weld material. Materials such as wood have fibers oriented in specific directions depending on how the pieces are fabricated or cut, and these materials often behave very differently under a tension load as opposed to a compression load with respect to that orientation.

In design, you often need to determine the state of stress at a point and orientation angle other than what may be shown on your basic stress elements. The state of stress on an inclined plane is dependent on only two variables: the normal stress of the original orientation and the orientation angle. Finding these stresses at unique orientations is known as *stress transformation.*

A common type of connection is called the *scarf splice,* in which the end of a material is *coped* (or cut at an angle) and connected to a matching piece with an adhesive such as glue. To design this connection, you need to be able to determine the state of stress along the inclined plane of the splice.

The simplest stress transformation you work with is for a uniaxially loaded member with an applied force P such as the one shown in Figure 7-4a. In this figure, if a tension member is sliced at an angle θ as shown, two internal forces must develop along the inclined surface to keep the object in equilibrium. The first is a normal force N that acts perpendicular to the cut plane (assumed as a tension force in this example), and the second is a shear force V that acts parallel to the cut plane (which is assumed up the shear plane for this example).

Remember, you can only write an equilibrium equation for a free-body diagram that contains forces. If you're working with the stresses on a stress element, you must first convert those stresses to forces before you can use static equilibrium equations.

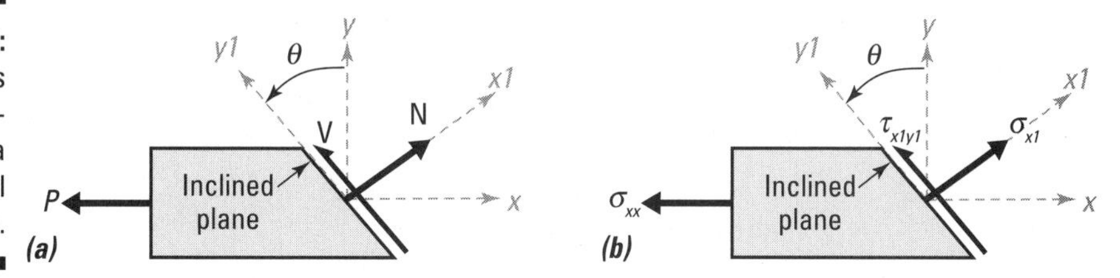

Figure 7-4: Stress transformation for a uniaxial stress state.

To calculate a transformed stress, you just need to know a couple of factors:

- **Orientation angle**: The *orientation angle* (θ) defines the orientation of the inclined or sloped surface and is measured as positive in a counterclockwise direction from the vertical *y*-axis. This angle is also the same angle from a horizontal reference (such as an *x*-axis) to an axis that is perpendicular to the inclined plane.
- **Cross-sectional area:** The *cross-sectional area* (A) is the area of the member that is oriented perpendicular to the longitudinal axis of the member. It's *not* the same as the area of the inclined plane.

However, the area of the sloped surface A_o is necessary in order to compute the average stresses and can be related to the cross-sectional area A of the member by the relationship

$$A_o = \frac{A}{\cos\theta}$$

The orientation angle θ is measured with respect to the vertical *y*-axis (or more specifically, the plane of the cross-sectional area A and not the slope of the inclined plane).

With this area now determined, you can then use the following formula to compute the corresponding *transformed normal stress* (σ_{x1}) — or the average normal stress acting on that inclined plane:

$$\sigma_{x1} = \frac{N}{A_o} = \frac{P\cos\theta}{\left(\frac{A}{\cos\theta}\right)} = \left(\frac{P}{A}\right)\cdot\cos^2\theta = \sigma_{xx}\cdot\cos^2\theta$$

The $x1$-axis is the axis that acts perpendicular to the slope of the interface, or the inclined cross-sectional area A_o. From geometry, you can then show that this $x1$-axis is also oriented an angle θ from the original x-axis. Likewise, the $y1$-axis is oriented parallel to the slope of the inclined plane, so it also creates an angle θ from the vertical y-axis.

The average shear stress on the inclined surface is given by

$$\tau_{x1y1} = \frac{V}{A_o} = \frac{P\sin\theta}{\left(\frac{A}{\cos\theta}\right)} = \left(\frac{P}{A}\right)\cdot\cos\theta\sin\theta = \sigma_{xx}\cdot\cos\theta\sin\theta$$

These equations illustrate my assertion earlier in the section that the two variables impacting the state of stress on an inclined plane are the normal stress of the original orientation (σ_{xx}) and the orientation angle (θ) of the inclined surface. I show both of these stresses in Figure 7-4b.

For example, if you know that the average normal stress for a bar is 10 ksi (C) — remember that the (C) means it's a compressive stress and therefore a negative value — and the orientation angle of a splice is 50 degrees counterclockwise from the cross-sectional area, you can compute the state of stress along that splice fairly simply:

$$\sigma_{x1} = (-10 \text{ ksi})\cdot(\cos 50°) = -4.13 \text{ ksi} = 4.13 \text{ ksi (C)}$$

$$\tau_{x1y1} = (-10 \text{ ksi})\cdot(\cos 50°)(\sin 50°) = -4.92 \text{ ksi}$$

Extending Stress Transformations to Plane Stress Conditions

Most objects and load scenarios that you cover are much more complex than the uniaxial loads in the preceding section, so you need to have a handy way to deal with stresses occurring in more than one direction, a situation known as a plane stress state.

Figure 7-5a shows a generic plane stress element subjected to two positive normal stresses σ_{xx} and σ_{yy} and a positive shear stress τ_{xy}. If you slice across this element at an angle θ from the vertical reference, you can see that the horizontal normal stress σ_{xx} contributes to the stresses on the inclined plane the same as the uniaxial case shown in Figure 7-4b in the preceding section. In addition, σ_{yy} becomes a second uniaxial condition that can add its effects to the stresses on the inclined plane. The difference is that the angle for the uniaxial state including σ_{yy} isn't the same angle (θ); instead, it's actually ($90° - \theta$).

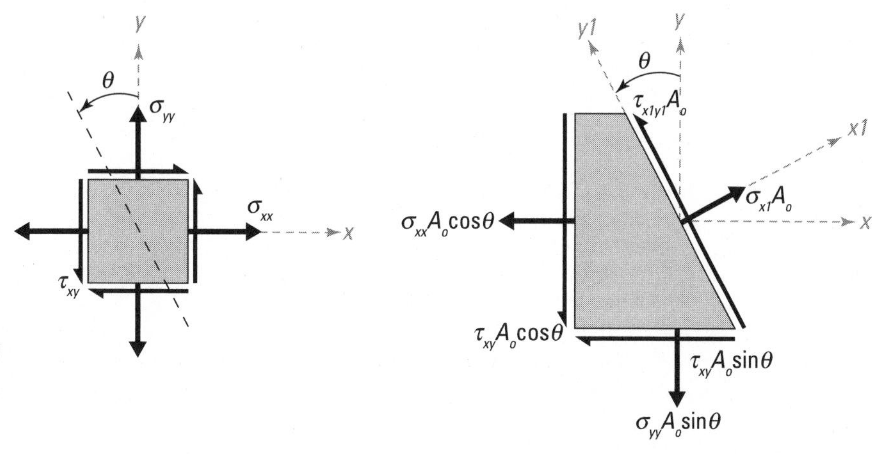

Figure 7-5: Plane-stress transformations.

If A_o is the area of the exposed inclined plane, you can apply the rules of equilibrium much like you do for uniaxial load cases. Using the free-body diagram shown in Figure 7-5b, you can then sum forces along the *x1*- and *y1*-axes.

$$\nearrow^{+}\sum F_{x1} = \sigma_{x1}A_o - \sigma_{xx}A_o\cos^2\theta - \tau_{xy}A_o\sin\theta\cos\theta - \tau_{xy}A_o\cos\theta\sin\theta - \sigma_{yy}A_o\sin^2\theta = 0$$

$$\Rightarrow \sigma_{x1} = \sigma_{xx}\cos^2\theta + \sigma_{yy}\sin^2\theta + 2\tau_{xy}\sin\theta\cos\theta$$

$$\nwarrow^{+}\sum F_{y1} = \tau_{x1y1}A_o + \sigma_{xx}A_o\cos\theta\sin\theta - \tau_{xy}A_o\cos^2\theta + \tau_{xy}A_o\sin^2\theta - \sigma_{yy}A_o\sin\theta\cos\theta = 0$$

$$\Rightarrow \tau_{x1y1} = -\left(\sigma_{xx} - \sigma_{yy}\right)\cos\theta\sin\theta + \tau_{xy}\left(\cos^2\theta - \sin^2\theta\right)$$

Just as with uniaxial cases, the transformed stresses on an inclined plane for plane stress problems become only a function of the original state of stress (σ_{xx}, σ_{yy}, and τ_{xy}) as well as the orientation angle (θ) of the inclined plane. Once again, the transformed stresses become completely independent of the cross-sectional area of the inclined plane A_o.

Most classic textbooks then apply a couple of algebraic substitutions and trigonometric identities to simplify the equations a bit further to look something like the following:

$$\sigma_{x1} = \frac{\sigma_{xx} + \sigma_{yy}}{2} + \frac{\sigma_{xx} - \sigma_{yy}}{2} \cos 2\theta + \tau_{xy} \sin 2\theta$$

$$\tau_{x1y1} = -\frac{\sigma_{xx} - \sigma_{yy}}{2} \sin 2\theta + \tau_{xy} \cos 2\theta$$

Regardless of which formula form you use, you can now quickly and easily transform the state of stresses from one orientation to another by simply plugging in the current state of stress and the orientation angle of the new inclined plane.

To determine the transformed normal stress that is oriented perpendicular to the $x1$-axis (which happens to be along the $y1$-axis) you can compute the normal stress by substituting $(\theta + 90°)$ for θ in the σ_{x1} equation — after all, the $y1$-axis is 90 degrees more in a counterclockwise direction than the rotated $x1$-axis. If you make this substitution, you can compute the normal stress along the $y1$-axis from the following:

$$\sigma_{y1} = \frac{\sigma_{xx} + \sigma_{yy}}{2} - \frac{\sigma_{xx} - \sigma_{yy}}{2} \cos 2\theta - \tau_{xy} \sin 2\theta$$

Consider the element shown in Figure 7-6, which is subjected to a stress in the x-direction of 10 ksi (C), a stress in the y-direction of 12 ksi (T), and a shear stress of –7 ksi. Suppose you want to determine the state of stress on a plane that is oriented 40° counterclockwise from the current orientation.

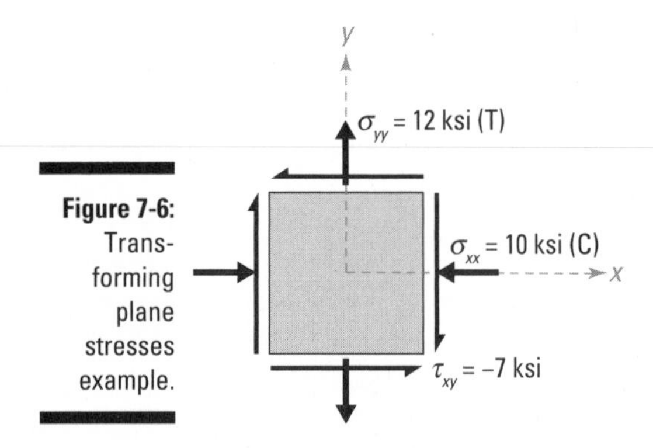

Figure 7-6: Transforming plane stresses example.

For this example, the current state of stresses and the orientation angle for the transformed plane are $\sigma_{xx} = -10$ ksi, $\sigma_{yy} = +12$ ksi, $\tau_{xy} = -7$ ksi, and $\theta = +40°$.

You can then calculate the transformed stresses on the new *x1*- and *y1*-axes as follows:

$$\sigma_{x1} = \frac{(-10+12)\text{ ksi}}{2} + \frac{(-10-(12))\text{ ksi}}{2}\cos(2(40°)) + (-7\text{ ksi})\sin(2(40°))$$

$$= -7.80\text{ ksi} = 7.80\text{ ksi (C)}$$

$$\sigma_{y1} = \frac{(-10+12)\text{ ksi}}{2} - \frac{(-10-(12))\text{ ksi}}{2}\cos(2(40°)) - (-7\text{ ksi})\sin(2(40°))$$

$$= +9.80\text{ ksi} = 9.80\text{ ksi (T)}$$

$$\tau_{x1y1} = -\frac{(-10-(12))\text{ ksi}}{2}\sin(2(40°)) + (-7\text{ ksi})\cos(2(40°)) = -12.05\text{ ksi}$$

These three stresses together define the transformed state of stress for an orientation of 40° counterclockwise from the original position.

Displaying the Effects of Transformed Stresses

Engineers and scientists are typically very visually oriented people and therefore like to see pictures of results (such as graphs, free-body diagrams, and other techy diagrams). To help illustrate your results, you can use two common types of diagrams to express the effects of transformed stresses: the stress wedge and the rotated stress element.

Wedging in on the action with stress wedges

The first method you can use to represent transformed stresses is with stress wedges. The *stress wedge* is useful for displaying the current state of stress while also displaying the normal and shear stress on the rotated plane. This basic technique is the method I use to derive the basic transformation stress equations earlier in the chapter.

I find stress wedges very useful when checking connection strengths on glue seams and welds on metal because you can easily align the inclined plane of the stress wedge to match the inclination angle of the connection or fiber plane.

Figure 7-7 shows the upper and lower stress wedge diagrams for Figure 7-6 in the preceding section.

To quickly sketch a lower stress wedge diagram, follow a few simple steps:

1. **Draw the original Cartesian reference axis with respect to the inclined plane.**

   In Figure 7-7a, the *x*-axis is oriented to the right, and the *y*-axis is oriented upward. (This setup follows the right-hand rule for Cartesian axes from Chapter 5.)

2. **Sketch the original *x*-direction normal stress σ_{xx} and the shear stress τ_{xy} on the vertical face of the stress wedge.**

   To sketch the lower stress wedge diagram for this example, I draw the original *x*-direction normal stresses on the left face. For a compressive stress in the *x*-direction, the arrow for the normal stress points into the block. The shear stress (which is a negative value) is applied upward on the left face per the sign convention that I lay out in "Establishing a sign convention for stresses" earlier in this chapter. This application causes the wedge element to want to rotate clockwise.

3. **Sketch the original *y*-direction normal stress σ_{yy} and the shear stress τ_{xy} on the horizontal face of the stress wedge.**

   The normal stress is a tension stress in this example, so the arrow for this stress pulls on the bottom horizontal face of the element. The shear stress on this element is negative, so it acts as a positive value on this face, which means that it acts horizontally to the right in order to rotate the wedge element counterclockwise.

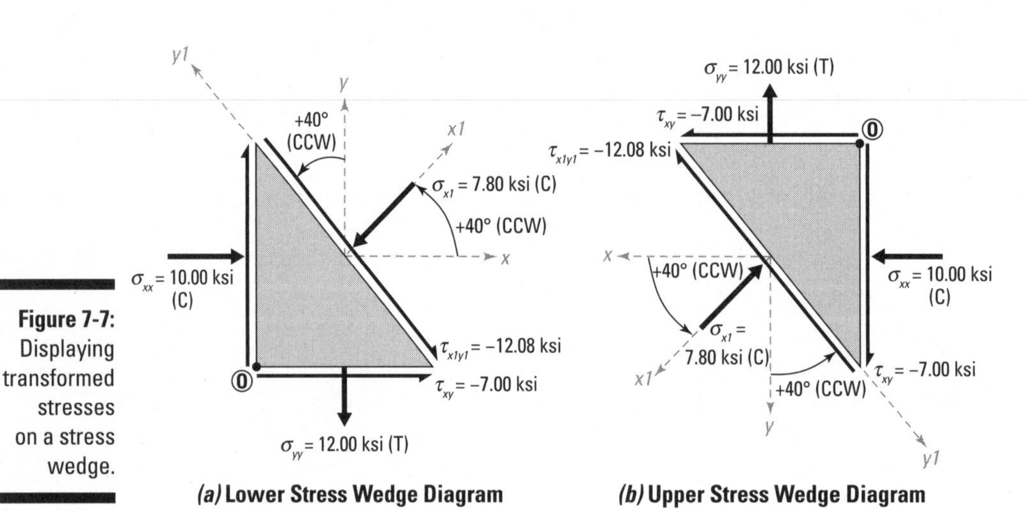

Figure 7-7: Displaying transformed stresses on a stress wedge.

4. **Draw the rotated $x1$-axis for the specified orientation angle θ outward and perpendicular (or normal) to the inclined plane.**

   In this example, the desired orientation angle is 40 degrees measured counterclockwise. Rotate the original x-axis by an angle of 40 degrees counterclockwise to locate the $x1$-axis, which is normal (or perpendicular) to the inclined plane. Likewise, rotate the original y-axis by an angle of 40 degrees counterclockwise to locate the $y1$-axis.

5. **Sketch the transformed normal stress σ_{x1} with respect to the $x1$-face and the transformed shear stress τ_{x1y1} with respect to the $y1$-axis.**

   You draw the transformed normal stress σ_{x1} parallel to the new $x1$-axis that you drew in Step 4. Because this stress was calculated to be a negative (or compressive) stress, the arrow along this line is pointing into the element. To sketch the transformed shear stress, choose an arbitrary point (such as the corner at Point O) and recall the basic sign convention for shear stresses (which I discuss in the earlier section "Establishing a sign convention for stresses"). Because the transformed shear stress is negative, it tends to rotate around Point O in a clockwise direction when it's applied to the inclined plane. For this reason, you sketch it on the inclined plane from upper left to lower right.

You can use a similar procedure to construct the upper stress wedge element shown in Figure 7-7b. The only difference is that your reference Cartesian axes are oriented in the opposite direction from the axes of the lower stress wedge element.

The stress wedge does a very good job of displaying information that acts parallel and normal to the inclined plane. However, notice that the σ_{y1} stress isn't shown on this element. The stress wedge only shows one transformed normal stress and one transformed shear stress at a time. The benefit of the stress wedge, however, is that it also displays the original state of stress prior to being transformed to the new orientation.

Rotating the basic stress element

The rotated stress element is another useful method for displaying transformed stresses and actually has a couple of additional benefits over the stress wedge I describe in the preceding section. *Rotated stress elements* simultaneously show the states of stress on two mutually perpendicular planes, which is very useful when you start calculating principal stress values. (I explain that topic more in the following section.) Figure 7-8 shows a rotated stress element for the same state of stress shown in Figure 7-7.

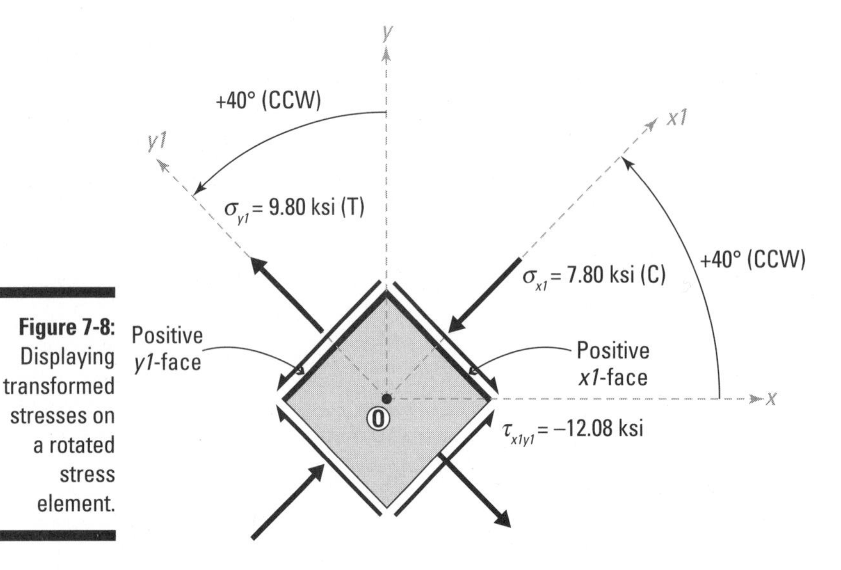

Figure 7-8: Displaying transformed stresses on a rotated stress element.

The biggest advantage of the rotated stress element is that you can display multiple stresses at the same time on an element. Unfortunately, though, you lose the original state of stress information from the figure. Constructing a rotated stress element is very similar to creating a stress wedge.

1. **Draw the original Cartesian axis with respect to the middle of an element.**
2. **Draw the rotated *x1*-axis for the specified orientation angle θ outward and normal to the exposed plane.**
3. **Draw the rotated *y1*-axis at an angle of θ from the original *y*-axis.**

   You can also locate this axis by rotating an angle $\theta + 90°$ from the original *x*-axis as well. Getting used to the *x*-axis version may help you when you deal with maximum and minimum stresses (see Chapter 8).
4. **Draw the rotated element such that the *x1*-face is perpendicular to the *x1*-axis and the *y1*-face is perpendicular to the *y1*-axis; complete the basic square shape of the plane stress element at this new rotated orientation.**
5. **Sketch the transformed normal stress σ_{x1} with respect to the *x1*-face and the transformed shear stress τ_{x1y1} perpendicular to the *y1*-axis.**

   You draw the transformed normal stress σ_{x1} parallel to the new *x1*-axis that you drew in Step 2. The stress of this example is a negative (or compressive) stress, so the arrow along this line must point into the element (acting normal to the *x1*-face).

To sketch the transformed shear stress acting on the $x1$-face, choose an arbitrary point (such as the center at Point O) and utilize the basic sign convention for shear stresses from earlier in this chapter. Because the transformed shear stress is negative, it should tend to rotate around Point O in a clockwise direction when it's applied to the inclined plane. For this reason, you sketch it on the inclined plane from upper left to lower right.

6. **Sketch the transformed normal stress σ_{y1} with respect to the $y1$-axis and the transformed shear stress $-\tau_{x1y1}$ perpendicular to the $y1$-axis.**

   You add the transformed normal stress σ_{y1} parallel to the new $y1$-axis that you drew in Step 3. This stress is a positive (or tensile) stress, which means the arrow along this line is pointing away from the $y1$-face. To sketch the transformed shear stress (acting on the $y1$-face), draw in the stress from the upper right to the lower left.

7. **Balance the remaining sides to ensure equilibrium.**

   After you draw the stresses on the positive $x1$- and $y1$-faces, you balance the opposite sides as I describe in Chapter 6.

When Transformed Stresses Aren't Big Enough: Principal Stresses

When engineers design a structural member to support a load, they must be sure to design for the worst case, which is typically for the maximum stress in an object. Even though this maximum stress may not be oriented at the same angle as the original stress element that you draw, you still must calculate these important values, known as the *principal stresses.*

Using maximum principal normal stresses as a failure criteria is only valid for certain materials such as ceramics. For other materials, such as metals, other established criteria may be more accurate. In fact, failure may not even be the result of peak (or maximum stresses) but rather the interaction between all stresses on the element. You can find more specifics about failure theories in most advanced mechanics of materials textbooks.

Regardless of the type of material you are working with, your ability to calculate the principal stresses and their angles is a fundamental skill in any basic mechanics of materials class.

Defining the principal normal stresses

The principal stresses represent the maximum and minimum states of stress for a given stress element. When calculating the principal stresses, you use the following formula:

$$\sigma_{P1,P2} = \frac{\sigma_{xx} + \sigma_{yy}}{2} \pm \sqrt{\left(\frac{\sigma_{xx} - \sigma_{yy}}{2}\right)^2 + \left(\tau_{xy}\right)^2}$$

When you evaluate this expression, you actually get two different values because of the ± in front of the second term. This operation is what determines the maximum and minimum value.

Consider an example where σ_{xx} = –10 ksi, σ_{yy} = +12 ksi, and τ_{xy} = –7 ksi. (If these figures look similar to the ones in examples I use earlier in the chapter, that's because they're the same example.)

You can calculate the principal normal stresses as

$$\sigma_{P1,P2} = \frac{(-10+12)\text{ ksi}}{2} \pm \sqrt{\left(\frac{(-10-12)\text{ ksi}}{2}\right)^2 + (-7\text{ ksi})^2} = (1.00 \pm 13.04)\text{ ksi}$$

$$\Rightarrow \sigma_{P1} = (1.00 + 13.04)\text{ ksi} = +14.04\text{ ksi} = 14.04\text{ ksi (T)}$$

$$\Rightarrow \sigma_{P2} = (1.00 - 13.04)\text{ ksi} = -12.04\text{ ksi} = 12.04\text{ ksi (C)}$$

So for this example, the maximum principal stress is the larger of the two values — 14.04 ksi (T) — and the minimum principal stress is the smaller of the two values: 12.04 ksi (C).

Orienting the angles for principal normal stresses

Finding the orientation angles of principal stresses is important to designers as well. For example, if you're working with a brittle material such as concrete or glass, you can get a cracking or crushing type of failure if the maximum principal stress exceeds the limits of the material. However, if you know the orientation of the maximum principal stress, you can actually design the structural object to better resist the applied stresses. In concrete members, this step may mean including reinforcing steel at the locations and orientations of high stresses.

To determine the principal angles, you use a simple relationship that relates the shear stress to the normal stresses of the original element:

$$\tan 2\theta_P = \frac{2\tau_{xy}}{\sigma_{xx} - \sigma_{yy}}$$

So for the example I lay out in the preceding section,

$$\tan 2\theta_{P1} = \frac{2(-7 \text{ ksi})}{(-10 \text{ ksi} - 12 \text{ ksi})} = +0.636$$
$$\Rightarrow 2\theta_{P1} = 32.47°$$
$$\Rightarrow \theta_{P1} = +16.23°$$

which indicates that one of the principal stresses occurs at an orientation angle of 16.23 degrees positive (or counterclockwise) from the original *x*-axis. The only problem now is whether that's the angle of the maximum or minimum principal stress.

With an angle of orientation, you can compute the normal stress at the new orientation by using the basic transformation equation earlier in the chapter. If you substitute the known current states of stress and this principal angle, your transformation equation automatically tells you which principal stress occurs at that orientation. For example,

$$\sigma_{x1} = \frac{(-10+12) \text{ ksi}}{2} + \frac{(-10-(12)) \text{ ksi}}{2}\cos(2(16.23°)) + (-7 \text{ ksi})\sin(2(16.23°))$$
$$= -12.04 \text{ ksi} = 12.04 \text{ ksi (C)}$$

Thus, you know now that at the angle of 16.23 degrees counterclockwise from the *x*-axis is the minimum principal stress of 12.04 ksi (C).

The maximum principal stress and the minimum principal stress occur at angles that are oriented 90 degrees apart. So if you can locate one of the principal orientation angles, you automatically know the second angle by using the following simple equation:

$$\theta_{P2} = 16.23° + 90° = 106.23°$$

The maximum principal stress of 14.04 ksi (T) lies on the *y1*-axis at an orientation of 106.23 degrees.

Next, you compute the corresponding shear stress that occurs at the principal orientation by using the basic transformation equations for shear stress as follows:

$$\tau_{x1y1} = -\frac{\left(-10-(12)\right)\text{ ksi}}{2}\sin\left(2(16.23°)\right)+(-7\text{ ksi})\cos\left(2(16.23°)\right) = 0.00\text{ ksi}$$

This result shows that no shear stress is present at the principal orientation for normal stresses.

If you know the principal normal stresses and their orientations, you can assume that the corresponding shear stress at that principal orientation is always zero.

You can then draw the final rotated stress element for the principal normal stresses by using the guidelines I describe in "Rotating the basic stress element" earlier in the chapter, resulting in the element shown in Figure 7-9.

As you can see from Figure 7-9, the stress element for this principal orientation consists of only normal stresses; no shear stresses exist at this orientation. But a shear stress is only zero at this orientation. At all other orientations, the shear stress has a different value — either positive or negative. And at one special orientation, the shear stress has a maximum value, which I show you how to determine in the following section.

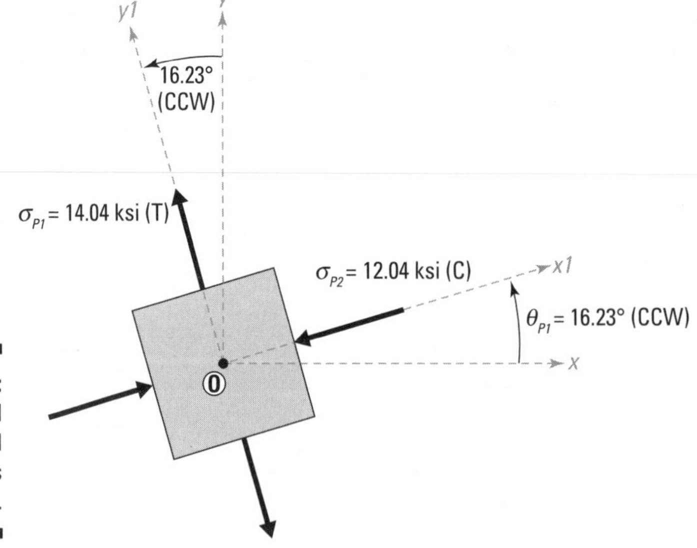

Figure 7-9: Principal normal stress element.

Defining a normal stress invariant rule

A special relationship known as a stress invariant exists within the transformation equations. A *stress invariant* relates normal stresses (σ_{xx} and σ_{yy}) on one set of mutually perpendicular planes to the normal stresses (σ_{x1} and σ_{y1}) on another set of mutually perpendicular planes:

$$\sigma_{xx} + \sigma_{yy} = \sigma_{x1} + \sigma_{y1}$$

Consider the example I often use in this chapter where $\sigma_{xx} = -10$ ksi, $\sigma_{yy} = +12$ ksi, and $\tau_{xy} = -7$ ksi.

In this example, the transformed stresses $\sigma_{x1} = -12.04$ ksi and $\sigma_{y1} = +14.04$ ksi. When you plug those numbers into the stress invariant, you get the following:

$$\sigma_{xx} + \sigma_{yy} = -10 \text{ ksi} + 12 \text{ ksi} = +2 \text{ ksi}$$

$$\sigma_{x1} + \sigma_{y1} = -12.04 \text{ ksi} + 14.04 \text{ ksi} = +2 \text{ ksi}$$

The values are the same!

Calculating principal shear stresses

Principal shear stresses are the maximum and minimum values of shear stress that can occur in a given plane. These stresses can become especially important in objects subjected to torsion and in the areas such as the *webs* (a part of an object that connects two stronger elements of a cross section) of many bending members.

However, you also have orientations for which the transformed shear stress is a nonzero value. Earlier in this chapter, I show an example of how to calculate the shear stress at an orientation of 40 degrees counterclockwise. For that example, the shear stress was –12.08 ksi, which proves that the shear stress can be a maximum value at some orientation.

To find the principal shear stress, you use the following formula, which is very similar to the second term in the principal normal stress equations (see the earlier section "Defining the principal normal stresses"):

$$\tau_P = \pm\sqrt{\left(\frac{\sigma_{xx} - \sigma_{yy}}{2}\right)^2 + \left(\tau_{xy}\right)^2}$$

For the example I lay out in that same section, you can compute the principal shear stresses:

$$\tau_P = \pm\sqrt{\left(\frac{(-10 - 12)\text{ ksi}}{2}\right)^2 + (-7\text{ ksi})^2} = \pm 13.04\text{ ksi}$$

Thus, the maximum shear stress is +13.04 ksi, and the minimum shear stress is –13.04 ksi.

Finding the principal shear stress orientation angle

After you know the principal shear stresses (covered in the preceding section), you need to determine the corresponding orientation angle. The basic equation for the principal shear stress angle is

$$\tan 2\theta_S = -\frac{\sigma_{xx} - \sigma_{yy}}{2\tau_{xy}}$$

For the example in "Defining the principal normal stresses" earlier in the chapter, you can compute the maximum shear stress angle as follows:

$$\tan 2\theta_{S1} = -\frac{(-10-12)\text{ ksi}}{2(-7\text{ ksi})} = -1.57$$
$$\Rightarrow 2\theta_{S1} = -57.52°$$
$$\Rightarrow \theta_{S1} = -28.76°$$

To determine whether this orientation angle contains the maximum (or positive) value or the minimum (or negative) value for principal shear stress, you plug this angle into the transformations equations along with the original state of stress for the element:

$$\tau_{x1y1} = -\frac{(-10-(12))\text{ ksi}}{2}\sin(2(-28.76°)) + (-7\text{ ksi})\cos(2(-28.76°)) = -13.04\text{ ksi}$$

From this test you can conclude that the minimum (or negative) value for the principal shear stress of –13.04 ksi appears on the face that is oriented at an angle of –28.76 degrees. The maximum (or positive) shear stress occurs at an orientation angle of 90 degrees from the minimum shear stress angle, or

$$\theta_{S2} = -28.76° + 90.00° = +61.24°$$

At an angle of +61.24 degrees, the maximum shear stress of +13.04 ksi appears.

You may notice that the minimum shear stress and the maximum shear stress have the same magnitude but opposite signs and that they occur 90 degrees apart from each other. For a plane stress element, these observations mean that both of the shear stresses occur on the same principal shear stress element; the maximum occurs on one face and the minimum occurs on the other. The negative sign simply means that the minimum shear stress components are acting in a clockwise rotation to help resist the counterclockwise rotation of the maximum principal stress components.

However, unlike the principal normal stresses, which have no corresponding shear stress, both the maximum and minimum shear stresses almost always have a corresponding normal stress to go with them. In fact, the corresponding normal stress is the same for both the maximum and minimum principal shear stresses. You can compute this value by substituting the original state of stress and the principal shear angle into the basic transformation equations:

$$\sigma_{x1} = \frac{(-10+12)\text{ ksi}}{2} + \frac{(-10-(12))\text{ ksi}}{2}\cos(2(-28.76°)) + (-7\text{ ksi})\sin(2(-28.76°))$$

$$= +1.00\text{ ksi} = 1.00\text{ ksi (T)}$$

$$\sigma_{y1} = \frac{(-10+12)\text{ ksi}}{2} + \frac{(-10-(12))\text{ ksi}}{2}\cos(2(61.24°)) + (-7\text{ ksi})\sin(2(61.24°))$$

$$= +1.00\text{ ksi} = 1.00\text{ ksi (T)}$$

At the orientation of the principal shear stress, the corresponding normal stresses along the $x1$- and $y1$-axes have the same value. In fact, with a few mathematical tricks, you can show that

$$\sigma_{AVG} = \sigma_{x1} = \sigma_{y1} = \frac{\sigma_{xx} + \sigma_{yy}}{2}$$

where σ_{AVG} corresponds to both of the normal stresses along the transformed $x1$- and $y1$-axes. In this example,

$$\sigma_{AVG} = \frac{(-10+12)\text{ ksi}}{2} = +1.00\text{ ksi} = 1.00\text{ ksi (T)}$$

Finally, you can draw the principal shear stress element as shown in Figure 7-10. For more on drawing this rotated stress element, refer to the section "Rotating the basic stress element" earlier in this chapter.

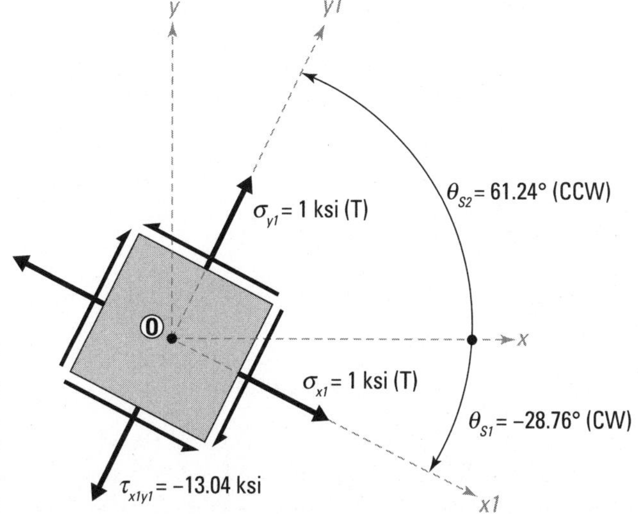

Figure 7-10: Principal shear stress element.

The principal shear stresses always occur at an angle of ±45 degrees from the orientation of the principal normal stresses. For this reason, if you can find the principal normal stresses and their orientations, you're well on your way to finding several of the most important stresses values (and orientations) need for design.

Distinguishing between in-plane and out-of-plane maximum shear stresses

For a plane stress element, the maximum shear stress τ_{MAX} can be either in or out of the plane of the element. You can determine this position by examining the principal normal stress values σ_{P1} and σ_{P2}.

- **Case 1: σ_{P1} and σ_{P2} have opposite signs.**

  The maximum shear stress in this case occurs within the plane of the element and is equal to the same value as the maximum shear stress in that plane (or the in-plane principal shear stress), τ_P:

  $$\tau_{MAX} = |\tau_P| = \frac{|\sigma_{P1} - \sigma_{P2}|}{2}$$

- **Case 2: σ_{P1} and σ_{P2} have the same signs (either both positive or both negative).**

  The maximum shear stress in this case occurs perpendicular to the plane of the element and has a value that is the larger of the two values:

  $$\tau_{MAX} = \frac{|\sigma_{P1}|}{2} \quad \text{or} \quad \tau_{MAX} = \frac{|\sigma_{P2}|}{2}$$

Utilizing Mohr's Circle for Plane Stress

Perhaps one of the most useful features of the rotated stress element is that it allows you to use a technique known as Mohr's circle for plane stress. (Flip to "Rotating the basic stress element" earlier in the chapter for more on rotated stress elements.)

Mohr's circle is a graphical technique for quickly computing the stress at any orientation, given that you know the state of stress (both normal and shear stresses) on any two perpendicular planes. Mohr's circle also tells you the orientation of the principal stresses without having to test the angle by plugging it into the transformation equations.

In this section, I explain the basic procedure for constructing a Mohr's circle for plane stress analysis and how to use the circle to find specific stress values.

Establishing basic assumptions and requirements for Mohr's circle

Analyzing Mohr's circle involves plotting values based on the current state of stress onto a set of Cartesian-type axes based on the state of stress. Keep the following points in mind when working with the Mohr's circle:

- **Normal stress plots on the horizontal axis.** Tension is measured to the right, and compression is measured to the left. Your stress elements may contain positive, negative, and even zero values!
- **Shear stress plots on the *y*-axis.** However, unlike most graphs from your algebra class, the positive shear stress (or a counterclockwise shear stress on the vertical faces) plots on the lower half of the graph.
- **All angles measured from Mohr's circle are twice their real value.** If you want to find the state of stress for an element rotated +30 degrees, the angle that you measure on Mohr's circle is 2(+30) = +60 degrees. The double angles are necessary to make the circle produce the same results as the transformation equations I mention earlier in this chapter.

Before you can apply Mohr's circle, you need a properly defined plane stress element with known normal stresses (both σ_{xx} and σ_{yy}) and shear stress (τ_{xy}). After you have the basic element established, you're ready to construct the circle.

Constructing the Mohr's circle

To illustrate the construction method, consider the state of stress for a member that is given as

$$\sigma_{xx} = -10 \text{ ksi} \qquad \sigma_{yy} = +12 \text{ ksi} \qquad \tau_{xy} = -7 \text{ ksi}$$

If you recognize these values, they're the same as the example I use earlier in the chapter to demonstrate the transformation equations (see Figure 7-6).

1. **Establish the Cartesian axes.**

   On the horizontal axis, you plot the normal stress with positive normal stresses (or tension) at the right end of the axis and negative normal stresses (or compression) at the left end. The vertical axis is reserved for the shear stress, and it crosses at a normal stress value of zero. The upper end of the vertical axis is reserved for shear stress pairs that cause a clockwise rotation (or negative shear stress), and the lower end of the vertical axis is for shear stress pairs that cause a counterclockwise rotation (or positive shear stress). See the "Establishing a sign convention for stresses" section earlier in this chapter for more on the sign convention for shear stresses.

2. **Determine the stress coordinates for the positive *x*-and *y*-axes of the current stress element.**

   The first point you plot is the state of stress on either of the vertical faces. I typically choose the axis on the positive side of the element (which is the vertical face on the right side of the base element in this example). I label this coordinate as Point V; it has stress coordinates V(–10, –7) ksi. The normal stress on the vertical face in the *x*-direction is a compressive stress, so its normal stress coordinate is a negative value; the shear stress on this face results in a clockwise rotation, so the shear stress coordinate is a negative value.

   The coordinates for the horizontal face (which I label as Point H) are given as H(+12, +7) ksi. The normal stress on the horizontal face in the *y*-direction is a tensile stress, so its normal stress coordinate is a positive value; the shear stress on this face results in a counterclockwise rotation, so the shear stress of the coordinate is also positive value.

3. **Draw a line connecting the two points of Step 2.**

   This line is a diameter of the Mohr's circle and represents the current position, or the current state of stress for your plane-stress element.

Computing coordinates and other important values on Mohr's circle

After you get the basic circle down (see the preceding section), you're ready to do a bit of geometry by using the properties of the circle and the points you plotted in the previous section. In this section, I show you how to find the center point of Mohr's circle and how you can compute the radius of the circle.

1. **Determine the coordinates of the center of the circle (located at Point C).**

   The normal stress coordinate of the centers σ_{CENTER} is actually located at the average value of σ_{xx} and σ_{yy}:

$$\sigma_{CENTER} = \frac{\sigma_{xx} + \sigma_{yy}}{2} = \frac{(-10+12)\text{ ksi}}{2} = +1.00\text{ ksi (T)}$$

   You also need to include the *y*-coordinate for the center point as well, but because this point is located on the horizontal axis, the shear stress (which is the *y*-direction) is zero ($\tau_{CENTER} = 0$ ksi).

   Therefore, the center of the circle (Point C) has coordinates (+1.00, 0.00) ksi.

2. **Draw a circle with the center located in Step 1 and through Points V and H of Step 2 in the preceding section.**

   Figure 7-11 shows the completed circle's construction.

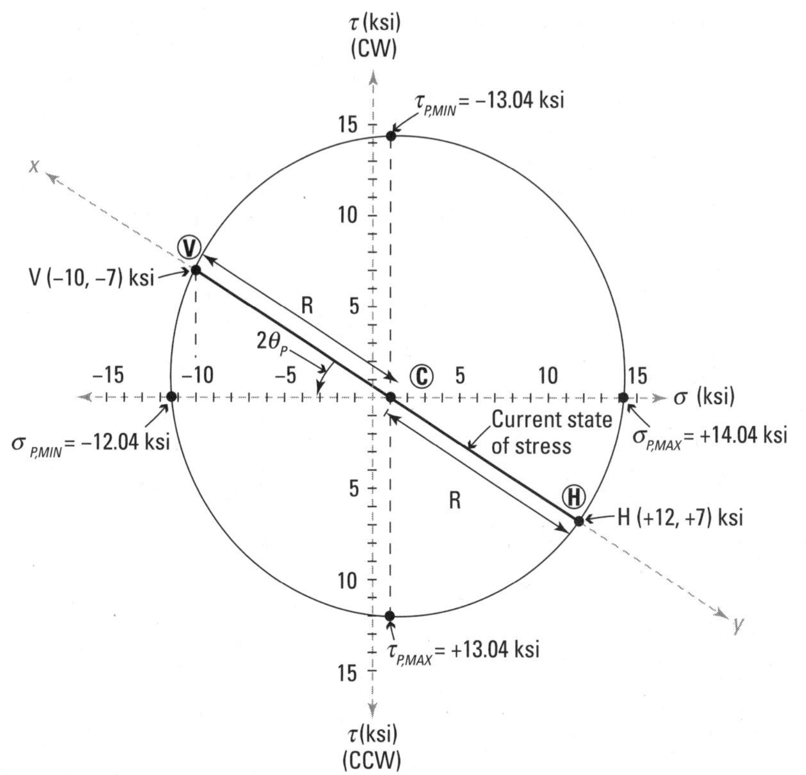

Figure 7-11: Mohr's circle example.

3. **Compute the radius *R* of the Mohr's circle.**

   Now that you have located the center of the circle, the next piece of information that you need to compute is the radius of the circle shown in Figure 7-11. To help you compute the radius, you can pull out the triangle shown in Figure 7-12.

   The triangle you examine is the triangle between the center of the circle (Point C), the coordinate of the vertical face stresses (Point V), and the horizontal axis of the circle. The radius *R* is the line that connects Point C with Point V. You can compute the vertical side of the triangle as $7 - 0 = 7$ ksi, which is the difference in the shear stress values of the coordinates. You calculate the horizontal side of the triangle between the center point (Point C) and the point on the *x*-axis directly below the normal stress on the vertical face: $1 - (-10) = 11$ ksi.

   After you have the sides of the triangle, you can use the Pythagorean theorem to compute the hypotenuse of this triangle (which coincidentally happens to be the radius *R* of the Mohr's circle).

$$R = \sqrt{(11\text{ ksi})^2 + (7\text{ ksi})^2} = 13.04\text{ ksi}$$

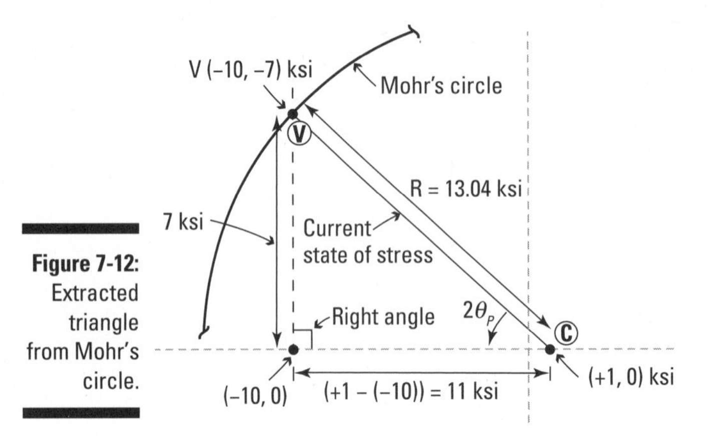

Figure 7-12: Extracted triangle from Mohr's circle.

Determining principal normal stresses and angles

Before you can do any further analysis, you must determine the relationship of the current state of stress with regards to the principal stresses. The principal stresses serve as a reference for all states of stress on the Mohr's circle. Just follow these steps:

1. **Compute the maximum and minimum normal stresses by adding the value of the radius *R* to the *x*-coordinate of the center point.**

   To get the maximum principal stress, you add the value of the radius to the *x*-coordinate of the center point (Point C).

   $$\sigma_{P,MAX} = \sigma_{CENTER} + R = +1 \text{ ksi} + 13.04 \text{ ksi} = +14.04 \text{ ksi} = 14.04 \text{ ksi (T)}$$

   To determine the minimum principal stress, you subtract the radius from the *x*-coordinate of the center point (Point C).

   $$\sigma_{P,MIN} = \sigma_{CENTER} - R = +1 \text{ ksi} - 13.04 \text{ ksi} = -12.04 \text{ ksi} = 12.04 \text{ ksi (C)}$$

2. **Compute the principal angle of the nearest principal stress.**

   Using the vertical face coordinate point (Point V), you can utilize the triangle of Figure 7-12 to compute one of the principal angles. You assign a positive or negative sign to this value depending on the direction that you must move from Point V to reach a particular principal stress. In this example, if you stand at Point V on the circle, you need to move counterclockwise around the circle to reach $\sigma_{P,MIN}$, so this direction results in a positive value for the angle $\theta_{P,MIN}$. From the triangle of Figure 7-12, you can show

$$\tan 2\theta_{P,MIN} = \frac{7 \text{ ksi}}{11 \text{ ksi}} = 0.636$$
$$\Rightarrow 2\theta_{P,MIN} = 32.46°$$
$$\Rightarrow \theta_{P,MIN} = 16.23°$$

On the Mohr's circle, and the triangle of Figure 7-12, you actually calculate a value for $2\theta_{P,MIN}$. To report this angle or include it on a rotated element sketch, you must divide the angle of your calculations by two.

3. **Compute the principal angle to the other principal stress.**

   Because $\sigma_{P,MIN}$ is oriented 180 degrees away from $\sigma_{xx'}$, you need to rotate Point V a distance of 212.46 degrees (180° + 32.46°) in a clockwise direction in order to land on $\sigma_{P,MAX}$. Remember to divide the angle in half to account for the double angle:

$$\theta_{P,MAX} = \frac{1}{2}(212.46°) = +106.23°$$

   or 106.23 degrees counterclockwise.

   Notice that these calculations from Mohr's circle match the calculations you perform with the basic transformation equations earlier in this chapter.

You can also measure this principal angle by using a clockwise rotation around Mohr's circle, in which case you need to rotate the *x*-face an angle of –73.77 degrees clockwise. Regardless of the direction you rotate (clockwise or counterclockwise), your stress element looks the same. At this point, you have the Mohr's circle fully established and can start to use it to find other transformed states of stress.

Calculating other items with Mohr's circle

If you're interested in using Mohr's circle to compute the principal shear stresses, you simply need to rotate Point V such that it lines up with the top or bottom of the circle (see Figure 7-11). If you want to find the maximum shear stress on a plane stress element, you rotate the line you draw for the diameter such that Point V is at the bottommost point on the circle, which makes the transformed Point V have stress coordinates that include the maximum shear stress. Likewise, if you want the minimum shear stress, the transformed Point V needs to be on the top of the circle.

For stresses on the face of the element that includes principal shear stress coordinates, you must remember to also include the corresponding normal stress coordinate as well. This normal stress is the same normal stress, σ_{CENTER} for the coordinates at Point C (or the center of the Mohr's circle).

As with the principal angles (see the preceding section), the direction you rotate from the original location of Point V determines the direction that you need to rotate the transformed stress element to achieve a particular shear stress on the *x1*-face.

For example, to get the shear stress on the *x1*-face to be the minimum (or negative) principal shear stress, you rotate the element 90 degrees clockwise along Mohr's circle from the minimum principal normal stress, $\sigma_{P,MIN}$ (which is located on the left edge of Mohr's circle). For the Mohr's circle in Figure 7-11, Point V is 32.46 degrees clockwise from the minimum principal normal stress, so you need to rotate the element first to the minimum principal stress (counterclockwise), and then back (clockwise) to the top of the circle. On Mohr's circle, you measure this angle as $2\theta_{S,MIN} = 32.46° - 90.00° = -57.54°$.

The minimum principal shear angle θ_S is one half the value on the Mohr's circle, or –28.77° (28.77 degrees clockwise). The stress coordinates at this location are then (+1 ksi, –7 ksi). The +1 ksi value is actually the normal stress coordinate (or the first term) of the coordinates of Point C — the center of the circle.

To get to the maximum value, you add 90 degrees counterclockwise from the minimum principal normal stress to get to the bottom of the circle. The state of stress on the vertical face of the transformed element then has stress coordinates of (+1 ksi, +7 ksi) at an angle of one half of (90.00° + 32.46°) = 122.46°, or 61.23°.

As with the principal angles for normal stresses, the principal shear angles on Mohr's circle are also recorded as $2\theta_S$ on Mohr's circle, and occur at 90 degrees from each other on the rotated stress element. If you forget to divide by 2, your maximum and minimum stress values wind up being on opposite faces of the element and your element won't be balanced properly.

Finding stress coordinates at arbitrary angles on Mohr's circle

After you have the principal stresses and the principal angles defined, you can easily find the state of stress at any orientation. Suppose you want to find the state of stress at an angle of 40 degrees counterclockwise for an element where $\sigma_{xx} = -10$ ksi, $\sigma_{yy} = +12$ ksi, and $\tau_{xy} = -7$ ksi. That means you need to rotate from Point V by an angle of 80 degrees on Mohr's circle in a counterclockwise direction. This new transformed Point V then forms a triangle with the horizontal axis and the center point on the circle (as shown in Figure 7-13a) such that

$$2\phi = 2(40°) - 2\theta_P = 80.00° - 32.46° = 47.54°$$

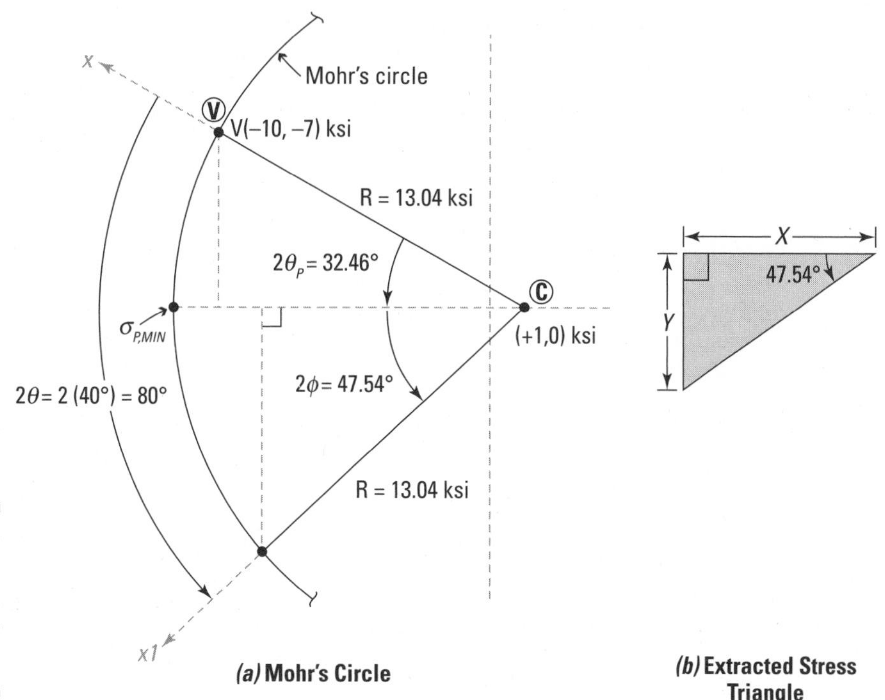

Figure 7-13: Finding coordinates at arbitrary angles.

After you have this angle measured relative to one of the principal normal stresses, you can use basic trigonometry and the triangle shown in Figure 7-13b to solve for X and Y and then calculate the transformed stresses:

$$X = R\cos(2\phi) = (13.04 \text{ ksi}) \cdot \cos(47.54°) = 8.79 \text{ ksi}$$

$$Y = R\sin(2\phi) = (13.04 \text{ ksi}) \cdot \sin(47.54°) = 9.61 \text{ ksi}$$

Because the new coordinate lies to the left of the center point (Point C), you calculate the normal stress coordinate from

$$\sigma_{x1} = \sigma_{CENTER} - X = (+1 \text{ ksi}) - 8.79 \text{ ksi} = -7.79 \text{ ksi} = 7.79 \text{ ksi (C)}$$

The shear stress coordinate is then equal to the value of the Y dimension of the triangle. Because the new stress coordinate is located below the horizontal axis, the shear is a positive value on the new vertical face.

$$\tau_{x1y1} = +Y = +9.61 \text{ ksi}$$

You now know the state of stress on the transformed *x*-face and can sketch a final rotated element.

You still need to compute the coordinates for the Point H, which lies on the opposite end of a diameter from the new Point V location. The procedure I outline here works just as well for that opposite point. Just remember that you need to add an additional 180 degrees (or subtract it) to get the new angle on Mohr's circle to that point.

Adding a third dimension to Mohr's circle

Drawing Mohr's circle for a three-dimensional state of stress (see "Identifying basic states of stress" earlier in the chapter) isn't possible, with one exception. If the state of stress is a principal state (or a triaxial state) in which no shear stress exists on any faces of the three-dimensional element, you can draw a special version of Mohr's circle by examining the normal stresses on any given plane of the element. That is, you can draw a circle for the XY plane, another circle for the YZ plane, and a third for the XZ plane. You simply plot the principal stresses σ_{P1}, σ_{P2}, and σ_{P3}. Assuming that σ_{P1} is the largest of the three values and σ_{P3} is the smallest of the three values, you can generate a figure of three circles such as Figure 7-14.

For Figure 7-14, you have one circle that connects σ_{P1} and σ_{P2}, another that connects σ_{P2} and σ_{P3} and a third that connects σ_{P1} and σ_{P3}. Beyond this, you can't do any transformation calculations on this version of the Mohr's circle.

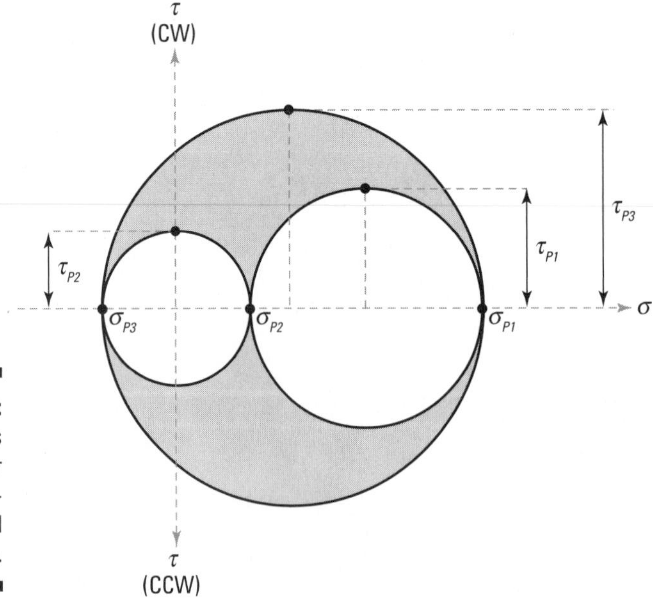

Figure 7-14: Mohr's circle for three-dimensional stresses.

However, you can identify the maximum shear stresses. You determine the maximum shear stresses for a triaxial element in a similar fashion to the plane-stress elements that I describe in the earlier section "Distinguishing between in-plane and out-of-plane maximum shear stresses."

You can compute the radius of each circle by finding the difference between the principal stresses of each circle; this difference is the diameter. As I note earlier in the chapter, the radius of a Mohr's circle is actually at the same value as the principal shear stress for a particular circle.

The maximum shear stress for a given triaxial stress element is the larger of the three principal shear stress values. In equation form, this fact means that

$$\tau_{P1} = \frac{|\sigma_{P1} - \sigma_{P2}|}{2} \qquad \tau_{P2} = \frac{|\sigma_{P2} - \sigma_{P3}|}{2} \qquad \tau_{P3} = \frac{|\sigma_{P1} - \sigma_{P3}|}{2}$$

Additionally, the maximum shear stress for the element is the largest of those three values:

$$\tau_{MAX} = \max(\tau_{P1}, \tau_{P2}, \text{ and } \tau_{P3})$$

For a plane stress problem in three dimensions, one of the principal stresses must be equal to zero. When this situation happens, the maximum shear stress may not be equal to the in-plane principal shear stress. In reality, even if you're working with a plane stress problem, the third, out-of-plane stress is a zero value.

If the zero-value principal stress is the middle point on the figure (that is, $\sigma_{P2} = 0$ ksi), the maximum shear stress is equal to the in-plane principal shear stress of the two dimensional problem. If the zero-value stress is either the maximum or minimum value, either

$$\tau_{MAX} = \frac{|\sigma_{P1}|}{2} \quad \text{or} \quad \tau_{MAX} = \frac{|\sigma_{P3}|}{2}$$

For this case, you choose the larger of the two values from these equations.

The maximum shear stress for the three-dimensional problem may not necessarily lie in the plane of the plane stress element. Instead, it may lie in one of the remaining *orthogonal* (or perpendicular) planes of the element. Thus, an in-plane principal shear stress may not necessarily be the maximum shear stress for a given problem.

Chapter 8

Lining Up Stress Along Axial Axes

In This Chapter

- Identifying axial stresses
- Working with bearing stresses
- Exploring pressure vessels
- Dealing with maximum stresses in geometry

In most mechanics of materials classes, often the simplest (and earliest introduced) stresses to compute are the normal stresses that act perpendicular to a member's cross section. Normal stresses result from different types of loads, such as through tension in a cable, compression of a very short column, or by the bending (or flexure) of a beam. Axial stresses are the most basic of the normal stresses, and they usually provide the jumping-off point into stress analysis.

In this chapter, I explain how to calculate several different types of axial stresses beginning with a basic bearing situation and then moving to multi-directional axial stresses caused by pressure vessels. Finally, I explain how sudden changes in geometry can greatly increase stresses in a member, and how you can account for those changes in your analysis.

Defining Axial Stress

An *axial stress* is a type of normal stress that acts parallel to an axis, such as the *longitudinal axis* (or an axis that runs down the length of a member), and acts perpendicularly (toward or away from) a cross section. Axial stresses can be a significant portion of the total stress that members such as columns and ropes or cables may experience.

An axial stress can occur in one of two orientations: tension or compression.

- **Tension:** A *tension* stress is a stress that pulls on a member, causing the member to experience an *elongation* (or increase in length). For the equations of this book, I treat a normal stress causing tension as a positive value.
- **Compression:** A *compression* stress is a stress that pushes on a member, causing the member to experience a *shortening* (or decrease in length). In this book's equations, a normal stress causing compression is a negative value.

Many textbooks use the term *axial stress* interchangeably with the term *normal stress*. However, this setup isn't strictly accurate. Although axial stresses are always normal stresses, not all normal stresses are axial stresses. I explain more about this discrepancy in Chapter 9 when I discuss stresses from bending, which are also normal stresses.

The units of an axial stress are measured in a force-per-area form such as psi or ksi in U.S. customary units and the Pascal in SI units. I discuss the basic units of stress in Chapter 6.

The basic equation for computing the axial stress in a tension member is

$$\sigma_{AXIAL} = \frac{P_{INT}}{A}$$

where P_{INT} is the internal axial force in the member, and A is the cross sectional area. For example, an axial tension rod that carries P_{INT} = 100 kip and has a cross-sectional area of A = 2 in^2 has an axial stress of σ_{AXIAL} = (100 kip)÷ (2 in^2) = 50.0 ksi.

Figure 8-1 shows typical stress elements in *pure tension* or *pure compression* — where stresses are applied either all in tension or all in compression. Figure 8-1a shows a uniaxial stress state (or a stress in only one direction), and Figure 8-1b shows a plane stress element where normal stresses (σ_{xx} and σ_{yy}) are acting in two directions simultaneously. In each of these figures, all arrows act either away from the object (tension) or toward the object (compression). In these figures, the only stresses acting on the element are normal stresses; you don't see any *shear stress* (or stresses that act parallel to the face on which they are acting) on the basic element.

In reality, you can also have a combination of the tension and compression cases, where axial stresses acting on the vertical faces are applying tension along one axis of the element and axial stresses on the horizontal faces are applying compression along the other axis of the element. You can also have shear stresses acting at the same time (which I discuss more in Chapter 10), but then that wouldn't be an axially loaded stress element, now would it?

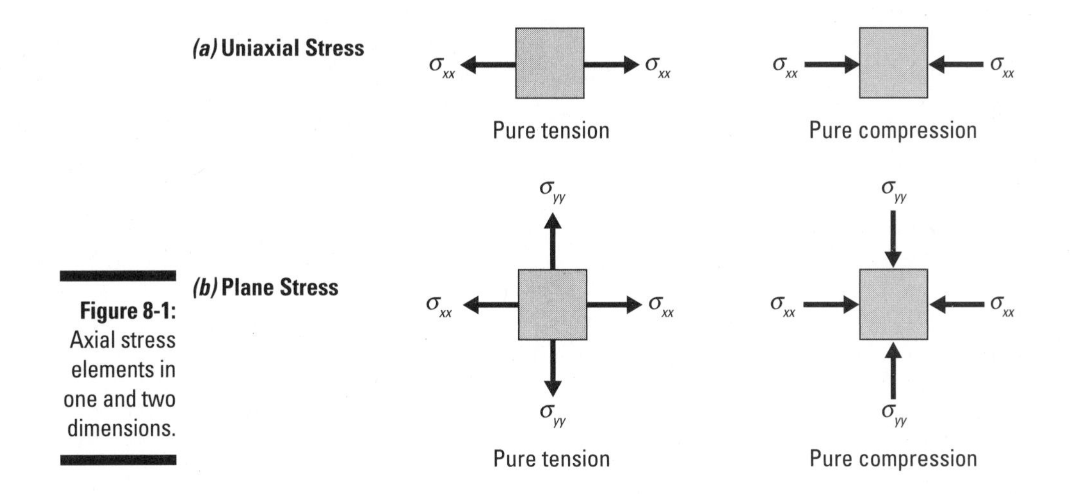

Figure 8-1: Axial stress elements in one and two dimensions.

If the only stresses on an element are in axial tension or axial compression, the corresponding element that you draw is automatically a *principal stress element,* which never has any in-plane shear stresses acting on it at the orientation, only normal stresses. For more on principal stresses and stress transformation, turn to Chapter 7.

As with all stress calculations, the biggest challenge you face for axial stresses is determining the appropriate internal loads (see Chapter 3) and the appropriate section property on which those loads are acting (see Chapters 4 and 5). After you have those items figured out, you're well on your way to completing your stress analysis.

Getting Your Bearings about Bearing Stresses

Bearing stresses, or the simple stresses that result from one object pushing onto another, are one of the most common types of axial stresses you work with in structures. Examples of bearing stress (sometimes referred to as *contact stress*) include the stress that develops from this book resting on the desk below or from the foundation of your house sitting on the ground.

If your house is sitting on loosely compacted sand, the settlement (sinking) you have on your hands (well, below your feet) is a direct result of bearing-type stresses — and more specifically, bearing stress failures.

Because a bearing stress is a type of normal stress, the symbol I use for bearing stresses in this text is σ with a subscript (such as *xx*, *yy*, *zz*, or other label) to indicate its direction of orientation. However, in examples in the coming sections, I simply refer to them as $\sigma_{BEARING}$ because I'm not actually concerned with the direction at this time.

Exploring bearing stresses on flat surfaces

For objects with flat contact surfaces, the resulting bearing stress is equal to the internal contact force between them (P_{INT}) and the common area at the contact location ($A_{CONTACT}$).

$$\sigma_{BEARING} = \frac{P_{INT}}{A_{CONTACT}}$$

The dimensions of the contact area $A_{CONTACT}$ usually come from the dimensions of the smaller object.

This basic calculation actually only gives an approximation to bearing stresses between two objects. In reality, the true bearing stress of deformable objects is a highly complex phenomenon governed by *Hertz contact theory,* which says that the contact stress between two objects becomes a function of the deformation of the two objects and the materials they're made of. However, for the purposes of a basic mechanics of materials discussion, such as this text, this theory is often neglected.

Consider the two unequal blocks shown in Figure 8-2, which illustrates how you analyze a problem for bearing stress.

Figure 8-2a shows a scenario where a smaller block is pushing against a larger block. Using statics, you can compute the internal contact force (or the force that the small block exerts on the large block, and vice versa) and show that $P_{INT} = P_{APPLIED}$. The contact area (shown in Figure 8-2b) is computed from the dimensions of the smaller object in contact with the larger, or $A_{CONTACT} = (hL)$ for this example. The average bearing stress is shown in Figure 8-2c; you can compute it as $\sigma_{BEARING} = (P_{INT}) \div (hL)$ for this example.

Bearing stress is the primary reason tools such as machine punches for making holes or indentations have as small of an area at the tip of the tool as possible. If the tip area is very small, the bearing stress increases as a result. If this stress exceeds the strength of the material being punched, the punch may create a dent, an impression, or even an actual hole in the object. To help the punch resist these high stresses, punches are often made from much stronger materials such as carbide alloys or coated in a stronger material such as diamond.

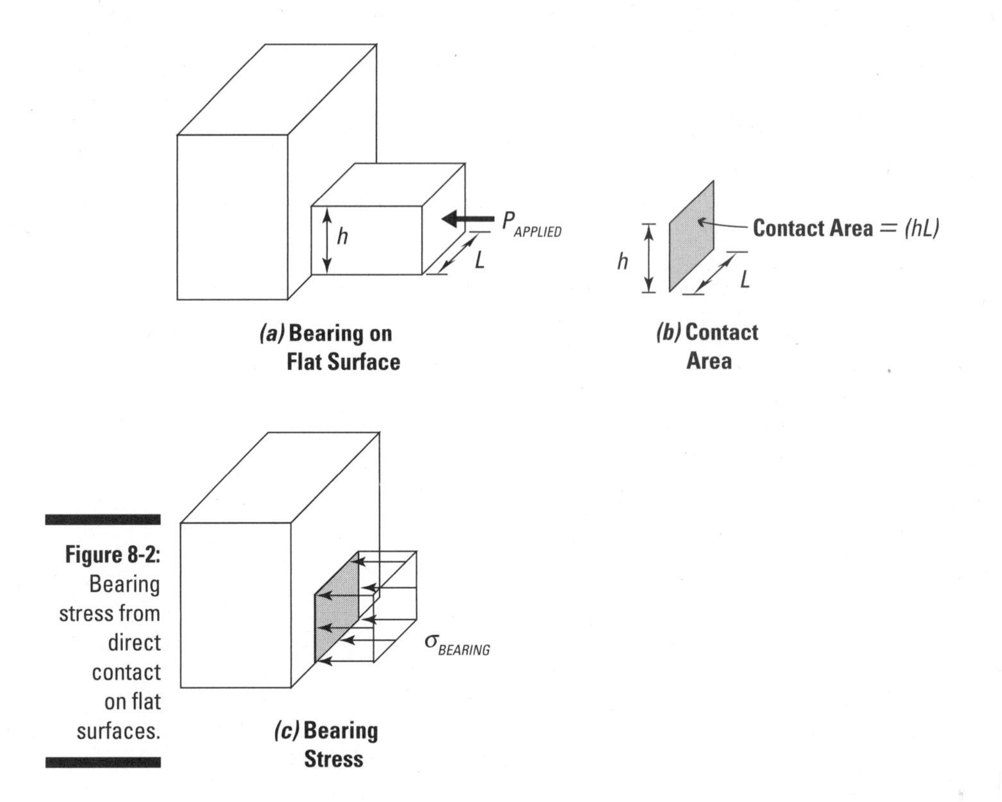

Figure 8-2: Bearing stress from direct contact on flat surfaces.

Perusing bearing stresses on projected planes

Some bearing situations aren't as simple as one flat object pushing on another (see the preceding section), so determining the actual area needed for calculating these bearing stresses is a bit more difficult. You encounter this situation a lot when you deal with bolts and shafts. A force applied to a bolt within a connection such as the one in Figure 8-3a is a common example.

In this example, the transmission of these forces ($P_{CONTACT}$) from the first plate to the bolt to the second plate occurs through bearing stresses between the plate (more specifically, the edge of the hole) and the area on the bolt's shaft that contacts the hole. For these situations, you don't actually calculate the area in direct contact because that's only one half of the circumference. Instead, you calculate the area that's perpendicular to the force being applied to the bolt, which is actually a projected area.

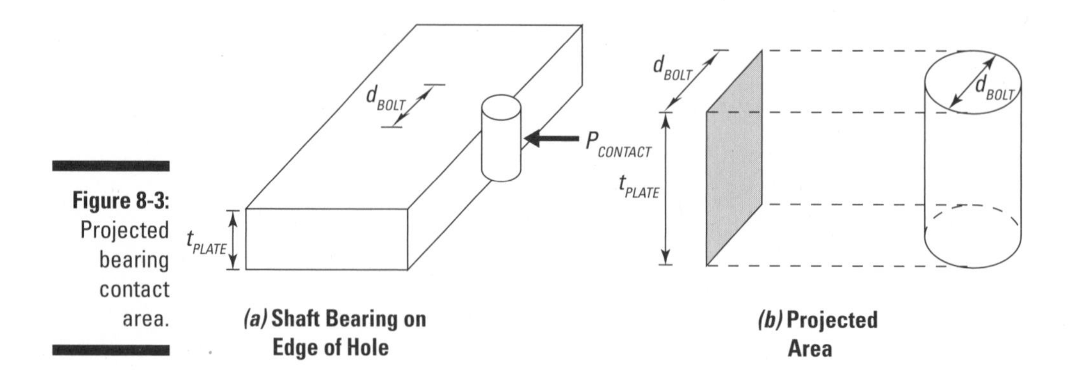

Figure 8-3: Projected bearing contact area.

You can compute the bearing stress on the projected area between the plate and bolt as shown in Figure 8-3b by using the following equation:

$$\sigma_{BEARING} = \frac{P_{CONTACT}}{A_{PROJECTED}} = \frac{P_{CONTACT}}{(d_{BOLT}) \cdot (t_{PLATE})}$$

where d_{BOLT} is the diameter of the bolt or shaft and t_{PLATE} is the thickness of the connecting plate. If the projected area is small for the same load, the bearing stress increases. This characteristic is one of the reasons why you can slice through a block of cheese with a small wire cutter much more easily than you can with a thick (or wide) butter knife.

The bearing stresses of bolts on the edges of holes are a very important concern in designing structural connections. If the design exceeds the bearing stresses, designers often either use larger diameter bolts (which increases d_{BOLT}) or increase the number of bolts in the connection (which decreases $P_{CONTACT}$ per bolt). Both of these methods result in a decrease in the bearing stress of the bolts in the connection.

Containing Pressure with Pressure Vessels

Pressure vessels are unique structures that typically have an internal pressure acting within a basic shell structure that differs from the exterior pressure (known as the *ambient pressure*). Examples of pressure vessels include storage tanks, basketballs, and even the pressurized cabin of an airplane.

The first pressure vessels were developed during the Industrial Revolution to serve as boilers or tanks to hold steam that powered steam engines. Design codes and certification procedures now stringently control the design of pressure vessels because of their propensity to suddenly explode upon failure.

Differentiating between thin- and thick-walled pressure vessels

Pressure vessels typically fall into two major categories that affect the equations you use for analysis:

- **Thin-walled pressure vessels:** *Thin-walled* pressure vessels are the most common type of pressure vessel you encounter in a basic mechanics class. A pressure vessel is typically considered to be thin-walled by design codes if it meets the criteria

  $$r \geq 10t$$

  where *r* is the inner radius of the pressure vessel and *t* is the wall thickness of the vessel. The pressure vessels I discuss in this chapter are all thin-walled pressure vessels.

- **Thick-walled pressure vessels:** *Thick-walled pressure vessels* are those vessels that have significant thicknesses with respect to their radii. Thick-walled pressure vessels are significantly more complex to analyze and are often beyond the scope of a basic mechanics of materials course. Check out the "When pressure vessels become thick-walled" sidebar if you're interested in more information on thick-walled pressure vessels.

When pressure vessels become thick-walled

Thick-walled pressure vessels become much more complex and require you to solve a differential equation in order to establish the basic equations for stress. The stresses for the radial direction *r*, the tangential direction θ, and the axial direction *z* have the basic form as shown:

$$\sigma_r = \frac{E}{1-\nu^2}\left(\frac{du}{dr}+\nu\frac{u}{r}\right) \quad \sigma_\theta = \frac{E}{1-\nu^2}\left(\frac{u}{r}+\nu\frac{du}{dr}\right) \quad \sigma_{zz} = \left(\frac{p_i r_o^2 - p_o r_o^2}{r_o^2 - r_i^2}\right)$$

where *E* and ν are material properties and *u* is a function that expresses the radial displacement of the wall and has a general form of

$$u = C_1 r + \frac{C_2}{r}$$

C_1 and C_2 are constants of integration based on the boundary conditions for the specific problem. When working with thin-walled pressure vessels, you typically neglect the radial stresses that act perpendicular to the thickness of the vessel, which are assumed to be very small in comparison to the other stresses present in a pressure vessel. However, to analyze a thick-walled pressure vessel, you must include the effects of those radial stresses making these vessels significantly more complex to analyze because they're no longer just plane stress problems.

Taking a closer look at thin-walled pressure vessels

When you're working with a thin-walled pressure vessel, you need to develop the basic equations for axial stress in pressure vessels. The following sections present some factors to consider and show you how to work with spherical and cylindrical pressure vessels.

Setting parameters

All thin-walled pressure vessel stress analysis centers on the same three parameters: internal pressure, inner radius, and wall thickness.

- **Internal pressure:** The *internal pressure* is the pressure inside the vessel, measured relative to the ambient exterior pressure. The pressure contained within the vessel is typical expressed in psi for U.S. customary units and N/m^2 (or Pa) for SI units.
- **Inner radius:** The *inner radius* is the distance measured from the central longitudinal axis radially outward to the inner edge of the outer wall of the pressure vessel. Its units are inches (U.S. customary) and meters (SI).
- **Wall thickness:** The wall thickness of the pressure vessel is the difference between the inner and outer radius of the pressure vessel and is usually measured in inches and meters.

You compute the actual normal stresses in thin-walled pressure vessels by calculating their magnitudes in two perpendicular directions, which I show in Figure 8-4. How you actually calculate these stresses depends entirely on the type of pressure vessel you're dealing with. I discuss more about the different types of pressure vessels in the later section, "Calculating spherical and cylindrical pressure vessels."

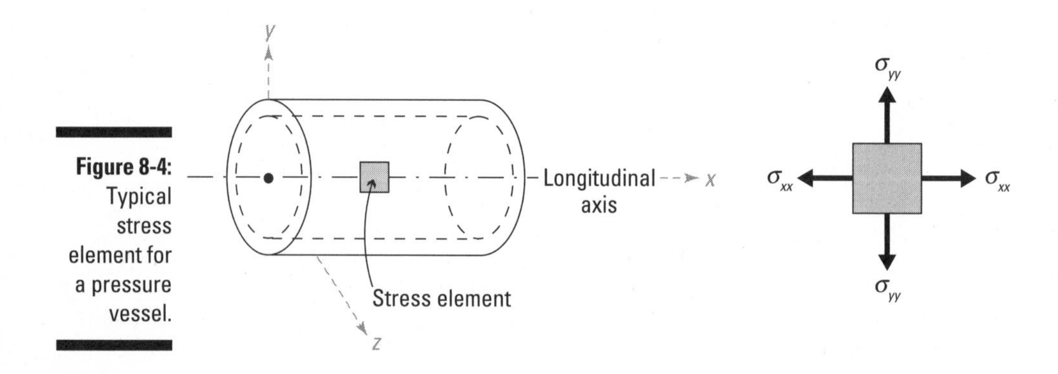

Figure 8-4: Typical stress element for a pressure vessel.

Calculating spherical and cylindrical pressure vessels

In mechanics of materials, thin-walled pressure vessels are divided into two basic categories, spherical and cylindrical. Although the procedure for analysis is fairly similar for both, you do need to consider a couple of simple differences in their respective stress equations.

A *spherical pressure vessel* is a pressure vessel that has a uniform radius r in all directions and a wall thickness t (refer to Figure 8-5). Examples of spherical pressure vessels include soccer balls and compressed air tanks.

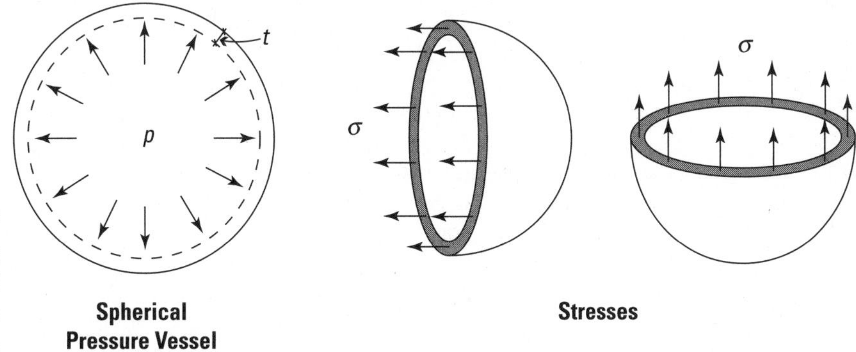

Figure 8-5: Spherical pressure vessels.

For a spherical pressure vessel, the stresses in each direction on the stress element are the same value and can be computed from the following:

$$\sigma = \sigma_{xx} = \sigma_{yy} = \frac{pr}{2t}$$

where p is the internal pressure measured relative to the exterior ambient pressure. For a spherical pressure vessel having a radius of 20 millimeters and a wall thickness of 1 millimeter, subjected to an internal pressure of 40 kPa, you can compute the stresses as shown here:

$$\sigma_{xx} = \sigma_{yy} = \frac{pr}{2t} = \frac{(40\text{ kPa})(0.02\text{ m})}{2(0.001\text{ m})} = 400\text{ kPa}$$

A *cylindrical pressure vessel* is a pressure vessel that has a circular cross section with radius r and wall thickness t. However, the vessel has a length L in the longitudinal direction (see Figure 8-6). Examples of cylindrical pressure vessels include cans of soda, pressurized air tanks, and airplane fuselages.

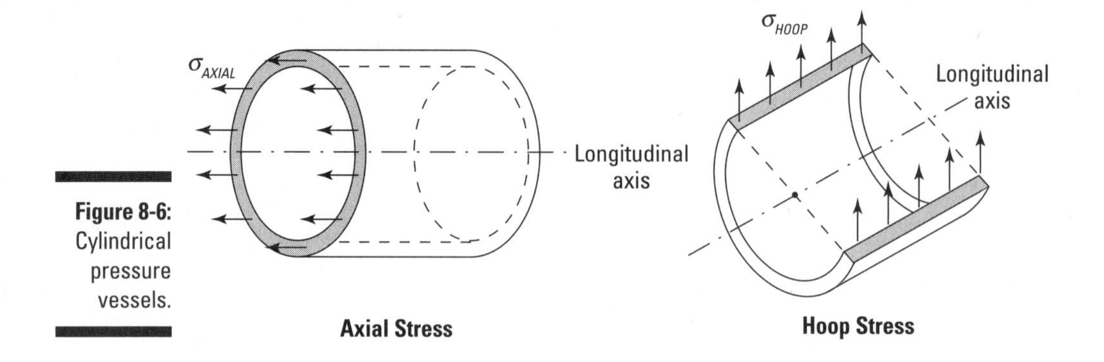

Figure 8-6: Cylindrical pressure vessels.

Unlike spherical pressure vessels, cylindrical pressure vessels have two distinctly different normal stresses acting in perpendicular directions. The first of these stresses is the axial stress, which acts parallel to the longitudinal axis of the vessel. In the calculations of this section, I refer to this stress as σ_{AXIAL}. The *hoop stress* (also known as the *circumferential stress*) acts tangentially to the radius of the cross section and is denoted as σ_{HOOP}.

Pay special attention to the direction of the longitudinal axis of the cylindrical pressure vessel. This direction is always the direction of the axial stress. For a cross section in the Cartesian YZ plane, the longitudinal axis (and thus the axial stress) is oriented in the Cartesian *x*-direction.

You can compute the axial and hoop stresses for a cylindrical pressure vessel by using the following equations:

$$\sigma_{xx} = \sigma_{AXIAL} = \frac{pr}{2t} \quad \text{and} \quad \sigma_{yy} = \sigma_{HOOP} = \frac{pr}{t}$$

where p is the internal pressure measured relative to the exterior ambient pressure. For the same cross section dimensions and internal pressure as the spherical pressure vessel example, you can find the stresses on a comparable cylindrical pressure vessel by using the following equations:

$$\sigma_{xx} = \sigma_{AXIAL} = \frac{pr}{2t} = \frac{(40\text{ kPa})(0.02\text{ m})}{2(0.001\text{ m})} = 400\text{ kPa}$$

$$\sigma_{yy} = \sigma_{HOOP} = \frac{pr}{t} = \frac{(40\text{ kPa})(0.02\text{ m})}{(0.001\text{ m})} = 800\text{ kPa}$$

The axial stress acts down the length of the member (down the longitudinal axis), and the hoop stress acts tangentially around the circumference of the cross section. These two stresses always act simultaneously and in their different directions.

The hoop stress is always double the stress of the axial stress for a cylindrical pressure vessel. That's why when you see a cylindrical tank rupture, the direction of the rupture is always parallel to the longitudinal axis of the tank. This parallel rupture happens because the hoop stress is the stress that acts in this direction, and it always has a larger magnitude than the corresponding axial stress.

If your pressure vessels are subjected to an internal vacuum, you need to include the effects of compression and special local stability issues that are beyond the scope of this text. The formulas in this section are for tension in pressure vessels due to an internal outward pressure.

After you have the basic stress element created, you can then begin your stress analysis by transforming the stresses, or determining their principal values, as I show in Chapter 7.

When Average Stresses Reach a Peak: Finding Maximum Stress

The calculation of an average normal stress (as I discuss in Chapter 6) requires that the resultant axial force be located at the centroid (see Chapter 4) of the cross section and that the member itself stay *prismatic* — or have a uniform cross section along the length — in order for the stress to remain uniform across a cross section. However, changes to geometry such as the presence of holes or notches cause that stress magnitude to vary from one point to another within the same cross section.

In this section, I show you how to compute the net area for average stresses and then how to determine the maximum stresses that can occur near holes or notches in a tension member.

Explaining gross versus net areas for average normal stress calculations

A consideration that you need to keep in mind is that the magnitude of a stress is directly related to the area on which the internal forces are acting. For the same internal load, the stress increases as the area decreases. Engineers frequently drill holes in members so that they can use bolts or pins to connect members. These holes reduce an area, which in turn can increase the stress. A cross section that isn't reduced because of holes is sometimes referred to as a *gross cross section,* whereas a cross section that contains holes or opening may be referred to as a *net cross section.*

Consider the bar example in Figure 8-7a. It's 6 inches tall x ¼ inch thick, with a 2-inch-diameter hole located in the midregion of the bar. As a designer, you may be interested in determining average normal stresses for an applied external axial tension of $F = +2{,}000$ lbs applied at the ends. As part of the design process, you'd calculate internal stresses at multiple locations in the bar, such as Location 1, Location 2, and Location 3 in this example. Locations 1 and 3 go through a gross cross section on either side of the hole, and Location 2 is on a net cross section passing through the center of the hole.

To calculate the average stress at Location 1, you must first slice the bar to expose the internal force acting on the cross section. From static equilibrium, you can find the internal axial force F_{INT} from the following:

$$\overset{+}{\rightarrow}\sum F_X = 0 \Rightarrow 2{,}000 \text{ lbs} - F_{INT} = 0 \Rightarrow F_{INT} = +2{,}000 \text{ lbs } (\text{T})$$

Next, you must compute the appropriate section property for calculating the stress at Location 1. Because you want to compute an average normal stress for an axially loaded bar, you calculate the area as depicted by the shaded region in Figure 8-7b. Because the cross section at Locations 1 and 3 contain the entire cross section (it has no holes or openings), you may see this shaded area referred to as a gross cross section.

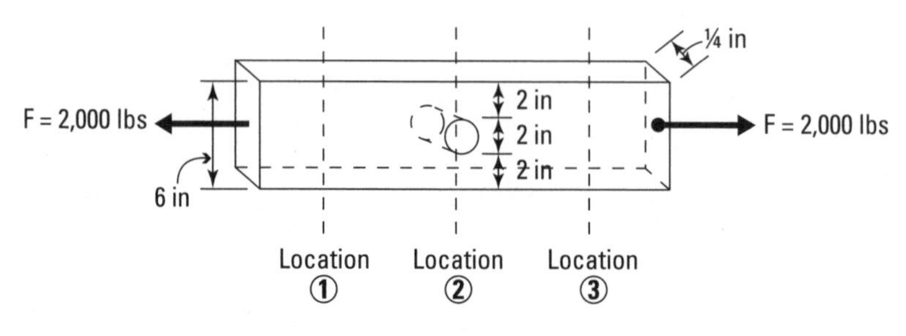

(a) **Real System**

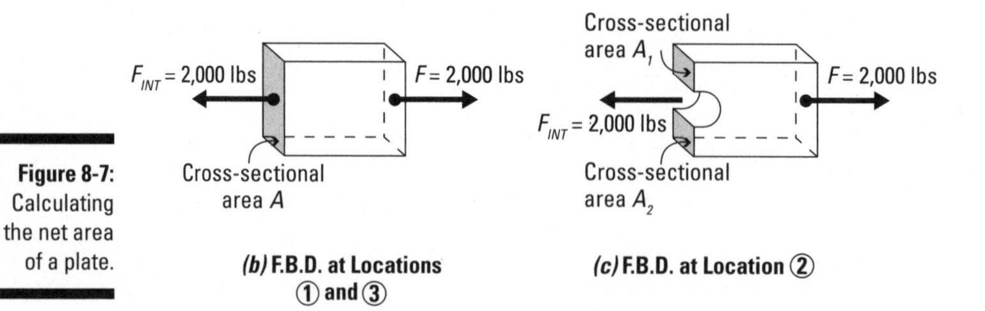

(b) **F.B.D. at Locations ① and ③**

(c) **F.B.D. at Location ②**

Figure 8-7: Calculating the net area of a plate.

As I show in Chapter 6, the gross cross-sectional area A_{GROSS} = (6 in) (0.25 in) = 1.5 in^2, and the average stress at these locations is σ_{AVG} = (+2,000 lbs) ÷ (1.5 in^2) = 1,333 psi.

If you're working with a very long, straight, uniform bar subjected to an axial tension force, the stress computed in the average stress equation is fairly accurate and consistent all along the length as long as the bar has no holes.

But wait, this example has a hole at Location 2! This hole actually removes material from the cross section, as shown by the shaded regions of Figure 8-7c, so this cross section is actually a net cross section.

For objects with holes or grooves, you can still compute an average stress as an approximation. However, you must reduce the total area at that cross section in order to calculate the average stress. You can calculate this reduced or area of the net cross section with the equation

$$A_{NET} = A_{GROSS} - A_H$$

where A_{GROSS} is the total area of the cross section not including holes, and A_H accounts for the area you're removing to account for the presence of holes or notches along the cross section.

When calculating net cross-sectional areas for stress computations, the area you're calculating is the area of solid material on which the force is physically acting. You must remember to subtract the holes.

To calculate the net area at Location 2, you can approach the problem in several ways.

- **Compute the area of the shaded region above the hole and simply add it to the shaded region below the hole.** This total summed area is the net area. The following equation shows you how to do this for the example in Figure 6-3:

  $$A = A_{NET} = A_1 + A_2 = (2\text{ in})\cdot(0.25\text{ in}) + (2\text{ in})\cdot(0.25\text{ in}) = 1.0\text{ in}^2$$

- **Find the gross cross-sectional area and subtract the area of the hole contained on the cross section.** Use the following equation:

  $$\begin{aligned} A = A_{NET} &= A_{GROSS} - A_H = A_{GROSS} - (d)\cdot(t) \\ &= (6\text{ in})\cdot(0.25\text{ in}) - (2\text{ in})\cdot(0.25\text{ in}) = 1.0\text{ in}^2 \end{aligned}$$

  where d is the diameter of the hole and t is the thickness of the bar (or the depth of the hole). This equation produces the same net area calculation.

For any given cross section containing multiple holes or cutouts, you often find that using the gross area and subtracting the holes is a much simpler calculation.

You can then calculate the average stress at Location 2 by using the following formula:

$$\sigma_{AVG} = \frac{+2{,}000 \text{ lbs}}{1.0 \text{ in}^2} = +2{,}000 \text{ psi}$$

Notice that at Locations 1 and 3 on either side of the hole (where the gross cross-sectional area is used), the average normal stress is +1,333.3 psi, while at Location 2 (where the presence of the hole requires you to calculate a net cross-sectional area A_{NET}), the average normal stress is significantly higher at +2,000 psi. This discrepancy occurs because the load is spread evenly across the entire gross cross section at the ends of the bar. But at Location 2, the same load is carried by a smaller area.

To reduce the average normal stress for a given axial load, you simply need to increase the cross-sectional area. An easy way to accomplish this task is to make the bar thicker at the hole locations. Doing so increases the cross-sectional area without affecting the width of the bar. Designers use this simple trick frequently in their work with bolted mechanical assemblies.

I should point out that the average stress condition is actually a very poor assumption near holes and openings because these reductions in cross-sectional area can actually cause a significant localized stress increase adjacent to the hole and openings. That is, the stresses along the cross section are no longer constant, which is a requirement in the definition of an average stress.

Using the force lines to locate maximum stress

Consider the object in Figure 8-8a in the preceding section, which shows an axially loaded member. The axial load is transmitted from one end of the object to the other by the *force flow lines,* which display the flow of forces through a member.

You can picture the approximate force flow lines as a series of strings stretched tightly and equally spaced, representing the flow of force from one end of the axial member to the other. If you place an object in the middle of these strings, displacing the strings around the hole or notch, you get an approximation of the flow of forces (or force flow lines) from one side of the opening to the other.

For the member shown in Figure 8-8a, the force lines remain constant (or evenly spaced) at a large distance away from the hole (such as Location 1). However, near the hole or opening (such as at Location 2), you see that the force flow lines tend to compress (or become closer together) as they travel past a hole or notch, which results in larger localized internal forces. As these forces increase, the corresponding stresses at these point locations must also increase.

To illustrate the effect of these compressed forces on a plate with a single hole, the corresponding normal stresses at Location 2 vary as shown in Figure 8-8b.

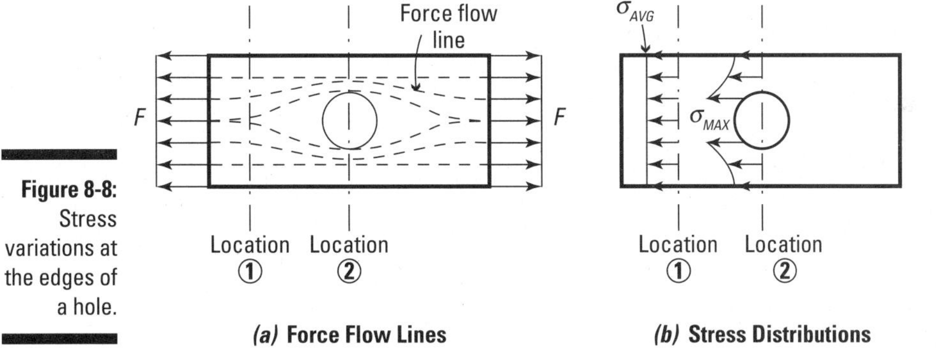

Figure 8-8: Stress variations at the edges of a hole.

As the figure shows, the stress at Location 2 is no longer uniformly distributed but rather increases to maximum values at the edge of the hole and gradually reduces as you move away from the edge of the hole.

Concentrating on normal stress concentrations

Accurately calculating the maximum stress near a hole or notch becomes a complex issue. In fact, stresses in members with holes or notches become directly related to the ratio of the dimensions of the hole or notch to the width of the bar. By contrast, the average normal stress σ_{AVG} is almost always fairly straightforward to calculate, as I show in the preceding section.

To simplify the calculation of the maximum stress, engineers and scientists use *stress concentration factors* to relate the maximum normal stress σ_{MAX} to the average normal stress σ_{AVG} by the following relationship:

$$\sigma_{MAX} = K \cdot \sigma_{AVG} = K \cdot \frac{F}{A}$$

In most classic texts, the stress concentration factor is represented by a constant *K* and is typically available in tables or figures. Figure 8-9 contains an example of one of these curves for a hole in the middle of a flat bar.

Originally, these concentration factors were determined based on experimental analysis results from physical testing, but now they can be determined numerically with computer models.

Fortunately for engineers, many design handbooks and textbooks contain stress concentration factor diagrams for a wide variety of geometric variations. Just remember, you must pay special attention to the input parameters (the geometric ratios) for each diagram because these ratios are defined differently from book to book and diagram to diagram.

Consider the 6-inch-wide bar with the 2-inch-diameter hole in Figure 8-7a earlier in the chapter. From Figure 8-9, you can compute the input ratio of $d \div w$ = 2 in ÷ 6 in = 0.333. Finding this value on the figure's horizontal axis (or the ratio) and reading up to the displayed curve, you can see that this point on the curve corresponds to a stress concentration factor *K* of approximately 2.30. This number indicates that the maximum normal stress at the edge of the hole is 2.30 times larger than the average normal stress on the same cross section.

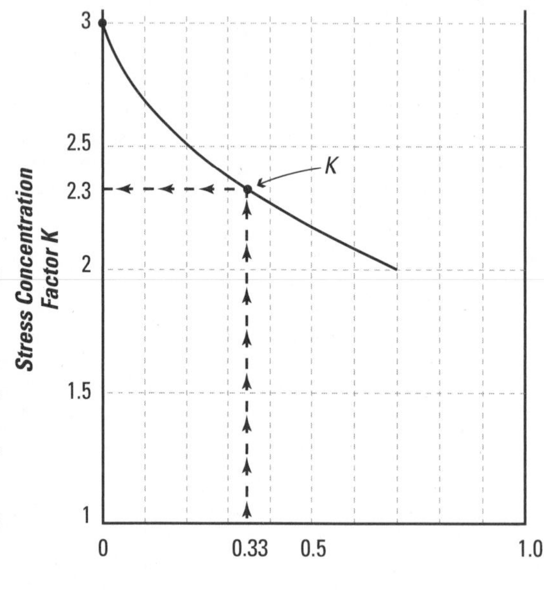

Figure 8-9: Stress concentration factor of hole in middle of flat bar.

That means that for the bar with the hole, the maximum normal stress σ_{MAX} (which is located at the edge of the hole) is actually

$$\sigma_{MAX} = K \cdot \sigma_{AVG} = (2.30) \cdot (2{,}000 \text{ psi}) = 4{,}600 \text{ psi}$$

Now here's where the situation gets a little scary. Suppose your bar is only capable of sustaining a normal stress of 2,000 psi (sometimes called the *capacity*). The bar only had an average stress of 1,333 psi in the full cross section and an average stress of 2,000 psi at the cross section containing the hole. However, because the hole is associated with a stress concentration, the maximum stress spikes up to 4,600 psi at a point immediately adjacent to the edge of the hole, a stress that exceeds the capacity of the material at that location. This can lead to serious problems, including failure of the member at the hole location.

Fortunately, design codes and other safety factors are included to help ensure that your tension member never reaches those stresses. Design codes and safety factors often require a member to be designed for limiting these stresses; that is, you may have to use a bar with a larger cross-sectional area, which reduces the stresses in the bar.

Design deals with a lot more than the considerations I mention here, and in reality, the fact that the maximum stress at a single point exceeds the capacity of the material doesn't necessarily mean that your bar will break from being overstressed. In fact, surprisingly, design engineers may not always design for the maximum stress but rather use an average stress because many materials simply deform and redistribute load from one force line to another.

For example, consider the case of a plate that is infinitely wide with the same 2-inch hole in the middle. That same ratio calculation for this scenario is now computed as

$$\frac{d}{w} = \frac{2}{\infty} \approx 0$$

which results in a corresponding stress concentration factor K that reaches its maximum value of 3.0 as indicated on the left side of the chart. For the example of a simple bar with a single hole, the stress concentration factor is never higher than 3.0.

Chapter 9

Bending Stress Is Only Normal: Analyzing Bending Members

In This Chapter

- Identifying bending assumptions
- Working with pure bending of symmetrical sections
- Calculating stresses for non-prismatic cases

Like axial stresses (see Chapter 8), bending stresses are a subset of normal stresses. *Bending* is the phenomenon that occurs when an object is loaded perpendicular to a longitudinal axis. You can describe bending as a result of an applied load or moment that causes an object to curve from its original shape; think of a fishing pole that has just caught a large fish. If you're fishing for a 200-pound sailfish to mount on your wall, you don't take a thin pole made to catch river trout; you need something more substantial to withstand the large bending stresses that creature can apply to your fishing pole. There are actually a lot of parameters that should guide your selection of fishing poles (most related to length, cross section dimensions, and material properties, though cost is probably a factor as well). That's where the concepts of this chapter come in very handy.

In this chapter, I show you how to actually compute normal stresses from bending, and I discuss several different types of bending that you encounter in mechanics of materials. Although you won't become a professional angler on the pro fishing circuit by reading this chapter (though you can take this book along to read when the fish aren't biting), you should be able to calculate the normal stresses within your fishing pole when you finally hook a big one.

Explaining Bending Stress

A *bending stress* is a normal stress that acts on a cross-sectional area as the result of an internal bending moment about an axis in the plane of the cross section. For example, if a member is oriented such that its cross section lies in the XZ plane and its length (defined by the *longitudinal axis*) is in the *y*-direction, the bending moments are about either the *x*-axis or the *z*-axis (or, in some cases, both). (A moment acting about this example's longitudinal *y*-axis isn't actually a bending moment; it's a torsional moment, which I cover in Chapter 11). The units for bending stress are the same as the basic force-per-area units of stress that I discuss in Chapter 6.

Members subjected to bending moments can also be referred to as *flexural members* because these objects tend to flex or curve instead of becoming longer or shorter in length when they deform. Bending moments (and therefore the bending stresses they cause) can be caused by an *eccentric axial load* (a load that doesn't act at the centroid of a cross section), distributed loads, concentrated moments, or *transverse forces* (forces acting perpendicular to a longitudinal axis).

Although all stresses from bending are normal stresses, all normal stresses aren't necessarily caused by bending moments. I explain more about other types of normal stresses in Chapter 8.

Handling Stresses in Bending

Bending moments actually cause both a tensile normal stress and a compressive normal stress. The maximum tension stress occurs at one end of the cross section, and the maximum compression ends at the other. Figure 9-1 shows a comparison of normal stresses caused by axial loading and bending for the same cross section.

Under an axial load, as shown in Figure 9-1a, the axial load causes a uniform stress across the entire cross section; that is, the normal stresses at Points A, B, and C all have the same numeric values because the internal axial load is concentric (or passing through the centroid of the section).

However, for the same member subjected to a positive bending moment (as shown in Figure 9-1b), the stresses at Points A, B, and C all have different values. At the top of the cross section (Point A), the bending stresses are a maximum negative compressive normal stress, and at the bottom (Point C), the stresses are a maximum positive tensile stress. At Point B (at the centroid of the cross section), the normal stresses from bending are actually zero.

For most bending members that are straight and prismatic, you can connect the stresses at these three points by a straight line. So if you know two of the stresses, you can always find the third. However, you can't do this if the member is made from multiple materials or if it's not straight or prismatic.

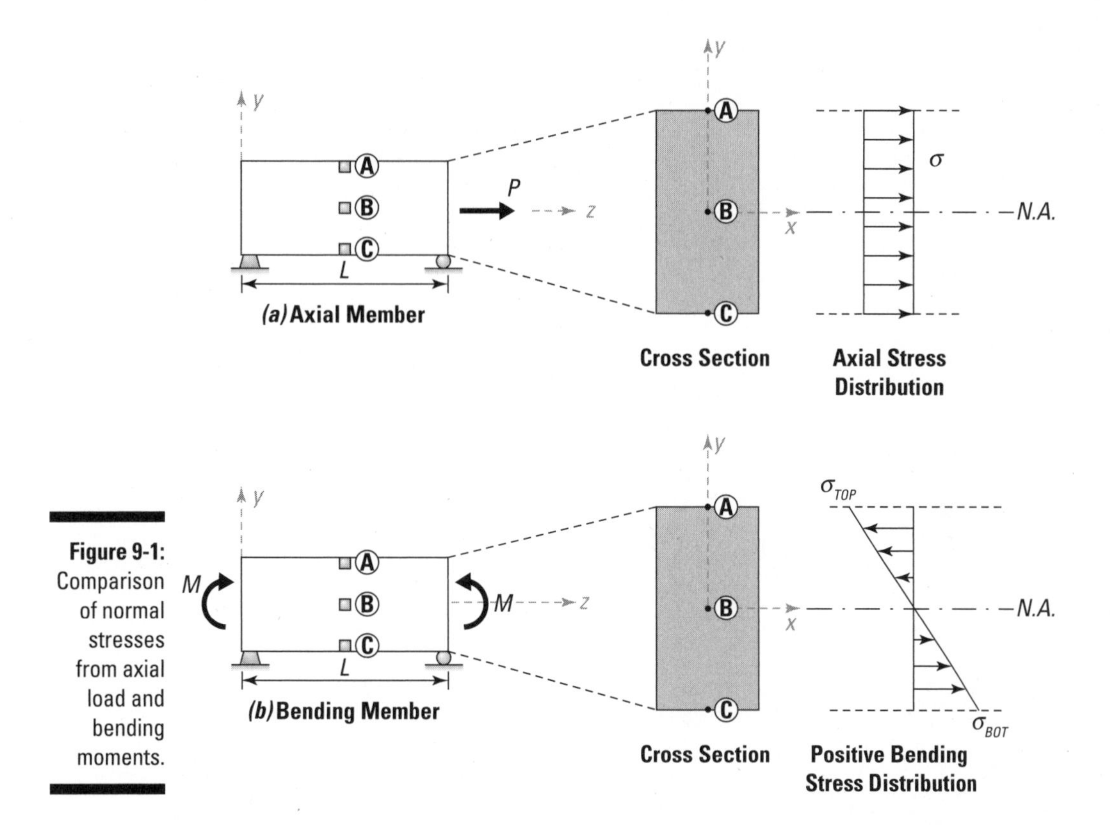

Figure 9-1: Comparison of normal stresses from axial load and bending moments.

Depending on the direction of the applied moment, the maximum tensile and compressive stress can occur in either edge of the cross section with respect to the bending axis. So be very mindful of the signs of the internal moment. However, in all cases, you have a maximum stress value at one edge of the cross section, and a minimum at the other. In between these points, the stress varies linearly.

At one point along the cross section, the signs of the stresses actually switch from being positive to negative (or vice versa). The point at which this occurs — where $\sigma_{zz,B} = 0$ — is called the *neutral axis* of the cross section. For symmetric cross sections of straight beams, the neutral axis occurs at the geometric centroid (for more on centroids and symmetry, turn to Chapter 4). The neutral axis represents the location of a *neutral surface* of a beam (where no deformation of the beam occurs; I talk more about deformation beginning in Chapter 12).

You also can display the state of stress at Points A, B, and C individually on stress elements. Figure 9-2a shows the state of stress at these points for the axial load example of Figure 9-1a. Figure 9-2b shows the state of stress at these points for the bending moment example of Figure 9-1b.

Along the length of a beam, stresses at the same position within a cross section may vary dramatically.

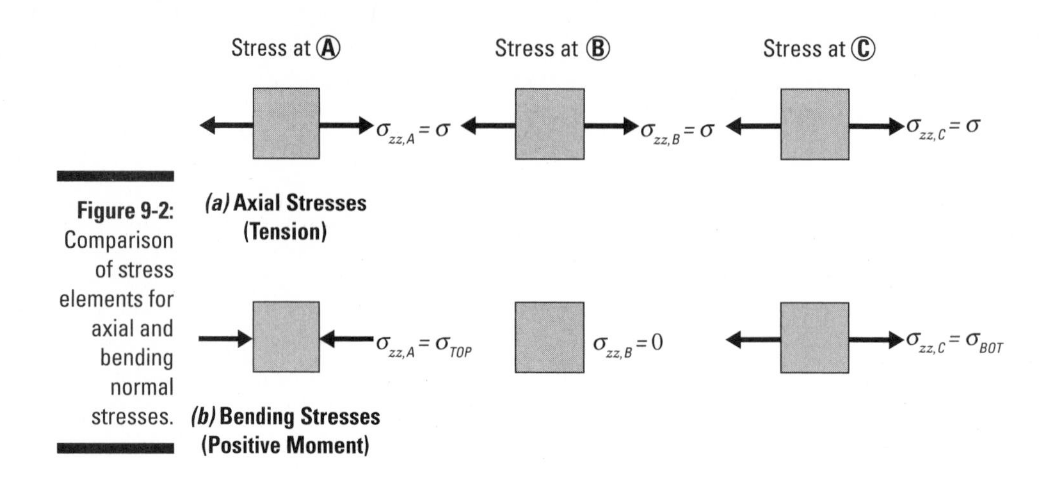

Figure 9-2: Comparison of stress elements for axial and bending normal stresses.

In the coming sections, I present you with the equations for normal stresses from bending and define some key assumptions to keep in mind when you actually start calculating bending stresses.

Solving Pure Bending Cases

Pure bending is a special condition in which a structural member is subjected to an internal moment without the presence of axial or shear forces or torsional moments. Pure bending can occur when a beam is subjected to the same uniform moment at each cross-sectional location as a result of either a concentrated effect or due to eccentric axial loads.

Establishing basic assumptions

Several basic assumptions affect the calculation of normal stresses from pure bending:

- **Prismatic members:** A *prismatic member* is a member that has the same perpendicular cross section along the length of the longitudinal axis. This characteristic means that for a given beam element, the cross section has the same area and moment of inertia at all locations along the longitudinal axis.
- **Isotropic and homogenous material properties:** *Homogeneous materials* are materials that have the same material properties throughout. *Isotropic materials* are materials that have the same material properties in a given direction, but may vary from one direction to another. In basic mechanics of materials, you typically assume that a material is both isotropic and homogeneous.

Although isotropic and homogeneous may seem to be saying the same thing, they're actually different. For example, a material like steel is considered to be both isotropic and homogeneous because the atoms that make it up are generally considered to be fairly uniformly distributed and to have the same general properties in all directions and throughout the member. Even a material such as concrete is often assumed to obey these properties despite being made from particles of distinctly different sizes (sand, aggregate, and cement). The assumption here is that the materials that make up a concrete mix are generally much smaller than the geometric dimensions of the member itself.

- **Plane-sections-remain plane:** This concept is part of Euler Bernoulli bending theory, which requires that all points within a given cross section remain in the same cross section when a beam is loaded and undergoes deformation. When displacements become large or certain loads are applied to certain geometric shapes, *warping* (out-of-plane deformations) can also occur, which results in additional stresses developing. In addition, transverse deformations are also affected by the presence of shear; for this text, I neglect these shear deformations unless otherwise noted.
- **Small displacements theory:** The object experiences small-enough displacements such that deformations don't induce additional stresses due to the eccentricity of internal loads. When deformations of beams remain very small, the analysis can (and often does) neglect axial effects. Small displacement theory affects the expressions for deformation by allowing trigonometric identities to be simplified, which influences internal force component calculations from statics.
- **Elastic behavior:** *Elastic behavior* means that even though a beam or member deforms when you subject it to a load, the deformation disappears and the object returns to its original position when you remove the load. If any portion of the deformation of an object remains after the load has been removed, the behavior is referred to as a *plastic behavior.* Plastic behavior is a complicated issue that is covered in many advanced mechanics classes. In this text, all behavior is considered elastic.

Computing stresses in pure-bending applications

The normal stress due to bending can be computed from the following basic relationship:

$$\sigma_{BENDING} = \frac{\text{moment about bending axis} \times \text{perpendicular distance from bending axis}}{\text{moment of inertia about bending axis}}$$

In this relationship, the *bending axis* is the centroidal axis about which the internal moment is acting. Although components of the moment can actually act about any three of the Cartesian axes, the bending axes are always contained within the cross section of the member. The third axis (which is normal to the cross section) is a longitudinal axis. Moments about this third axis are known as *torsional moments* and cause twisting (not bending), which I discuss in more detail in Chapter 11.

This relationship shows that the stresses for linear, elastic objects at a given location within a cross section are directly proportional to the distance of that cross section from the neutral axis. Normal stresses from bending can be the result of bending about either of the centroidal axes, or a combination of both.

In calculating bending stresses, which moment of inertia you use is especially important because you can actually have multiple moments acting simultaneously about different bending axes. This phenomenon is known as *biaxial bending,* and I discuss it in more detail in Chapter 15. You must use the moment of inertia with respect to the same axis as the internal moment.

Use the subscripts to help you determine which moment goes with which moment of inertia; they're always the same. If a bending moment is about the *x*-axis, the moment of inertia you need to use also needs to be about the *x*-axis. You also know that the perpendicular distance in the equation is always perpendicular to this same axis and must be measured within the plane of the cross section, so you'd measure this distance in the y-direction.

Consider a flexural member with a cross section contained in the XY plane. To calculate the normal stress σ_{zz} (which is a normal stress in the *z*-direction) due to bending about the centroidal *x*-axis, the basic relationship for symmetric bending is given by the following flexure formula:

$$\sigma_{zz} = -\frac{M_x y}{I_{xx}}$$

where M_x is the internal moment that you compute from statics at the cross section of interest about the *bending axis* (or the axis about which the internal moment is acting). In this example, the bending axis is about the centroidal *x*-axis. I_{xx} is the moment of inertia (which I discuss in Chapter 5) about the axis of bending, and *y* is the perpendicular distance from the neutral axis to the location of interest within the cross section. Positive values of *y* are on the compression side (above the neutral axis for positive moments about a horizontal axis) and negative on the tension side for the same case.

You can also have a second normal stress in the same direction due to bending about the centroidal *y*-axis:

$$\sigma_{zz} = -\frac{M_y x}{I_{yy}}$$

Time traveling to basic flexural theory's origins

Though some academics debate about who was first, the general consensus is that Galileo Galilei (1564–1642) and Leonardo da Vinci (1452–1519) were the first to formulate a general flexure theory for beams. da Vinci was more qualitative in his approach, focusing mainly on construction applications, while Galileo attempted to relate applied loads to stresses. Despite a fundamental error in his assumptions, Galileo's principles stood well into the 19th century.

In 1668, Robert Hooke (1635–1703) developed a relationship between forces and deformation that is perhaps the most widely known relationship: Hooke's law, which engineers still use today in mechanics of materials. In 1687, Sir Isaac Newton (1642–1727) applied his basic laws of motion to define equilibrium and deformation to flexural members. Edme Mariotte (1620–1684) and Parent (1666–1716) corrected Galileo's error, completing the basis for flexural calculations.

Later, mathematicians and scientists expanded basic flexural theory into many different areas, such as Euler (1707–1783) and the buckling of columns, Coulomb (1736–1806) and his final flexural formulation, and Navier (1785–1836) and his basic theory of elasticity. In the 20th century, Timoshenko (1878–1972) incorporated shear deformations into basic flexural calculations (among countless other mechanics contributions).

where M_y is the internal moment about a bending axis in the *y*-direction; I_{yy} is the moment of inertia about the bending axis in the *y*-direction; and x is the perpendicular distance from the neutral axis (which is parallel to the bending axis) to the location of interest within the cross section.

If the moment varies along the length of the beam, you need to compute the stresses independently at every cross section, or use generalized moment equations (see Chapter 3) to express the stress as both a function of location within a cross section and position along the length of the beam.

Looking at pure bending of symmetrical cross sections

Consider the symmetrical T-shaped cross section located in the XY plane as shown in Figure 9-3a. This section is subjected to an internal bending moment M_x of +60 kip-ft about the *x*-centroidal axis.

For a positive bending moment M_x, the region of the cross section below the neutral axis is always in tension. Similarly, the region above the neutral axis is always in compression.

The cross section has the dimensions and neutral axis location shown. You calculate the moment of inertia about the horizontal neutral axis I_{xx} to be 164.6 in^4. For guidance on computing the moment of inertia, flip to Chapter 5.

To compute the normal stress at Point A (at the bottom edge of the cross section), or $\sigma_{zz,A}$, the perpendicular distance to this location (sometimes referred to as the *fiber*) is taken as y = –4.18 in. This distance is negative because it occurs below the horizontal neutral axis:

$$\sigma_{zz,A} = -\frac{(+60\text{ kip-ft})(-4.18\text{ in})}{(124.6\text{ in}^4)}\left(\frac{12\text{ in}}{1\text{ ft}}\right) = +24.15\text{ ksi} = 24.15\text{ ksi (T)}$$

Similarly, you determine the normal stress at Point B at the top of the cross section, or $\sigma_{zz,B}$, at a distance of y = +2.82 in as follows:

$$\sigma_{zz,B} = -\frac{(+60\text{ kip-ft})(+2.82\text{ in})}{(124.6\text{ in}^4)}\left(\frac{12\text{ in}}{1\text{ ft}}\right) = -16.30\text{ ksi} = 16.30\text{ ksi (C)}$$

Together, these two stress values can help you fully describe the normal stress distribution due to bending, which you can see in Figure 9-3b.

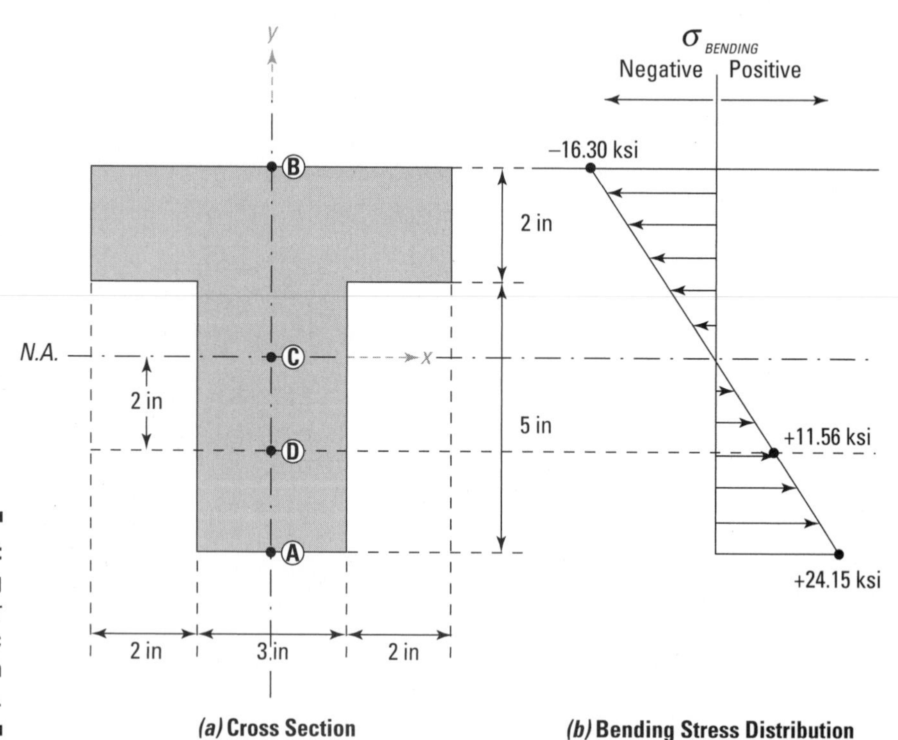

Figure 9-3: Calculating stresses for a symmetric T-beam section.

You must be careful about the units of the internal moment. In many cases, (especially in larger structures), you compute the moment with units of kip-ft when using U.S. customary units. Although the kip isn't an issue, you definitely must convert the feet to inches in order for your units to cancel out and leave you with the proper ksi units.

Employing similar triangles and a linear stress distribution to find unknown stresses

Because the normal stress distribution due to bending is always linear, if you compute one stress value, you can use the known stress at the neutral axis (which is 0 ksi by definition, as I explain in the earlier section "Handling Stresses in Bending") as your second point. After you find one of the stresses in the cross section, you can use its distance and proportions to determine the values elsewhere.

Suppose you want to calculate the stress at Point D, which is located at a distance of 2 inches below the neutral axis for the cross section shown in Figure 9-3a. The stress at Point B is $\sigma_{zz,B} = -16.30$ ksi (C) (see the preceding section for that calculation). Using proportions, you can create the following relationship between this stress (without its sign) and its distance from the neutral axis:

$$\frac{|\sigma_{zz,B}|}{2.82 \text{ in}} = \frac{|\sigma_{zz,D}|}{2.00 \text{ in}} \qquad |\sigma_{zz,D}| = \left(\frac{2.00 \text{ in}}{2.82 \text{ in}}\right)(16.30 \text{ ksi}) = 11.56 \text{ ksi}$$

The final step is to determine the sign associated with the stress. Because the location of Point D is on the opposite side from the known stress $\sigma_{zz,B}$, it must also be opposite in sense (have an opposite sign). The stress at Point B is compressive, so the stress at Point D must be in tension, or $\sigma_{zz,D} = +11.56$ ksi (T) which is also shown in Figure 9-3b.

Using an elastic section modulus to figure normal stresses from bending

Another method for quickly determining maximum normal stresses due to bending is to compute a hybrid section property known as the *elastic section modulus,* which helps to quickly identify the maximum and minimum stresses in a cross section. Coincidentally, these stresses appear at the extreme fibers of the cross section (the minimum is at one edge, and the maximum is at the other). For a cross section in the XY plane where the bending axis is the x-centroidal axis (or the horizontal neutral axis), you can compute S_x as follows:

$$S_{x,TOP} = \frac{I_{xx}}{c_{TOP}} \text{ and } S_{x,BOT} = \frac{I_{xx}}{c_{BOT}}$$

where $S_{x,TOP}$ is the elastic section modulus for the top of the cross section and $S_{x,BOT}$ is the elastic section modulus for the bottom of the cross section. In this expression, c_{TOP} is the perpendicular distance from the neutral axis to the topmost fiber in the cross section, and c_{BOT} is the distance from the neutral axis to the bottommost fiber.

The units of the elastic section modulus S are expressed as a volumetric unit, meaning it can be in^3 in U.S. customary and m^3 in SI units. You may see different units used here, depending on the dimensions of the cross section. After you have the elastic section modulus for a given cross section, you can directly relate the internal moment about the horizontal neutral axis M_x to the magnitude of the stress $|\sigma_{zz}|$ at the top and bottom by the relationships

$$\left|\sigma_{zz,TOP}\right| = \frac{M_x}{S_{x,TOP}} \text{ and } \left|\sigma_{zz,BOT}\right| = \frac{M_x}{S_{x,BOT}}$$

For rectangular cross sections, the elastic section modulus for opposite edges of the cross section have the same value; that is, $S_{x,TOP} = S_{x,BOT}$.

The elastic section modulus can only give you the stress at the extreme fibers of the cross section. You can't use this relationship to find normal stresses due to bending at any other location. However, you can use it to determine a stress at an extreme fiber and then use the proportional triangle method I explain in the preceding section to find stresses at other locations in the cross section.

You can also have an elastic section modulus about the vertical neutral axis. You just replace I_{xx} with I_{yy} and c_{TOP} and c_{BOT} with respective distances to the left and right edges. This section modulus is usually labeled as S_y.

Although this substitution may seem trivial, it's very handy in design, which I discuss in Chapter 19. If you know the limiting stress of a material and the applied internal moment from statics, you can quickly compute the required elastic section modulus that you need to support this load. Many design references even have tables of common shapes that contain section modulus values already tabulated for you.

Bending of Non-Prismatic Beams

Non-prismatic beams require you to perform the same basic calculation as in the preceding section, but you need to work with generalized equations for moment and the section modulus because you must account for variations in dimensions of the cross section based on position along the length of the beam (or the z variable in this case). In fact, in tapered beams, the maximum stress no longer necessarily occurs at the location of maximum moment. Consider the example of Figure 9-4a, which shows a non-prismatic tapered beam with a constant width of 1 inch lying in the XY plane. The beam is subjected to a concentrated load of 100 pounds and a concentrated moment of 200 lb-ft acting in the directions shown.

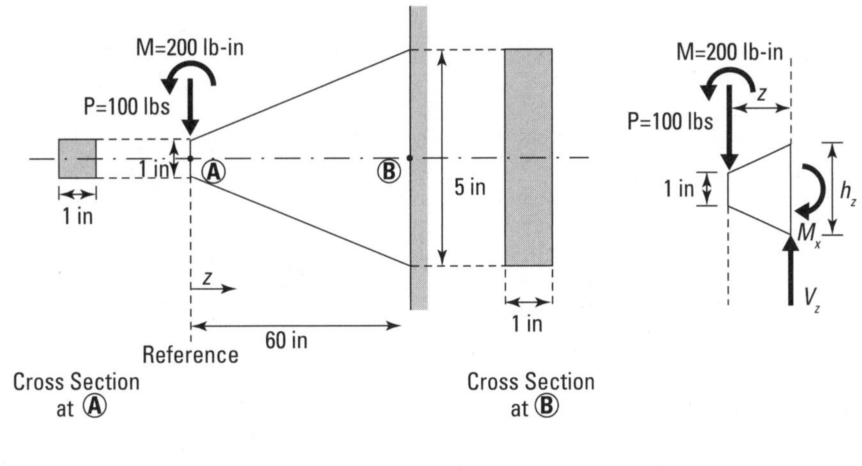

Figure 9-4: Tapered beam example.

***(a)* Tapered Beam**

(b)* Generalized Free-Body Diagram at Position *z

The formulas for bending that I present in the previous section are based on the assumption that a beam is prismatic and has the same uniform cross section along the longitudinal axis of the member. For a *tapered beam* (having a gradually changing cross section along the length) or a *stepped beam* (having multiple prismatic sections along the longitudinal axis), those basic formulas can still work, but you must break the beam into finite segments that are all prismatic. The accuracy of your solution becomes dependent on the number of segments that you use.

To solve this problem, you cut a general section at a distance z from the left end. To determine the internal moment M_x at this location, you simply apply basic statics (as I show in Chapter 3) to the free-body diagram (F.B.D.) shown in Figure 9-4b. The internal moment M_x at a position z is found to be $M_x = 100z + 200$ (lb-ft).

Likewise, the basic section modulus for the rectangular cross section at a location z from the left end is

$$S_z = S_{z,TOP} = S_{z,BOT} = \frac{1}{6}\left(1 + \frac{z}{14}\right)^2$$

You can find the normal stress at any arbitrary location x by substituting into the basic stress equation I show in the preceding section:

$$|\sigma_{zz}| = \frac{M_x}{S_z} = \frac{117{,}600z + 940{,}800}{(14 + z)^2}$$

Finally, all you need to do is plug in the desired value for z at your location of interest and then quickly compute the maximum stresses in a specific cross section at any location along the length of the beam.

At the location $z = 0$, you find that $|\sigma_{zz}| = 4{,}800$ psi; at $z = L = 60$ in (at the support at Point B), you discover that $|\sigma_{zz}| = 4{,}116$ psi. Notice that the maximum stress now occurs at a point different from the location of maximum moment.

For a beam with a rectangular cross section, the magnitude of the compressive stress at one edge of the cross section is equal to the tensile stress at the other end. In the Figure 9-4 example, the equation for the moment M_x at location z is clockwise on the right end of the F.B.D., which is negative, indicating the top of the beam is in tension and the bottom of the beam is in compression.

At this point, you currently have a function in terms of the position z that fully defines the stress, σ_{zz} at all locations along the length of the beam. You can now use basic calculus to take the derivative of this function and set it equal to zero:

$$\frac{d\sigma_{zz}}{dz} = 0$$

If you solve this function for z, you've then found the location of either a maximum or minimum value. After you have this location determined, you can substitute it back into the function to actually determine the maximum stress.

Chapter 10

Shear Madness: Surveying Shear Stress

In This Chapter

- Understanding shear stresses
- Performing calculations with different types of shear stresses
- Using shear flow diagrams to determine horizontal shear stresses

Imagine you're sitting at your desk and you reach for a piece of paper to jot down a quick note before you forget that brief moment of insight you gained while studying mechanics of materials. Out of habit, you quickly bend the piece of paper over the edge of the desk and rip it in half. You have just sheared this piece of paper.

In this chapter, I explore shear in more detail by building on the concept of average shear stress that I discuss in Chapter 6 and applying it to several basic applications. I then illustrate how flexural loads (from transverse loads and even concentrated moments) on beams can induce internal shear forces and how you can compute shear stresses once you know these shear forces. Finally, I show you how you can use shear flow to find shear forces at various locations within a cross section.

So, time to get ripping . . . err, shearing.

It's Not Sheer Folly: Examining Shear Stress

A *shear behavior* (or internal shear force) occurs when an object is loaded such that it results in a motion where one part of the object tries to move parallel past another. If you've ever heard clippers or scissors referred to as *shears,* that's because those tools' basic purpose is to shear objects. Scissors and hedge shears are devices intended to load an object in shear, and they cause shear stresses at localized points in order to exceed the material's capacity and ultimately cause separation (or cut).

As I first discuss in Chapter 6, a *shear stress* is a type of stress that acts on a cross-sectional area as the result of an internal shear load applied parallel to that cross section. Figure 10-1a shows an example of a basic shearing condition where a shear force V on one side of the vertical cut line is balanced by a matching force in the opposite direction on the other side. The shear stress element for this example is aligned such that one of the vertical edges is aligned with the cut line of the section.

The shear stress on the element (shown in Figure 10-1b) at the cut line (on the right side of this element) is developed in response to the applied shear loads acting parallel to the cross section. This shear stress is then balanced by the vertical shear stress on the left side of the element and the two horizontal shear stresses (one on the top and one on the bottom) in order to ensure that the element maintains equilibrium.

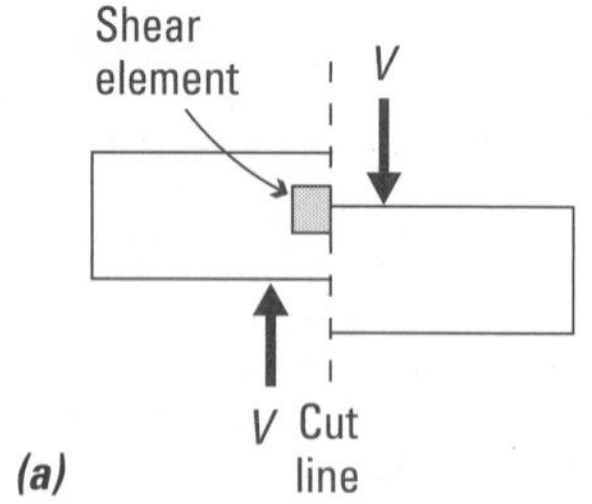

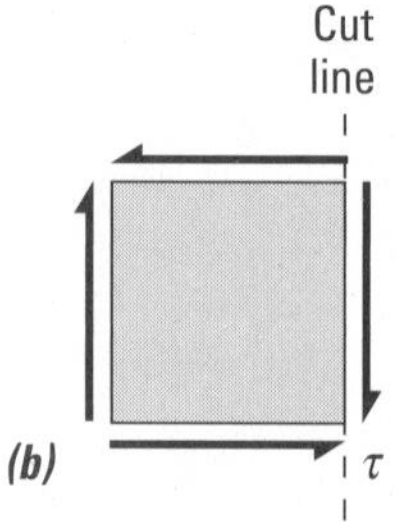

Figure 10-1: Shear deformation and shear stress element.

Stresses become especially important in glue seams and nails between objects in members that are built up from multiple smaller objects. Causes of shear stresses include direct shear loads on an object and flexural loads in bending members. Pure shear can also be caused by *torsional moments,* or moments around a longitudinal axis. I discuss these moments and their effects more in Chapter 11.

You measure shearing stress as you do all stresses: in force per area. These units are the same as the basic units of stress I cover in Chapter 6. The symbol for shear stresses in most classic textbooks (and this text as well) is the Greek symbol tau (τ) with some sort of subscript to denote a direction or plane on which the stress is acting.

Working with Average Shear Stresses

Unlike internal bending moments, which can cause normal stresses in multiple directions on a cross section, average shear stresses are more like axial stresses in that they generally are assumed to act in the same direction as the internal loads that create them.

Average stresses are very useful in determining shear stresses across the thickness of a plate or the diameter of a bolt because you calculate them under the assumption that the shear stresses are uniform (or constant) along the entire cross section. However, in some cases, this assumption isn't completely accurate, so I show you how to handle those cases in "Exploring Shear Stresses from Flexural Loads" later in the chapter.

Shear on glue or contact surfaces

Gluing two pieces of paper together is a simple example of an average shear stress situation. For example, consider the single glue seam connecting two plates as shown in Figure 10-2a. To compute the average shear stress, all you need to know is V_{INT}, the internal load acting on the contact surface, and A_{GLUE}, the area of the adhesive between the two plates. By drawing a free-body diagram of either plate (I show the lower plate in Figure 10-2b) and applying equilibrium equations, you can determine the internal shear on the glue seam, $V_{INT} = 10\ lbs$.

The dimensions needed to compute the area of the glue seam A_{GLUE} are revealed on the same free-body diagram — in this case, the area of the seam is 6 inches x 3 inches. Even though the plates being connected are larger, you can count only the dimensions of the actual adhesive seam.

With these values determined, you can now compute the average shear stress of this glue example: τ_{AVG} = (10 lbs) ÷ (6 in × 3 in) = 0.56 psi.

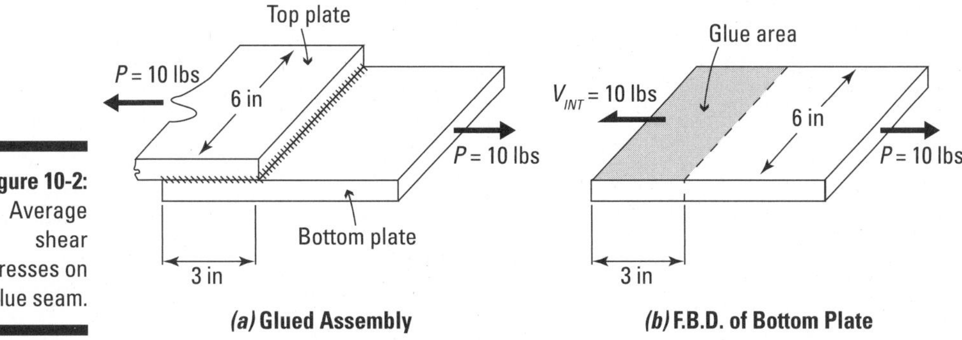

Figure 10-2: Average shear stresses on a glue seam.

Shear for bolts and shafts

Designers often use bolts and shafts in connections in order to transfer forces from one member to another.

When analyzing a connection to determine the internal forces applied, you must imagine what the connection looks like when it's separated and then picture the fracture path that would be required to separate the two parts in the connection.

Two common fracture paths are the single and double shear planes across shafts or bolts. For two plates bolted together, the fracture path is a *single shear plane* — it runs across a bolt at a single location. If you have three plates bolted together by a common bolt, the fracture path needs to cross the bolt at two locations (a *double shear plane*) in order to separate the plates. You can actually have many, many shear planes if you use the same bolt to connect all those plates. In the coming sections, I show you how to calculate the average shear stresses for single and double shear.

Single shear situations

The simplest connection, such as the one shown in Figure 10-3, is created when you connect two plates with a bolt, shaft, or rivet. If you look at the free-body diagram in Figure 10-3b, you can conclude that in order to separate the two connections, you need to slice across the bolt. If this cross section (known as a shear plane) fails and is unable to support the load, the connection can be separated. A connection where only one shear plane is stressed is known as a *single shear connection.*

For example, consider the connection shown in Figure 10-3a that shows a tension connection consisting of two plates subjected to 500 Newton of tension and connected by a 25-millimeter-diameter round bolt.

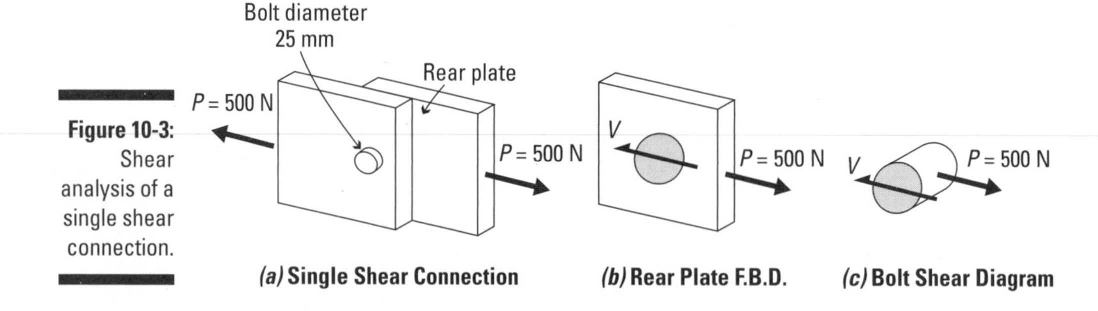

Figure 10-3: Shear analysis of a single shear connection.

If you want to determine the stress in the bolt due to this load, you first need to calculate the internal force. To find this force, you separate the connection into its two parts. By examining one of the plates, such as the rear plate, you can create a free-body diagram as shown in Figure 10-3b. To keep the rear plate in equilibrium, the force V across the bolt can be computed from statics as

$$\overset{+}{\rightarrow}\sum F_x = 0 \Rightarrow -V + P = 0 \Rightarrow V = P = +500 \text{ N}$$

where the plus sign indicates that the assumed direction of the force *V* on the free-body diagram was correct. (For more on static equilibrium, turn to Chapter 3.) This fact means that the force that must be transmitted through the bolt is 500 Newton.

Next, you need to find the shear plane on which the shear force is acting — in this case, the shaded area of Figure 10-3b and Figure 10-3c, which shows the cross section of the bolt. Use the following formula:

$$A_{BOLT} = \frac{\pi}{4}\left(d_{BOLT}\right)^2 = \frac{\pi}{4}\left(25 \text{ mm}\right)^2 = 490.87 \text{ mm}^2 = 4.90 \times 10^{-4} \text{m}^2$$

You can then compute the average stress felt by the bolt:

$$\tau_{AVG} = \frac{V}{A_{BOLT}} = \frac{500 \text{ N}}{4.90 \text{ x } 10^{-4} \text{ m}^2} = 1.02 \text{ MPa}$$

Double shear situations

A *double shear connection* has two stressed planes. You can find such connections in a clevis assembly, which is a common design for connections in building structures. In order to separate this connection, you have to cut the bolt in two places (along two different shear planes).

A *clevis assembly* is a connection consisting of a U-shaped member with a hole across each leg (or prong) that connects to another member with a hole by a single pin or bolt passing through each object.

The example in Figure 10-4a shows much the same connection as Figure 10-3 in the preceding section. Here, you have a single plate that is sandwiched between two other plates, connected with the same 25-millimeter-diameter bolt that is loaded by the same 500-Newton force.

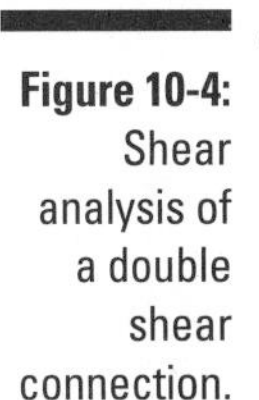
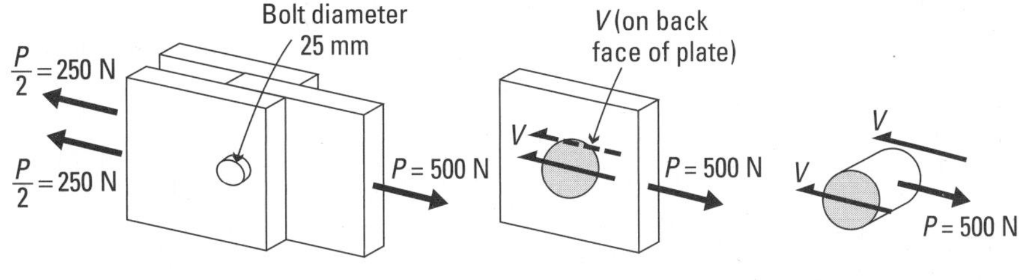

Figure 10-4: Shear analysis of a double shear connection.

Just as in the preceding section, you must use equilibrium to determine the internal shear force acting on each plane. In this example, if you apply equilibrium to the free-body diagram of Figure 10-4b, you see that $P = 2V$, which means that $V = 250$ Newton on each of the failure shear planes (the shaded regions shown on both ends of the bolt in Figure 10-4c). You can then determine the shear stress on each plane with the following equation:

$$\tau_{AVG} = \frac{V}{2A_{BOLT}} = \frac{500\text{ N}}{2\left(4.90 \text{ x } 10^{-4}\text{ m}^2\right)} = 0.51\text{ MPa}$$

Changing from a single shear connection to a double shear connection for the same load reduces the shear stress in the bolt or shaft by half. That means you can actually use a smaller bolt to hold the same load.

One way to increase the amount of shear load a bolt can carry is to increase the number of shear planes required to fully separate the shear connection. Generally speaking for a given load, the more shear planes a connection has, the lower the average shear stress is.

The general relationship for multiple shear planes on bolts and shafts is given by

$$\tau_{AVG} = \frac{P}{nA_{BOLT}}$$

where P is the applied load, A_{BOLT} is the cross-section area of one shear plane (not all of them), and n is the number of shear planes required to separate the connection.

Punching shear

Take the end of a paper clip in one hand and start pushing it against a sheet of paper in your other hand. Under small loads, the force may not be strong enough to cause the end to poke through the paper, but under larger loads, it may well break through. This type of loading is known as *punching shear,* and it's an average shear stress situation that occurs when one object is pushed through another (often bigger) object.

In some cases, such as in fabrication or tooling, this result may be an intentional consequence of the process. But in other situations, such as when a concrete slab is supported by a concrete column, if the loads are too big, or the physical dimensions of the slab or column are too small, punching shear may be the unfortunate result.

Consider the application of a round mechanical punch to a thin plate as shown in Figure 10-5a. The ⅛-inch-diameter punch exerts a force of 5,000 pounds on a 1⁄16-inch-thick plate.

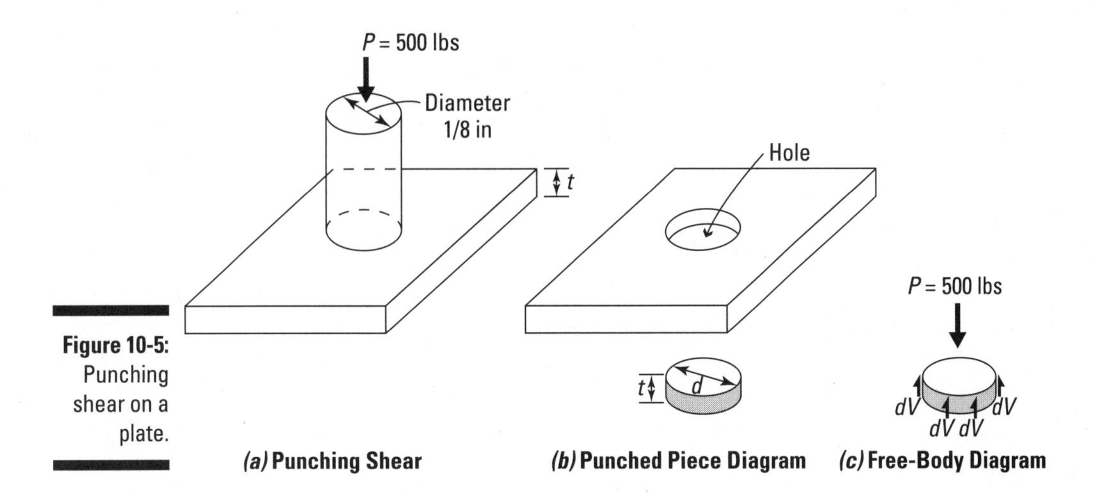

Figure 10-5: Punching shear on a plate.

To determine the area on which the internal shear forces are acting, consider the piece that must be removed in order to allow the punch to push through the plate (such as the one shown in Figure 10-5b). In this case, this removed piece is a disc (or a circular prism) that has the same diameter as the punch and the same thickness as the plate.

After you have this punched piece determined, you can then draw a free-body diagram showing the punching force P acting on the disc and the resulting internal shear forces dV that must act around the perimeter of the shape (or the shaded region of the disc in Figure 10-5c). Assuming that the force is shared equally around the entire shaded region, you then know that the total internal shear force $V = \Sigma dV$ and must be the same as the applied punch force $P = V = 500$ lbs. You can then compute the shaded area on which this shear force is acting as

$$A_{PUNCH} = \text{perimeter of punch} \cdot \text{thickness of punched object}$$

In this example, you calculate the perimeter as πd and the thickness of the plate as t. So the area of this punch is $A_{PUNCH} = \pi dt = (3.14)(0.125 \text{ in})(0.0625 \text{ in}) = 0.0245 \text{ in}^2$. Now that you know both the shear force and the shear area on which this force is acting, you can compute the average punching shear stress:

$$\tau_{AVG} = \frac{V}{A_{PUNCH}} = \frac{500 \text{ lbs}}{0.0245 \text{ in}^2} = 20{,}408 \text{ psi} = 20.4 \text{ ksi}$$

Exploring Shear Stresses from Flexural Loads

The average shear stress formulas I show in "Working with Average Shear Stresses" earlier in the chapter are highly inaccurate for flexural members. To calculate shear stress from flexure, you must use a different calculation. The fact is, shear stress in flexural members varies depending on the location of the point of interest within a given cross section. The basic relationship for flexural shear stress calculations is given by the following:

$$\tau = \frac{VQ}{It}$$

where V is the internal shear force on the cross-section at the location of interest, I is the moment of inertia (see Chapter 5), t is the width of the cross section at the position of interest, and Q is the first moment of area (see Chapter 5) of the position in the cross section about the neutral axis.

To change the shear stress distribution in a symmetric section, you must establish the shear stresses at various points within the cross section.

These points typically include the following:

- **Topmost point in a cross section.**
- **Neutral axis:** The neutral axis that you choose must be the neutral axis that's perpendicular to the applied shear load.
- **Bottommost point in a cross section.**
- **Locations where width changes values**: This location is particularly important in cross sections for I-sections, T-sections, channels, and angles because the changes in width actually cause shear stresses to change in magnitude.

Determining the shear stress distribution in uniform cross sections

To illustrate how you compute the shear stresses at specific points within a rectangular cross section, consider the cantilever beam of Figure 10-6a.

In Figure 10-6a, a cantilever beam supports a load of P = 10,000 lbs and has a rectangular cross section of dimensions 3 inches (width) x 12 inches (height) as shown in Figure 10-6b. Based on these dimensions, you can compute the moment of inertia as I_{xx} = 432 in^4. (For more on computing moments of inertia, turn to Chapter 5.)

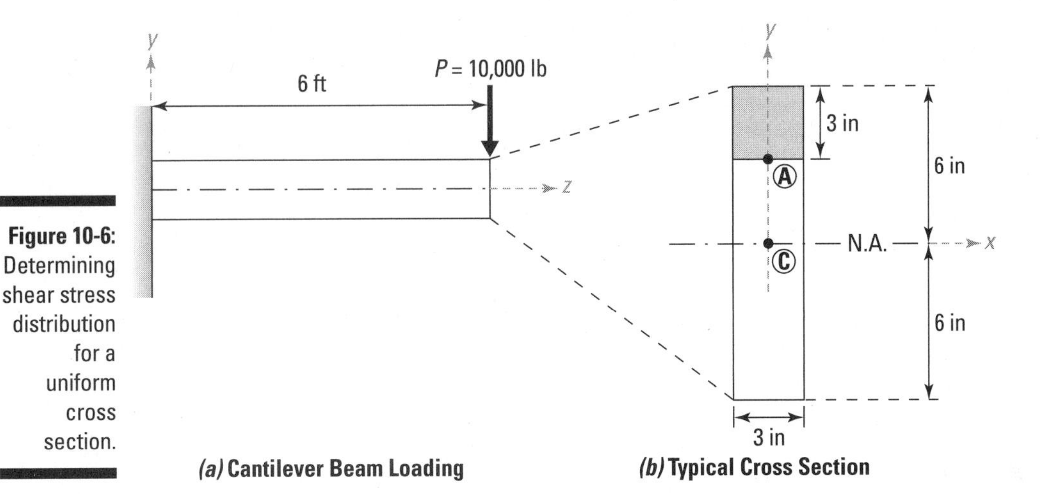

Figure 10-6: Determining shear stress distribution for a uniform cross section.

As I note in Chapter 5, the first moment of area Q of a point at the extreme top and bottom of a cross section must be equal to zero. That means that the corresponding shear stress for a vertical applied load at these locations is also zero. Thus, for this example, $\tau_{TOP} = \tau_{BOT} = 0$ psi.

The maximum shear stress actually occurs at the neutral axis (or Point C) because at this value, Q_{NA} is a maximum value. You compute the first moment of area about the x-neutral axis as $Q_{NA} = (6\text{in})(3\text{in})(3\text{in}) = 54\ \text{in}^3$.

You can now plug these figures into the flexural shear stress calculation as follows:

$$\tau_{MAX} = \frac{VQ_{NA}}{I_{xx}t} = \frac{(10{,}000\ \text{lbs})(54\ \text{in}^3)}{(432\ \text{in}^4)(3\ \text{in})} = 416.7\ \text{psi}$$

Similarly, you can find the stress at another point such as Point A, which is located 3 inches from the top of the cross-section (see the shaded region of Figure 10-6b). All you need to do is compute the first moment of area about the x-neutral axis:

$$Q_A = (3\text{in})(3\text{in})(4.5\text{in}) = 40.5\ \text{in}^3$$

and then incorporate the appropriate width at this location.

$$\tau_A = \frac{VQ_A}{I_{xx}t} = \frac{(10{,}000\ \text{lbs})(40.5\ \text{in}^3)}{(432\ \text{in}^4)(3\ \text{in})} = 312.5\ \text{psi}$$

Repeating this process at several more points, you can find the stresses at different locations and plot the shear stress for this example as shown in Figure 10-7.

As you can see from this stress distribution, the shear stress reaches its maximum value at the neutral axis and is zero at both the top and bottom. Between these points, the shear stress distribution is a second order curve (or a parabolic shape).

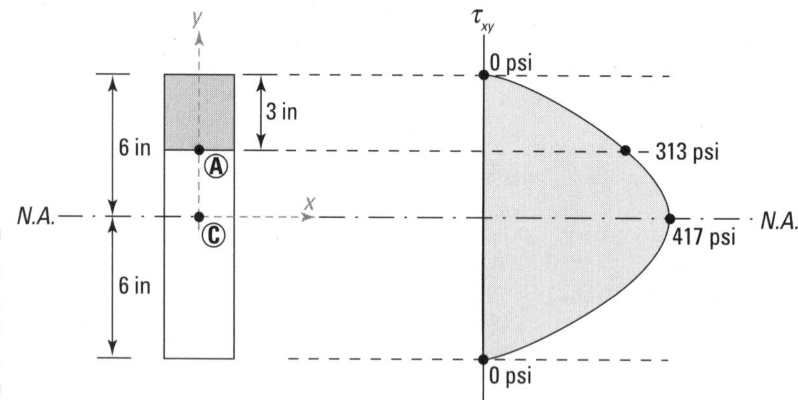

Figure 10-7: Drawing shear stress distribution.

Handling shear stresses in nonuniform cross sections

Determining the stress distribution in a nonuniform cross section (a cross section that has varying widths) is much like the procedure for rectangular cross sections (see the preceding section). However, you need to make one additional modification to the basic formulas for shapes where the cross section has a sudden change in thickness.

Consider the T-section I show in Figure 10-8 with a cross section that changes width at Point A.

The width above Point A in the cross section is greater than the width immediately below Point A. Q_A remains unchanged because you haven't changed position within the cross section. However, immediately above Point A, the shear stress is significantly less (because the width is greater) than immediately below Point A where the width is smaller. The shear stress experiences an instantaneous increase from one side of Point A to the other.

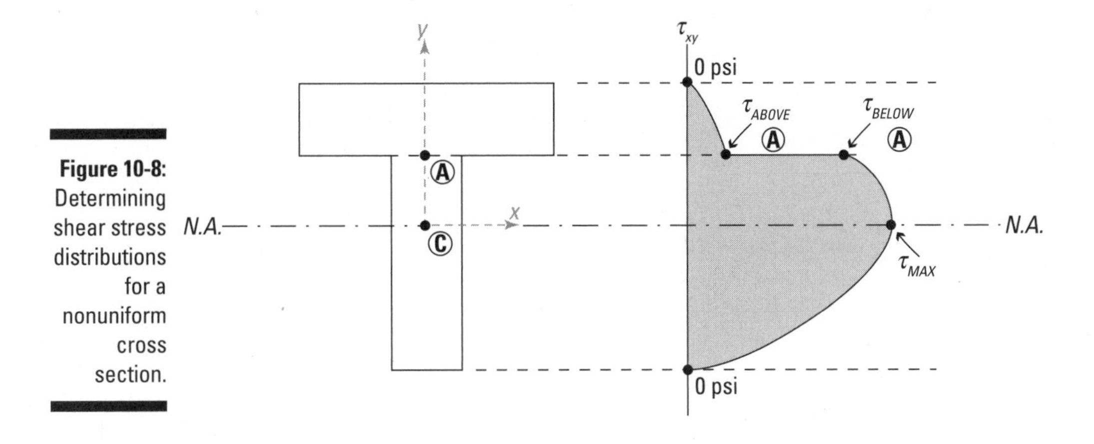

Figure 10-8: Determining shear stress distributions for a nonuniform cross section.

Calculating Shear Stresses by Using Shear Flow

As the earlier "Exploring Shear Stresses from Flexural Loads" section illustrates, sometimes working with shear stress diagrams can be a bit awkward when the stress jumps in magnitude, particularly if your cross section has sudden changes in thickness (such as with flanged cross sections like in I-sections or channels).

To eliminate this problem, you can modify the basic shear stress formula such that the thickness is initially eliminated from your calculations. In doing so, you can then determine shear effects in horizontal directions more conveniently, even when the applied shear force is acting vertically.

Going with the shear flow

The basic shear stress calculation I describe earlier in the chapter (where $\tau = VQ/It$) works best for determining shear stress in cross sections without openings or at connection locations where contact is continuous over an entire surface, such as a glue interface. For connections such as nails and screws where fastening ability is concentrated at a single location, the shear stress formula doesn't work as well as you may like.

Instead, you need to calculate the shear flow in a section along the shear plane of the fastener. *Shear flow* (usually depicted as a variable q) tells you how a shear stress accumulates as you move through the cross section of a member or along a cross section and is always located at the center line of the thickness of the member at a specific location.

Shear flow is zero at the ends of the cross section, while the maximum shear flow value occurs at the neutral axis.

Be careful that the units on your shear flow calculations are in force per length (N/m for SI units and lb/in in U.S. customary). Most shear calculations work with shear force or shear stress, which both have different units than shear flow.

Sketching shear flow diagrams

To illustrate how to create a shear flow diagram, consider the wide-flange (or I-section) shown in Figure 10-9a subjected to a shear force V acting downward across the cross section.

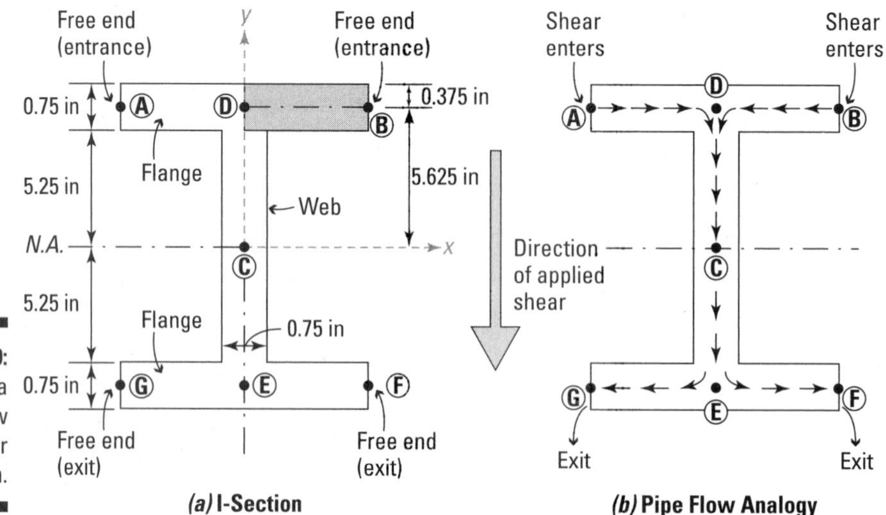

Figure 10-9: Sketching a shear flow diagram for an I-section.

You can think of this cross section as being a system of pipes as shown in Figure 10-9b. When you pour the shear in from the top (assuming it's acting downward), it enters at the free ends of the cross section and flows into the middle area and then out the free ends at the bottom. The farther the shear flow travels from the free end at the entrance, the more it increases, finally reaching a maximum value when it crosses the neutral axis. After crossing the neutral axis, it then begins to decrease as it moves toward the exit. In general, the direction of the shear flow follows the direction of the implied internal force.

When you start to create a shear flow diagram, you need to keep a couple of things in mind as you move through the process:

- **Determine the direction of flow based on the direction of internal shear along the web of the beam.** The *web* of a cross section is the portion in the middle of the cross section that passes across the neutral axis and connects the topmost portion of a cross section with the bottommost portion.
- **Shear flow is always applied at the center line of the region in which it is acting.**

To create a shear flow diagram for the cross section of Figure 10-9a (as shown in Figure 10-9b), you follow a few basic steps.

1. **Trace vertical stress requirements through the web of the member.**

   Usually, the web is oriented vertically, but in some sections it may actually be inclined. The vertical component of the shear flow arrow through the web is always in the same direction as the internal shear force, which in this example is from top to bottom (or downward) — from Point D to Point C to Point E.

2. **Draw flow lines (or arrows) from the free ends (entrance) at the top to the top of the web location.**

   As I indicate earlier, the shear flow at the free ends of the cross section must be equal to zero. For this example, one flow path goes from Point A to Point D. In some cross sections, you may have multiple free ends where the shear flow enters the cross section. Figure 10-9 has two entrance points; the second flow path runs from Point B to Point D.

3. **Draw flow lines from the bottom of the web(s) to the free ends (exit) of the cross section.**

   Just as with the entrance points, the shear flow at the exit points must also be equal to zero, so the shear flows from the bottom of the web to each of the exit points. That is, for this example, one path flows from Point E to Point G, and the other flows from Point E to Point F.

After you have the path determined, you can actually begin to calculate values of shear flow at important points, such as the ends of the member, the top and bottom of the web section, and the neutral axis. You also want to compute the values of shear flow at any corners or bends that may occur along the cross section.

Figuring shear flow quantities

When you have the directions of the shear flow established, you're ready to actually begin calculating the numerical values.

The computation required for computing the shear flow q in a cross section is given by the basic relationship

$$q = \frac{VQ}{I}$$

where V is the applied vertical shear force on the cross section, Q is the first moment of area for the cross section, and I is the moment of inertia of the cross section. (Head to Chapter 5 for more on moments of area and moments of inertia.)

This relationship is basically the same as the shear stress equation earlier in the chapter except that it removes the thickness from the calculation, making $q = (\tau)(t)$.

Shear flow is actually a gradient of the shear stress and is independent of thickness, whereas a shear stress actually changes instantly when a cross section thickness varies.

To compute the shear flow q_1, which is the value of the shear flow as it moves from Point A to Point D, you simply take a rectangular region — for this region, it's the entire upper-left flange — and calculate the first moment of area Q for this region: Q_1 = (4 in)(0.75 in)(5.625 in) = 16.875 in^3. The shear flow from q_1 at Point A is then given as

$$q_1 = 16.875\frac{V}{I}$$

Because the flange is oriented parallel to the neutral axis, the shear flow increases linearly from zero at Point A to the value of q_1 at Point D.

Similarly, you calculate the shear flow q_2 in the same fashion. Because the dimensions of the flange between Point B and Point D are the same as the flange from Point A to Point D, q_2 must also be equal to q_1.

At Point D, where the flows from each of the entrance points combine into one shear flow value moving down the web of the cross section, you can compute the flow at the top of the web by adding the flows from the ends of each of the flanges:

$$q_3 = q_1 + q_2 = (16.875 + 16.875)\frac{V}{I} = 33.75\frac{V}{I}$$

Finally, you can compute the maximum shear flow q_4, which also occurs at the neutral axis. To compute the first moment of area, Q_4 = (2)(16.875 in^3) + (5.25 in)(0.75 in)(2.625 in) = 44.09 in^3. With this value determined, you now know that the shear flow at the neutral axis is

$$q_4 = 44.09\frac{V}{I}$$

Because the shear flow from Point D to Point C is moving toward the neutral axis (as opposed to parallel to it), the shear flow distribution is parabolic. The lower portion of the shear flow diagram is similar to the upper because of the symmetry of this shape about the horizontal neutral axis. Figure 10-10 shows this final shear stress distribution.

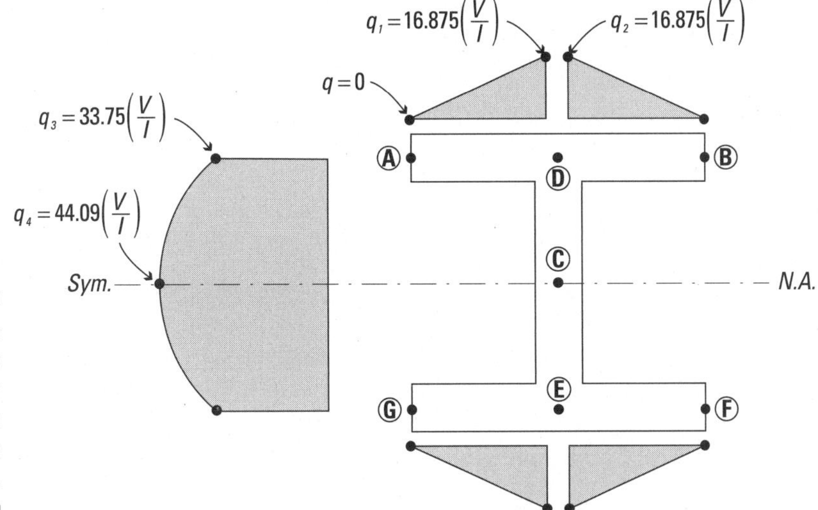

Figure 10-10: Shear flow diagram for I-section example problem.

Finding shear centers from shear flow diagrams

You can use a completed shear flow diagram to actually calculate the equivalent shear forces in different parts of a cross section, such as horizontal forces in the flanges or vertical forces in the vertical web of an I-section. The force in a particular part of the object is equal to the area under the shear flow diagram and can be computed from the basic integral

$$F = \int q(x) \cdot dx$$

where q is the shear flow in that part. With these shear forces determined, you can then determine a very special point in a cross section by summing moments to find the point where all of the shear forces are balanced rotationally. This point, known as the *shear center,* is the point through which you can apply an applied shear force without causing a twisting or rotational effect on the cross section. The shear center is always located on an *axis of symmetry* (see Chapter 4) if the object has one. If an object has two axes of symmetry, the shear center occurs at the intersection of those two axes. If a shear force is applied to a cross section at any location other than the shear center, it creates a torsion effect (which I cover in Chapter 11), and the object twists about its longitudinal axis. This occurrence is a major problem for some sections, such as channel sections (C-shapes) as shown in this figure.

(continued)

(continued)

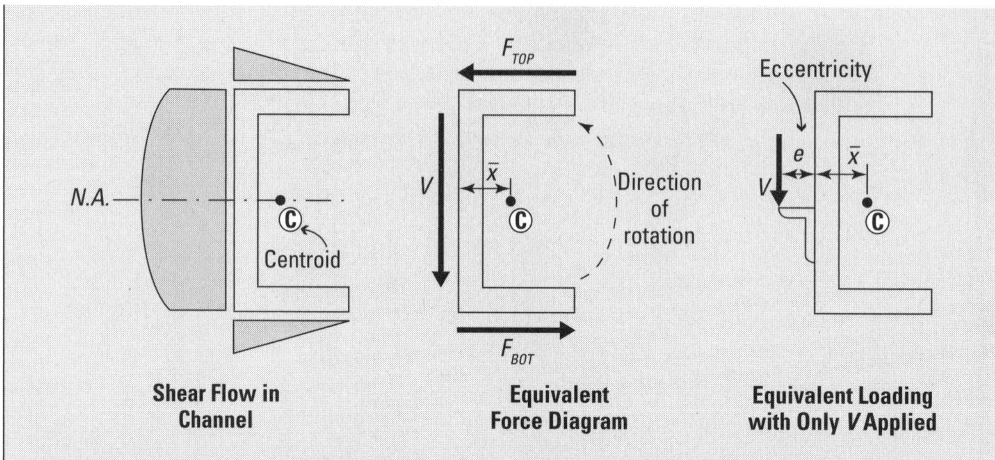

As you can see, the shear flow diagram and, consequently, the resulting forces F_{TOP}, F_{BOT}, and V induce a rotational effect about the centroid at Point C. These forces together want to rotate the shape in a counterclockwise direction. The equivalent location of just the vertical applied force V would have to be applied at Point E to provide an equivalent force-couple system. In these shapes, the shear center can be located outside of the cross section.

Chapter 11

Twisting the Night Away with Torsion

In This Chapter

- Developing the assumptions for torsion
- Exploring torsional stresses
- Determining shear stresses in various cross sections

Other chapters in Part II look at different types of loading and the stresses they cause on an object. In this chapter, I show you another type of loading: torsion, which creates shear stresses in the object on which it acts. From spinning shafts in a hydroelectric power generating station to the simple task of turning a doorknob or twisting the cap off a bottle of your favorite beverage, torsion is an everyday part of your life.

In this chapter, I show you some of the basic assumptions required for classical torsional analysis, and how you can use these assumptions to determine the shear stresses in circular and non-circular cross sections. I then explain a basic methodology for performing basic shear stress analysis of cross sections of multiple enclosed cells.

Considering Torsion Characteristics

A torsional moment, sometimes referred to as *torque*, is a moment that acts around a longitudinal axis of the object. Unlike bending moments, which cause a beam to curve, torsional moments cause an object to twist.

Torsional moments vary from bending moments in that they always rotate about an axis that isn't in the plane of the cross section. For an object with a cross section in the XY plane, a torsional moment acts around the *z*-axis. For the same cross section in the XY plane, moments about the *x*- and *y*-axes produce a bending result, which creates normal stresses on a stress element (as I explain in Chapter 9).

Defining the direction of torsion can sometimes be a bit difficult because you have to be thinking about which end of the object the torque is acting on. Consider the object shown in Figure 11-1a, which is subjected to a torque T acting in opposite directions on the end of the shaft. In the position shown, these torsional moments represent a positive torque situation as defined by the right-hand rule for moments in Chapter 3.

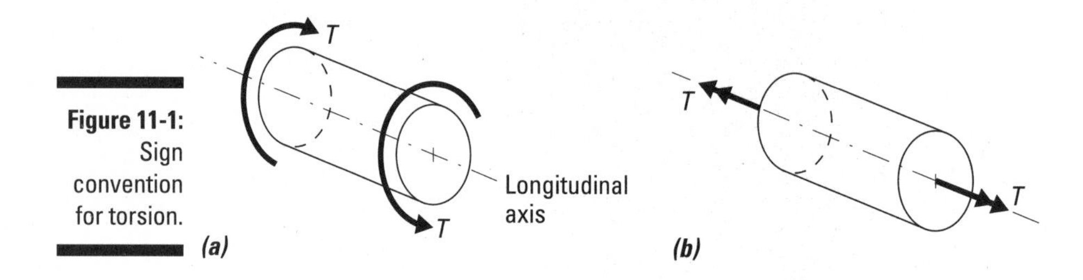

Figure 11-1: Sign convention for torsion.

You can also display this convention by using the double-headed vector notation that I describe in Chapter 3. For the bar shown in 11-1a, the corresponding double-headed notation has both of the double arrowheads pointing away from the object (as in Figure 11-1b), which indicates a positive torque.

As with all stresses, the units of a shearing stress from torsion are measured in a force-per-area form. These units are the same as the basic units of stress I discuss back in Chapter 6.

Working with Shear Stresses Due to Torsion

You need to keep a few factors that affect an object's response to torsion in mind when you start analyzing those twisting objects. The following factors play a key role in an object's response:

- **Internal torque:** Like all stress calculations, you must first determine the internal loads from statics that act on the cross section. For torsion problems, those loads are the internal torsional moments about the longitudinal axis of the member. If you increase the torque on a cross section of a shaft, you also increase the shear stress.
- **Position in cross sections:** Shear stresses from torsion are directly related to the distance from the centroid (or longitudinal axis) of the cross section to the point at which you're calculating the shear stress. You must measure this position radially outward from the centroid.

- **Torsional constant:** The *torsional constant* is a section property that is a measure of the resistance of a beam or shaft to applied torque and is a function of the cross section's geometric dimensions and shape. Different shapes of cross sections behave differently under torque. Round shapes are the simplest, and that's where I start the discussion later in the chapter.

An important assumption I make in this book is that any torque applied to the shaft doesn't cause permanent deformation after it's removed. The ability of an object to rebound to its initial position is known as *elasticity* (which I discuss further in Chapter 14).

Defining the shear stress element for torsion

Just as with axial stress in Chapter 8, the direction of the shear stress due to torsion on an exposed face is always in the same direction as the internal torsion on that face. Consider the object shown in Figure 11-2a.

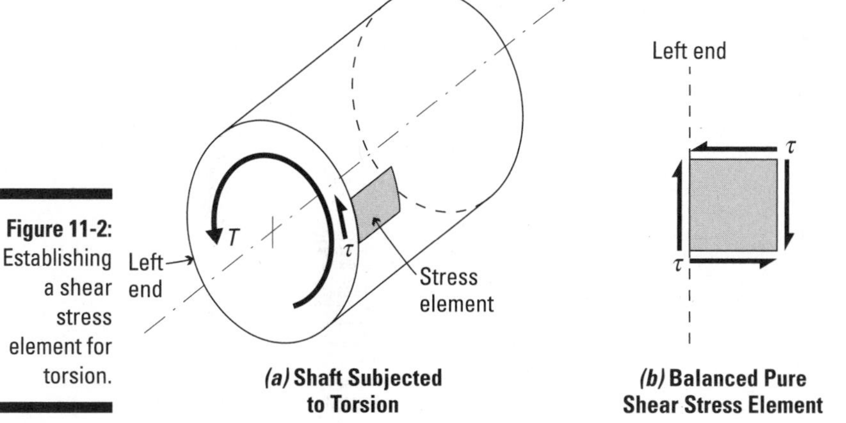

Figure 11-2: Establishing a shear stress element for torsion.

Torque applied to round or circular cross sections (such as many mechanical shafts), creates stress on an element in a state of *pure shear* — that is, a state of stress with no normal stress, only shear stresses.

If torsion combines with other effects, such as bending or axial loads, you can actually have an element with both normal and shear stresses on it as well. Don't worry too much about combined stress effects here; I clarify more about them in Chapter 15.

For the shaft of this example (which is subjected to positive torsion), if you draw an element such that the left face of the element lies on the left end of the bar, the shear stress that appears on the left end of the shaft must be acting in the same direction as the internal torque T on the cross section. Using the methodology for balancing a stress element (see Chapter 6), you can develop the pure shear stress element shown in Figure 11-2b.

Computing the torsional constant

The *torsional constant* is a section property that describes an object's resistance to a torsional moment. In classic mechanics, the torsional constant is the same constant that St. Venant helped to develop and is represented by the variable J. The larger this constant is numerically, the more torque you need to apply to make an object twist a given amount.

The most common (and perhaps the simplest) cross sections to analyze for torsion are circular shafts. Solid shafts and hollow circular tubes behave very similarly under torsion loads, and consequently have very similar torsional constants. For a circular sections

$$J = \frac{\pi}{4}c^4$$

and for hollow circular tube sections

$$J = \frac{\pi}{4}\left(c_o^4 - c_i^4\right)$$

where c is the radius of a solid circular shaft; c_o is the outer radius of a hollow circular shaft; and c_i is the inner radius of the hollow circular shaft (see Figure 11-4 later in the chapter for reference).

You may recognize that the values for the torsional constants for round cross sections are actually identical to their polar moments of inertia (see Chapter 5). Although mathematically this observation makes little difference in your calculations, it's a very important yet subtle difference.

Circular cross sections don't experience any unusual response to torsion. The behavior of circular shafts is fairly uniform because as a round shaft twists, the cross section doesn't experience any distortion, also known as *warping*.

The torsional constant is vastly different for non-circular shapes. For non-circular cross sections, torsion causes the cross section to distort, so you need a different approach to estimate their torsion constant. In fact, in many cases you can't compute this constant directly; you have to experimentally obtain it. Fortunately, scientists and physicists have already done this task for many cases and have tabulated the coefficients that can help you approximate this value in many references and textbooks.

For a single rectangular cross section, the torsional constant is $J = \beta ab^3$, where a is the long dimension of the rectangle and b is the short dimension (usually the thickness). The coefficient β is determined based on the proportions of the two sides of the rectangle a/b. Table 11-1 shows several values for β.

Table 11-1 Warping Coefficient for Rectangular Sections

a/b	1.0	1.5	2.0	5.0	10.0	Infinity
β	0.141	0.196	0.229	0.291	0.312	0.333

For composite shapes consisting of multiple rectangular sections, you can get an approximation of the torsional constant by summing the torsional constants for each of the individual rectangular regions that make up the composite shape. For example, consider the T-section in Figure 11-3.

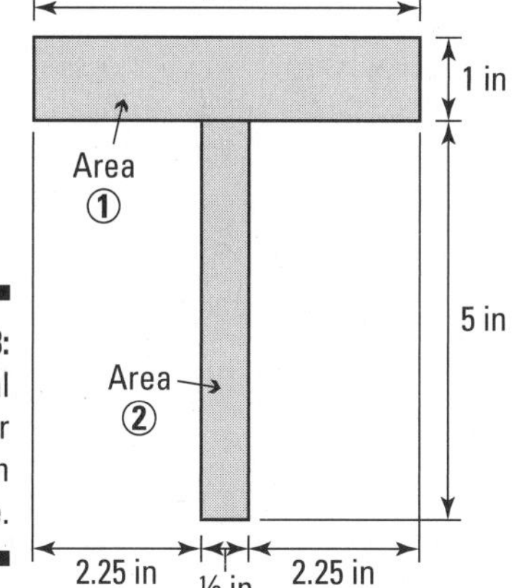

Figure 11-3: Torsional constant for T-section example.

For this shape, you can divide the region into two rectangular shapes and compute the torsional constant for each. With this torsion constant computed, you can approximate the shear stress of non-circular shafts, which I discuss a little later in the chapter.

$$\text{For Area 1: } \frac{a}{b} = 5 \Rightarrow \beta = 0.291 \Rightarrow J_1 = (0.291)(5\text{ in})(1\text{ in})^3 = 1.455\text{ in}^4$$

$$\text{For Area 2: } \frac{a}{b} = 10 \Rightarrow \beta = 0.312 \Rightarrow J_2 = (0.312)(5\text{ in})(0.5\text{ in})^3 = 0.195\text{ in}^4$$

$$J_{TOT} = \sum J_i = J_1 + J_2 = 1.455\text{ in}^4 + 0.195\text{ in}^4 = 1.650\text{ in}^4$$

Computing Shear Stress from Torsion

Most torsion examples in a basic mechanics of materials textbook require that your computations be conducted on solid round or hollow circular sections (which I highlight in the next section). However, you can also use other techniques for evaluating torsion on non-circular cross sections and cross sections of multiple cells, which I explain a little later.

Tackling torsion of circular shafts

Basic torsional stress formulations center on the assumption that deformation of a point in the cross section of a shaft is directly proportional to the distance of that point from the center of the shaft (or the longitudinal axis). To determine the shear stress τ at any point within a circle cross section

$$\tau = \frac{Tr}{J}$$

where T is the internal torque on the cross section; J is the torsional constant for the cross section (which I discuss in the preceding section); and r is the direct distance from the center (or longitudinal shaft) to the point of interest. You must measure r radially within the cross section.

If you plot this relationship (as I do in Figure 11-4a), you can see that for a solid circular shaft, the stress increases radially from the centroid, where the shear stress is zero, to its maximum value τ_{MAX} at the *outer fiber* (or the outer edge of the cross section) of the shaft.

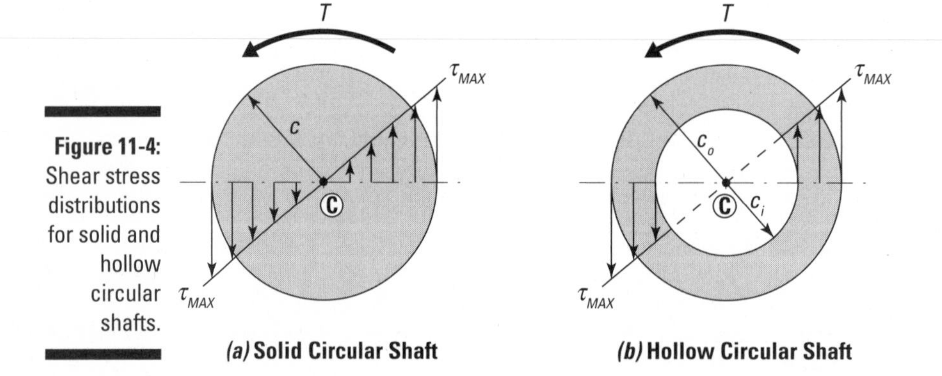

Figure 11-4: Shear stress distributions for solid and hollow circular shafts.

For a hollow shaft, the maximum stress is still located at the outer fiber, but the stress distribution is truncated such that no stress is acting on the empty region of the hollow shaft (see Figure 11-4b).

If you look at Figure 11-4, you may notice that for a hollow shaft, only a small portion of the total shear stress is actually missing from the middle of the shaft — that is, that the middle of the shaft is never fully stressed. That means that the outer portions of a circular cross section carry the majority of the torque. If designed properly, hollow tubes are generally more efficient when dealing with torsion because a hollow shaft typically weighs less but can support similar torsional moments.

To compute the maximum stress in a solid round shaft with a radius of 200 millimeters (or 0.2 meters) and an applied torque of 5,000 Newton-meters,

$$\tau_{MAX} = \frac{Tc}{J} = \frac{(5{,}000 \text{ N-m})(0.2 \text{ m})}{\frac{\pi}{4}(0.2 \text{ m})^4} = 796.2 \text{ kPa}$$

where T is the internal torque on a cross section; J is the torsional constant; and c is the outer radius of the cross section.

Determining torsion of non-circular cross sections

For non-circular sections (such as the T-section shown in Figure 11-3 earlier in the chapter), the largest shear stresses occur at the centers of the cross section's faces. For rectangular sections the maximum shear stress is

$$\tau_{MAX} = \frac{Tt}{J}$$

where T is the internal torque on the cross section; t is the thickness of the cross section perpendicular to the longest side; and J is the torsional constant that I show in "Computing the torsional constant" earlier in this chapter.

For example, a rectangular beam with dimensions of 5 inches (height) x 1 inch (thickness) is subjected to an internal torque of 1,200 pound-inches. The maximum shear stress in the 5-inch sidewall is computed as

$$\tau_{MAX} = \frac{Tt_{MAX}}{J} = \frac{(1{,}200 \text{ lb-in})(1.0 \text{ in})}{0.291(5.0 \text{ in})(1.0 \text{ in})^3} = 824.7 \text{ psi}$$

Note: In this equation, 0.291 = β as given by Table 11-1 (for a/b = 5.0) earlier in the chapter.

St. Venant: Jack of all (engineering) trades

Adhémar Jean Claude Barré de Saint-Venant (1797–1886) was a mathematician who did considerable work in the areas of mathematics and fluid and solid mechanics. In fluid mechanics, he was the first to publish the correct derivation for the Navier-Stokes equation, which describes the flow of viscous fluids. In mathematics, he developed a version of vector calculus known as *exterior differential forms.* And in solid mechanics, he was one of the first to recognize that every cross section, except for circular shapes, experiences some degree of warping when subjected to torsion loading. His principle is what shows that circular cross sections have the greatest torsional stiffness. Accounting for the effects of warping requires experimental data, some of which you can see in the section "Computing the torsional constant" in this chapter.

Applying shear flow to torsion problems in thin-walled sections

A very useful method known as the *Bredt-Batho theory* can greatly simplify torsion problems of thin-walled sections. In fact, for this theory to work, the shape no longer even has to be circular, and you don't need to compute those pesky torsional constants (J) that I describe earlier in this chapter.

The theory utilizes the idea that in a thin-walled object, the shear flow q resulting from torsion is constant at all locations around the cross section. Further, it states that you can express the average shear stress at any location as the product of the shear flow q and the thickness t at that location: $\tau_{AVG} = (q) \div (t)$.

This theory says that the average stress at a given location can be given by

$$\tau = \frac{T}{2tA_M}$$

where T is the internal torque; t is the thickness at a given location; and A_M is the area of the cross section bounded by the median centerline of the outer wall thickness (see Figure 11-5).

However, as with other torsion methods, you need to keep some limitations in mind:

- **Sections are thin walled.** A *thin-walled section* is a cross section where thickness of the object is significantly smaller than the dimensions of the cross sections.
- **Sections are closed sections.** A *closed section* is a section, such as a tube or hollow section, with an outer perimeter that closes on itself. Objects with slits don't count as closed sections.

- **Shear flow is constant.** The shear flow (which I introduce in Chapter 10), is assumed to be constant around the cross section that's subjected to torsion.

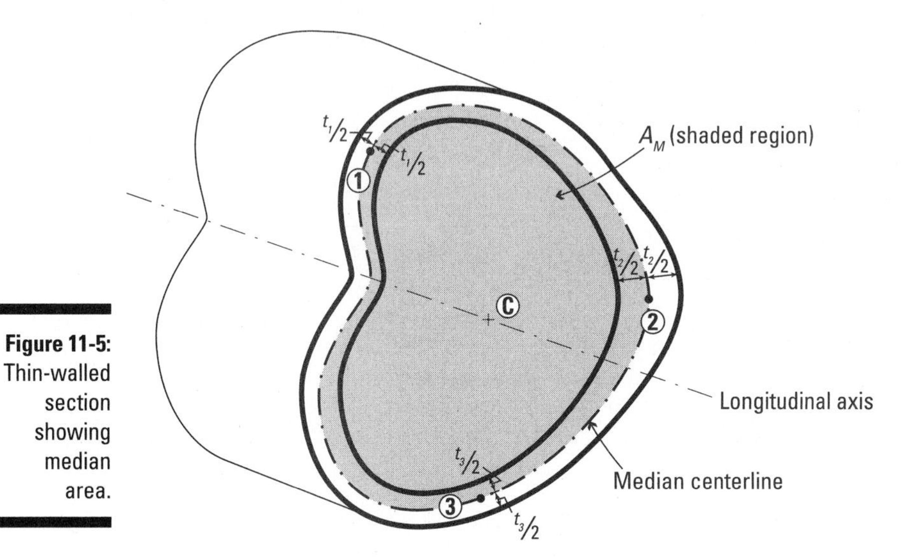

Figure 11-5: Thin-walled section showing median area.

Consider the rectangular tube shown in Figure 11-6a, which has a height of 200 millimeters, a width of 90 millimeters, and a wall thickness of 10 millimeters.

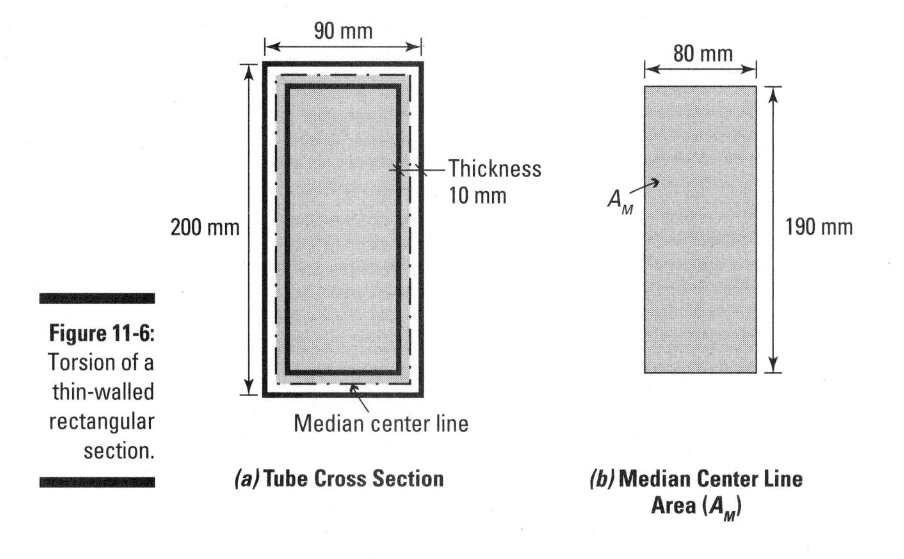

Figure 11-6: Torsion of a thin-walled rectangular section.

You can find the median center line area in Figure 11-6b with the equation A_M = (80 mm)(190 mm) = 15,200 mm². If the applied torque is 5,000 Newton-meters, you can determine the shear stress as

$$\tau = \frac{T}{2tA_M} = \frac{(5{,}000\text{ N-m})}{2(15{,}200\text{ mm}^2)(10\text{ mm})}\left(\frac{1{,}000\text{ mm}}{1\text{ m}}\right)^3 = 16.45\text{ MPa}$$

Using shear flow to analyze torsion of multicell cross sections

In some structures, a cross section is subdivided into several small compartments that are attached to an outer shell or skin. This type of construction is common in ship hulls and airplane wings, where intermediate stiffeners help transmit forces from an inner region of a cross section to an outer skin. These types of systems are referred to as *multicell cross sections* (see Figure 11-7.)

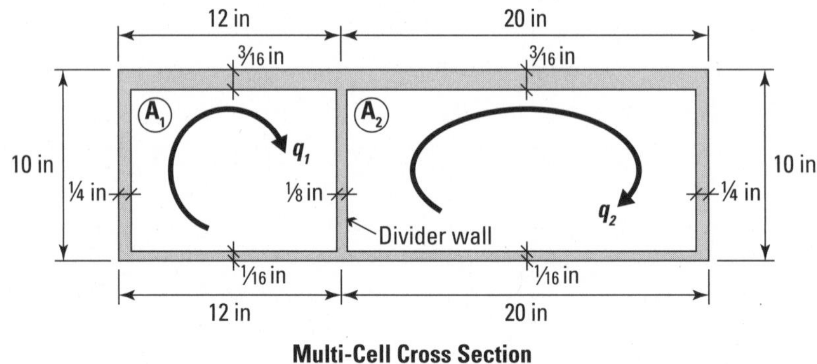

Figure 11-7: Multicell cross section example.

In multicell systems subjected to torsion, each of the individual compartments works to resist applied torsion. Using statics, you can show that the total torque T that can be applied to a multicell region is

$$T = 2q_1A_1 + 2q_2A_2$$

where q_1 and q_2 are the shear flows of each of the cells, and A_1 and A_2 are areas of each of the cells. From this equation, you can see that the system is statically indeterminate (see Chapter 3) because at this point you don't know the individual values of the cell's shear flow at the start. To provide the extra information, you have to look at the deformation behavior equations of the element, known as *compatability equations* (which I explore in more detail in Chapter 16). For now, I just mention the relationships that you use:

$$w_{11}q_1 + w_{12}q_2 = 2A_1\Psi$$

$$w_{21}q_1 + w_{22}q_2 = 2A_2\Psi$$

where Ψ is the twist per unit length of the member, and w_{11}, w_{12}, w_{21}, and w_{22} are equations that correlate the deformation between each of the cells. These formulas are of the following basic relationship:

$$w = \frac{1}{G}\sum\left(\frac{\text{side length}}{\text{wall thickness}}\right)$$

w_{11} is for Cell 1; w_{22} is for Cell 2; and $w_{12} = w_{21}$ is for the portion that connects the two individual cells — the divider wall. w_{12} and w_{21} are also a negative value.

The variable G that you see in the equation is a material constant for shear that I discuss in Chapter 14. As long as the whole cross section has the same G value, this shear constant doesn't ultimately affect the calculations. For now, you can just leave this value in its variable form. With these basic relationships, you can then set up a system of equations related to the parameter Ψ, which then allows you to compute the shear flow in each cell, and ultimately the shear stress.

To determine the stress in the walls of a multicell section (such as Figure 11-7), just follow these steps.

1. **Compute w_{11}, w_{12}, w_{21}, and w_{22}.**

   For Figure 11-7, use the following equation. Remember that when you calculate w_{12} and w_{21} for the divider wall, you need to make these values negative.

   $$\text{Cell 1: } w_{11} = \frac{1}{G}\left(\frac{10}{0.25} + \frac{12}{0.1875} + \frac{10}{0.125} + \frac{12}{0.0625}\right) = \frac{376.0}{G}$$

   $$\text{Cell 2: } w_{22} = \frac{1}{G}\left(\frac{10}{0.25} + \frac{20}{0.1875} + \frac{10}{0.125} + \frac{20}{0.0625}\right) = \frac{546.7}{G}$$

   $$\text{Divider: } w_{12} = w_{21} = -\frac{1}{G}\left(\frac{10}{.125}\right) = -\frac{80.0}{G}$$

2. **Substitute the correlations from Step 1 into the compatibility equations and solve for the twist per unit length, or Ψ.**

   If A_1 = (12 in)(10 in) = 120.0 in^2 and A_2 = (20 in)(10 in) = 200.0 in^2, you can write the compatibility equations as follows:

   $$w_{11}q_1 + w_{12}q_2 = 2A_1\psi \Rightarrow (376.0)q_1 - (80.0)q_2 = 2(120.0)G\psi = (240.0)G\psi$$

   $$w_{21}q_1 + w_{22}q_2 = 2A_2\psi \Rightarrow (-80.0)q_1 + (546.7)q_2 = 2(200.0)G\psi = (400.0)G\psi$$

3. **Use basic algebra to solve these two equations simultaneously for q_1 and q_2.**

   For this example, $q_1 = 0.819G\Psi$ and $q_2 = 0.852G\Psi$.

4. **Substitute the two expressions from Step 3 into the original equilibrium equation to solve for $G\Psi$.**

   Suppose the multicell cross section of Figure 11-7 is subjected to an internal torque T of 200,000 lb-in. You can solve for $G\Psi$ as follows:

$$T = 2q_1A_1 + 2q_2A_2 = 2(0.819G\psi)(120.0) + 2(0.852G\psi)(200.0) = 200{,}000$$
$$\Rightarrow (537.4)G\psi = 200{,}000$$
$$\Rightarrow G\psi = 372.2$$

5. **Determine the shear flow in each section by substituting the value of $G\Psi$ back into the q_1 and q_2 equations.**

   For Figure 11-7, that math looks like this:

$$q_1 = 0.819G\psi = 0.819(372.2) = 304.8\frac{\text{lb}}{\text{in}}$$
$$q_2 = 0.852G\psi = 0.852(372.2) = 317.1\frac{\text{lb}}{\text{in}}$$

6. **Determine the shear stresses in the outer walls.**

   As Chapter 10 states, the shear flow $q = \tau t$. The shear stresses are maximum in the thinnest walls of each section. For the outer walls of cell #1, $\tau_{MAX} = (304.8 \text{ lb/in}) \div (0.0625 \text{ in}) = 4{,}877$ psi. For the outer walls of cell #2, $\tau_{MAX} = 317.1 \text{ lb/in} \div (0.0625 \text{ in}) = 5{,}074$ psi.

7. **Calculate the shear stresses in the divider walls.**

   The shear stresses in the divider walls are a little different. Notice that for Cell 1, a clockwise shear flow q_1 acts downward through the divider wall. On the other hand, for Cell 2, a clockwise shear flow q_2 acts upward through the divider wall. This discrepancy means that the two shear flows are fighting against each other. Thus, the shear stress in the divider wall can be expressed as the difference of the shear stress from each of the shear flows of the adjacent cells:

$$\tau_{\substack{MAX \\ DIVIDER}} = \frac{q_1}{t_{DIVIDER}} - \frac{q_2}{t_{DIVIDER}} = \frac{(304.8 - 317.1)\frac{\text{lb}}{\text{in}}}{(0.125 \text{ in})} = -98.4 \text{ psi}$$

Although this example deals with shear stresses for rectangular sections, the shape of the section really makes no difference in the calculations as long as you know the length of each wall (or an arc-length if it's curved) and the corresponding uniform thickness for that segment. The only other major requirement is that the wall's thickness must be much smaller than the dimensions of the wall's length. You can also extend this methodology to cells of any combination of shapes and size. Just remember that each cell within the cross section has its own unique value for shear flow.

Part III
Investigating Strain

"We don't use a strain gauge. Usually, we just push Phil out the window, and if we don't hear a thump, we know our calculations were right."

In this part . . .

Strains are the second major fundamental area of mechanics of materials, so in this part I explain the basic concept of strain and how to calculate both longitudinal and shear strain values. I show how you can use these values to determine maximum and minimum strain values (known as *principal strains*) and their orientations by using strain transformation techniques. I also define the different material properties that let you relate load to deformation and, more importantly, stress to strain.

Chapter 12

Don't Strain Yourself: Exploring Strain and Deformation

In This Chapter

- Defining the basics of strain
- Computing normal and shear strains
- Analyzing objects under thermal loads
- Introducing plane strains

When you're working in your yard on Saturdays, strain is what you feel when you overstuff that bag of leaves and then try to move it. It starts as a pain in your lower back and can eventually lead to you blacking out on the front lawn or experiencing a loss of feeling in your toes.

Fortunately for you (and your insurance provider), strain in mechanics of materials is usually a bit less painful. When an object is trying to support a given load, *strain* is the effort with which it's trying to resist; you measure it in terms of a deformation.

Assumptions regarding deformation vary a bit from statics to mechanics of materials. Statics equations are limited to *rigid bodies* (bodies that don't experience deformation under a load). However, in the real world, this scenario can't be entirely true, simply because all objects are actually *deformable bodies* — they experience deformation when they're loaded.

In this chapter, I describe one of the most important links between statics and deformation: the concept of strain in objects. I also introduce you to several different types of strain that an object can experience under a load. Finally, I cover thermal strains, which are caused by changes in temperature.

Looking at Deformation to Find Strain

Strain is a measure of the deformation of an object in response to internal loads, so the more you strain an object, the more it deforms.

Although you can technically only use the equations of equilibrium in Chapter 3 on rigid bodies, you usually find that the deformations of an object under normal loads typically remain so small that you can reasonably assume that their actual magnitude is almost zero. For this reason, you can usually say that a deformable object with very small deformations is basically the same as a rigid body, which then allows you to use the basic statics relationships of equilibrium.

Unlike stress in an object (see Chapter 6), which you can't actually see, deformation is a visible and measurable quantity. When you pull on a tension rod, you can see the rod physically increase in length (or elongate). When you bend a beam, you see it curve. Deformations are a direct indicator of strain. And the best part is that unlike stress, you can physically measure deformation, which lets you compute strain. (I explain more about how to do that in Chapter 13.)

In mechanics of materials, you work with two basic types of strain:

- **Normal strains:** A *normal strain* is a strain computed from relative displacements that are measured perpendicular to two reference planes. Normal strains measure the relative perpendicular movement of one reference plane with respect to another. The symbol for normal strain is usually the lowercase Greek symbol epsilon (ε).
- **Shear strains:** A *shear strain* is a strain computed from relative displacements that are measured parallel to two reference planes. Shear strains measure the relative parallel movement of one reference plane with respect to another. The symbol for shear strain is usually the lowercase Greek symbol gamma (γ).

In the following sections, I introduce the basic relationships for strain and then show you how to calculate both normal strains and shear strains.

Strained relationships: Comparing lengths

All strain calculations are affected by two primary factors, reference length and deformation:

- **Reference length:** The *reference length* (sometimes referred to as a *gauge length*) is the length prior to the deformation occurring and is measured in specific directions depending on the type of strain you're calculating. In experimental testing, this gauge length is typically specified to be a particular dimension.

- **Deformation:** The *deformation* is a measure of how much an object deforms from its original dimensions or size in a given direction. Depending on which deformation you measure, you can calculate different types of strain, which I describe later in this chapter.

The basic relationship for all types of strain is given as

$$\text{strain} = \frac{\text{deformation}}{\text{reference length}}$$

Examining units of strain

Strain is actually a unitless quantity, even though units are often reported as meter per meter in SI units and inch per inch in U.S. customary units. Algebraically speaking, these units cancel each other out. However, engineers often leave them in this form as a reminder of which units the calculation involved.

Because deformations of objects are generally much smaller than the original length, in most engineering applications the strains you actually calculate are usually on the order of 10^{-5} or 10^{-6} in size. When they aren't expressed as meter per meter or inch per inch, you commonly see the units written as the Greek lowercase symbol mu (μ), which is actually the SI prefix for micro- (see Chapter 2). However, you can use this symbol with strains calculated from either SI or U.S. customary units.

The SI prefix micro is actually 10^{-6}. By using this unit, you can convert a really small number such as 0.00001 in/in or 10.0×10^{-6} in/in to a much more simplified representation of 10μ. This setup simplifies the numerical expressions significantly and can help you avoid costly errors in writing strain values or miscounting all of those extra zeroes in the decimal answers (don't worry, it's an easy mistake to make)!

Using formulas for engineering and true strains

In mechanics of materials, you actually find two variations of normal strain: engineering strain and true strain. Each of these strain values serves a unique purpose.

Engineering strain is a strain calculation that you typically use when deformations in objects remain really small (usually less than about 5 percent of the original reference length). In this case the engineering strain equation looks like the same basic strain relationship I describe in the earlier section

"Strained relationships: Comparing lengths." For a normal strain calculation, this expression may look like the following:

$$\varepsilon = \frac{\Delta L}{L_o}$$

where ΔL is the total change in length (or the deformation) and L_o is the original length measured in the direction of the deformation.

True strain, on the other hand, is used for larger deformations because in reality, strain increases exponentially based on the force applied to the object because the reference length is constantly changing due to the deformation. For a normal strain calculation, you can express the true strain by using the following equation:

$$\varepsilon = \ln\left(\frac{L}{L_o}\right)$$

where L_o is the initial reference length and L is the current length when you calculate the strain value.

For very small strains, the true strain and engineering strain yield similar values.

Normal and Shear: Seeking Some Direction on the Types of Strain

Depending on the type of stress you put on it, an object can experience a normal strain, a shear strain, or a combination of both. These strains are a direct result of the stresses applied to the object; normal stresses cause normal strains and shear stresses cause shear strains.

In the following sections, I show you how to actually identify normal and shearing strains, how to represent this information on a basic graphic, and how to perform their basic calculation.

Getting it right with normal strain

Just as with normal stress components (see Chapter 6), normal strains are usually measured parallel to a longitudinal axis or perpendicular to the cross section on which the normal stresses are acting. The member in Figure 12-1 illustrates a simple element that shows the normal strains that can develop from a uniaxial (or single direction) load in the x-direction.

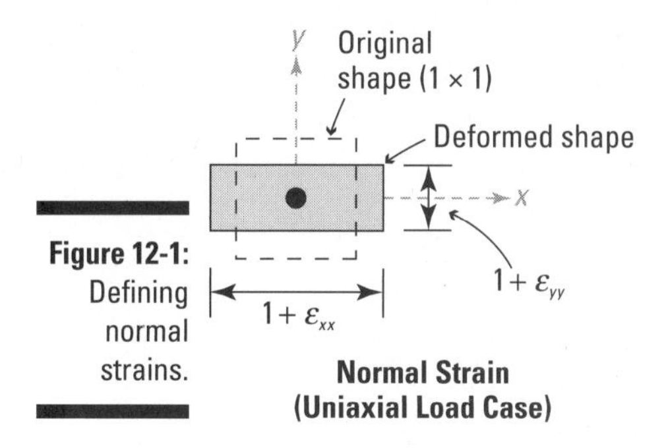

Figure 12-1: Defining normal strains.

The strain in the *x*-direction (ε_{xx}) is an elongation, but the applied stress also causes a shortening in the transverse (or *y*-direction for this problem), which causes a transverse normal strain, ε_{yy}. (I discuss these transverse normal strains in more detail in Chapter 14.)

Normal strains are often caused by axial or bending effects (though other effects such as temperature can also cause normal strains — I talk more about these effects later in this chapter). In two dimensions, you can have two normal strains. In three dimensions, you can have as many as three normal strains — just like stresses!

Establishing a sign convention for normal strains

The sign convention for normal strains is similar to the sign convention for normal stress. The sign convention I use in this book is fairly standard among most classic textbooks (see Figure 12-1). I define normal strains as being positive if they result in an elongation of the strain element. In Figure 12-1, ε_{xx} is a positive strain. Likewise, I consider normal strains negative if they cause a shortening of the strain element, like ε_{yy} in Figure 12-1 does.

Like normal stresses in two and three dimensions, you can have stresses that are both positive and negative acting on the same element. That is, the state of strain on an element may be such that it has an elongation in one direction and a shortening in another direction.

Computing average normal strains

For normal strains, you measure the percent elongation in the same direction as both the basic length and the deformation.

For example, consider a 2.0-meter-long bar that experiences an axial elongation of 0.1 meters. The normal strain in the longitudinal direction is

$$\varepsilon_{xx} = \frac{0.01\text{ m}}{2.0\text{ m}} = 0.005\ \frac{\text{m}}{\text{m}} = 5{,}000\mu$$

Compare this example to a 20.0-meter-long bar that experiences an axial elongation of 0.1 meters. The normal strain for this bar is

$$\varepsilon_{xx} = \frac{0.1\text{ m}}{20.0\text{ m}} = 0.005\ \frac{\text{m}}{\text{m}} = 5{,}000\mu$$

Although strain is important, deformation is also a concern for design. The bar of the second example is much longer and has deformations nearly ten times as large, but the strain in the bar is still the same value. Consider the 6-inch-round, 1-inch-diameter cylinder in Figure 12-2. Under a compressive load, the axial deformation (in the *z*-direction) is 0.02 inches. At the same time, this load causes an increase in the radius of 0.00074 inches (or 7.4×10^{-4} inches).

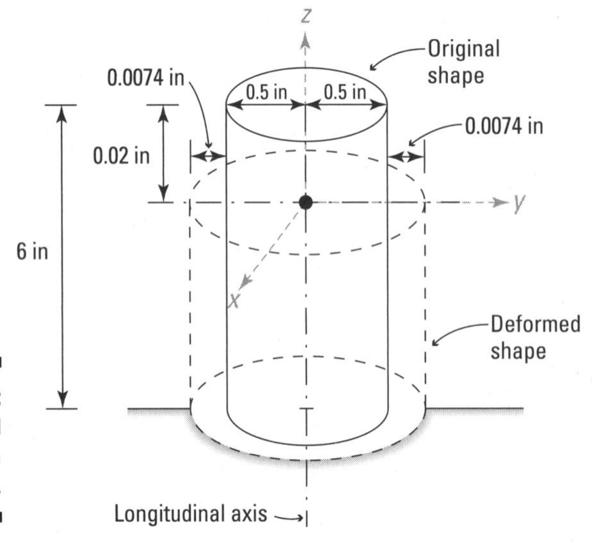

Figure 12-2: Normal strains on a cylinder.

You can compute the strain in the axial direction (along the *z*-axis) as

$$\varepsilon_{zz} = \frac{-0.02\text{ in}}{6.0\text{ in}} = -0.003333\ \frac{\text{in}}{\text{in}} = -3{,}333\mu$$

And in the radial direction (which happens to also include the *x*- and *y*-directions),

$$\varepsilon_{xx} = \varepsilon_{yy} = \frac{0.00072\text{ in}}{1.0\text{ in}} = 0.00072\ \frac{\text{in}}{\text{in}} = +720\mu$$

Finding a new angle with shear strain

Instead of causing a member (or element) to become longer or shorter, shear strain defines the behavior of opposite faces of an element (see Figure 12-3).

Under a shear strain, a square cross section may become a rhombus as opposite faces move parallel with respect to each other.

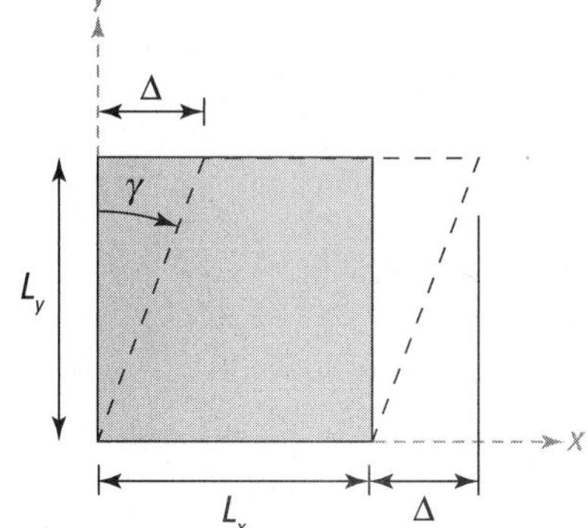

Figure 12-3: Defining shear strains.

You encounter shear strains in problems that include internal shear forces from bending (which I define in Chapter 10) and in problems involving torsion, which I mention in Chapter 11.

Establishing a sign convention for shear strains

The sign convention for shear strains is somewhat different than the convention for shear stresses in Chapter 6. To define shear strain, you have to look at the deformation of the strain element — and more specifically, the angles of the horizontal and vertical faces.

Because shear stresses are equal on horizontal and vertical faces of a two-dimensional element, the shear strains that result from these stresses are also equal. For this reason, the shear strain of the horizontal face is one-half the total shear strain on the element; the shear strain on the vertical face is also one-half (as shown in Figure 12-4a).

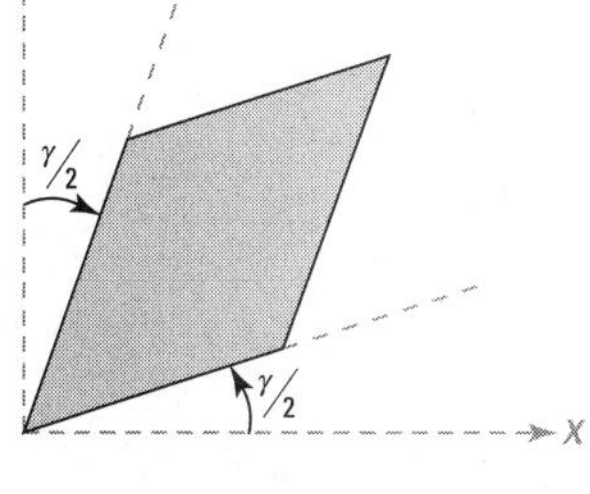

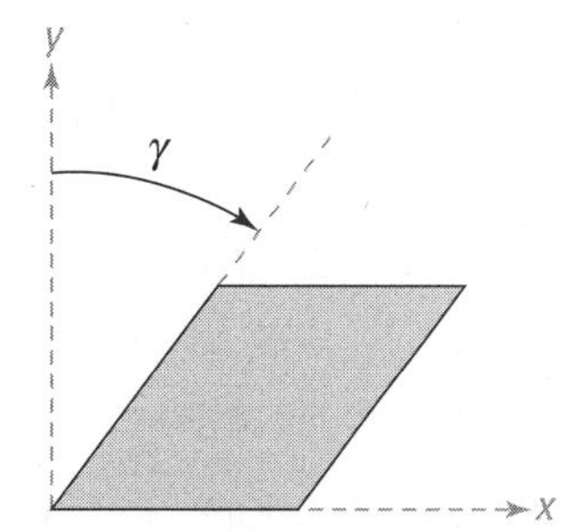

Figure 12-4: Sign convention for positive shear strain.

***(a)* Half Angles on Horizontal and Vertical Face**

***(b)* Full Angle on Vertical Face**

A counterclockwise rotation of the horizontal face and a clockwise rotation of the vertical face both correspond to a positive shear strain.

To help you keep these two competing sign conventions for shear strains straight, I often find lumping the shear strain completely onto one face (as I show in Figure 12-4b) more convenient. In this figure, the horizontal faces are parallel to the Cartesian *x*-axis, and the vertical faces are parallel to the Cartesian *y*-axis. If you keep the horizontal face on the bottom of the element in its original position, you can then state that a positive shear strain is a clockwise rotation of the vertical faces. The top horizontal face moves parallel to the bottom. As long as the shear strains remain small (and they usually are), this representation should be acceptable.

Determining average shear strains

Shear strains are often very small, and as a result, the shear strain is often approximately equal to the angle of the rotated face. Shear strain relates the movement of one parallel plane with respect to another and is also dependent on the perpendicular distance between those planes. For example, consider the strain element shown in Figure 12-3.

If the basic element has a length L_y=2 in, and the top plane moves horizontally (and parallel to the bottom) by an *x*-distance of $\Delta = +0.00005$ inches, you can compute the shear strain from the following:

$$\gamma_{xy} = \frac{+0.00005 \text{ in}}{2.0 \text{ in}} = +0.000025 \frac{\text{in}}{\text{in}} = +25\mu$$

The shear strain in this calculation is positive because the vertical faces (the faces that are parallel to the *y*-axis in this example) rotate in a clockwise direction. Remember that technically, both the vertical and horizontal faces will actually rotate by an amount of $\gamma_{xy} \div 2 = 12.5\mu$ on each face.

Expanding on Thermal Strains

In most applications, strains are usually caused by some sort of applied external stress. But another unique category of problems that cause strain exists within mechanics of materials: Thermal effects are often a significant source of normal strains. These *thermal strains* are a type of normal strain that occurs when changes in temperature cause changes in the dimensions of an object, and they always happen simultaneously in all directions (two directions for two dimensions and three directions for three dimensions). The thermal strain in each direction is the same, although the thermal deformations may be different.

The primary factors affecting thermal strains are

- **Change in temperature:** The more you heat or cool an object, the larger the magnitude of the thermal strains you experience.
- **Coefficient of thermal expansion:** The thermal coefficient, known as the *coefficient of thermal expansion,* is a material constant that describes how an object reacts when subjected to changes in temperature. The thermal coefficient is a measure of a material's change in volume due to temperature effects. This value is based on the behavior of a material's atoms, which begin to move faster under increased temperature, resulting in a larger average distance between the molecules.

The basic relationship for thermal strains is given as

$$\varepsilon_{THERMAL} = \alpha(\Delta T)$$

where ΔT is the change in temperature and α is the coefficient of thermal expansion for the material. You can also calculate the deformation due to temperature, $\Delta_{THERMAL}$ by remembering that strain is a function of original length and deformation:

$$\Delta_{THERMAL} = \alpha(\Delta T)L_O$$

where L_O is the original length of the object.

Although you typically find the coefficient of thermal expansion expressed in units of per degree Celsius (or /°C), some resources may express these values in degrees Fahrenheit or in Kelvin. The temperature change ΔT must be expressed in the same units as the thermal coefficient.

Table 12-1 shows you some example values of these coefficients under normal temperatures (usually taken as approximately 20 degrees Celsius).

Table 12-1 Approximate Coefficients of Thermal Expansion

Material	α *(× 10^{-6}/°F) (approximate)*	α *(× 10^{-6}/°C) (approximate)*
Steel	6.7	12
Brass	10.5	19
Aluminum	12.8	23

You can see that metals have a wide range of thermal coefficients. For the values shown in Table 12-1, an aluminum bar experiences a thermal deformation that is nearly twice as much as the same length of a bar made from steel.

Most mechanics of materials textbooks list typical values (or ranges of values) for common materials' coefficients of thermal expansion in a table or appendix. These values are based on results from experimental testing of materials, so if you can't find it in a table, you can always test for it.

Consider a steel cable in a suspension bridge support. If this cable is 500 meters long and experiences a change in temperature of 30 degrees Celsius, the corresponding axial deformation due to thermal effects is

$$\Delta_{THERMAL} = \alpha(\Delta T)L_O = (12 \times 10^{-6}/°C)(30°C)\ (500\ m) = 0.18\ m = 18\ cm$$

Although this deformation may seem small, thermal strains can produce tremendous forces in other parts of the structure if you don't account for them in the design process.

Considering Plane Strains

In two dimensions, the strain element has two normal strains and one shear strain. In three dimensions, you may encounter elements with as many as three normal strains and three shear strains. A special state of strain known as *plane strain* exists if all strains in or acting on a given plane have a zero value. You can often assume a state of plane strain in objects with one very long dimension (such as the length or thickness), such as dams or thick-walled pipes.

For example, if you have an element with a normal strain in the *z*-direction, a normal strain in the *y*-direction, and a shear strain in the YZ plane, this situation is a plane strain condition. Notice that none of the strains in this example has an *x* subscript, meaning that the element of this example has no normal strain acting in the *x*-direction and no shear strain acting in either the XY or XZ planes.

If you can define the strains and the planes they're acting on, as I show in Chapter 13, you can tell whether you have a plane strain condition by simply examining the subscripts of the strains. If none of the nonzero strain values has a particular *x*-, *y*-, or *z*-direction subscript, you know that you have a plane strain condition. For example, if you determine that an element has a state of strain defined as normal strains of $\varepsilon_{xx} = 100\mu$, $\varepsilon_{yy} = 0\mu$, and $\varepsilon_{zz} = -200\mu$ and shear strains of $\gamma_{xy} = 0\mu$, $\gamma_{xz} = +150\mu$, and $\gamma_{yz} = 0\mu$, you know this element is a plane strain element because all strains in the *y*-direction are all zero — that is, all strains that contain a *y* subscript have a zero value.

Chapter 13

Applying Transformation Concepts to Strain

In This Chapter

- Sketching strain elements
- Transforming strains with equations and Mohr's circle
- Working with strain gauges and strain rosettes

The ability to *transform* strains (calculate them at specific orientations within an object) becomes especially important in experimental testing of structures. Although you can't physically measure stress in an object, you can measure the deformations by using instruments such as strain gauges. However, to determine the state of strain, you must be able to apply strain transformation equations to the readings you get from those sensors.

In this chapter, I show you how to perform the basic strain transformations by using both equations and the graphical Mohr's circle, both of which are similar to the tools I use with stresses in Chapters 6 and 7. I conclude the chapter by showing you how to use strain gauges to measure the strains at a point in an object. After you master strain transformations, you're then ready for a wide array of application problems.

Extending Stress Transformations to Plane Strain Conditions

The strain transformation equations — like the stress transformation equations in Chapter 7 — are derived based on simple equilibrium equations and aren't affected by the material properties of the object. In fact, the relationships between stresses in two transformed states are almost identical to the relationships between strains.

Using the stress transformation equations, you can make the following substitutions to create the strain transformation equations:

- Normal strain ε_{xx} for any normal stress σ_{xx}
- Normal strain ε_{yy} for any normal stress σ_{yy}
- Half of the in-plane shear strain ($\gamma_{xy}/2$), for the in-plane shear stress τ_{xy}

Just as with stress transformations, strain transformations become very important when you start to analyze materials with imperfections or fibers oriented at specific angles (such as with wood products or along welded seams).

In this section, I show you how to perform the strain transformation calculations by using the transformation equations and then introduce you to a modified form of Mohr's circle for plane stress that you can apply to plane strain conditions.

Transforming strains

Figure 13-1 illustrates the strain deformation of an element, and from these figures you can derive the basic strain transformation equations. In Figure 13-1a, you see the effects of a normal strain in the *x*-direction, or ε_{xx}. Figure 13-1b shows the effects of normal strain in the *y*-direction, or ε_{yy}. And Figure 13-1c displays the effects of the shear strain in the XY plane, or γ_{xy}.

You can illustrate the orientation of a shear strain on a plane strain element in multiple ways, as I show in Chapter 12. For the examples of this chapter, I lump all shear strain into a single value reference from the vertical axis.

The basic equation for the normal strain transformation ε_{x1} for a plane strain condition in the XY plane is given by the following:

$$\varepsilon_{x1} = \varepsilon_{xx} \cos^2 \theta + \varepsilon_{yy} \sin^2 \theta + \gamma_{xy} \sin \theta \cos \theta$$

which relates a strain onto a rotated *x1* axis at an angle θ from the original *x*-axis.

Likewise, you can define the equation for the shear strain transformation (γ_{x1y1}) by using the same strain values in the following:

$$\gamma_{x1y1} = -\left(\varepsilon_{xx} - \varepsilon_{yy}\right)\sin 2\theta + \gamma_{xy} \cos 2\theta$$

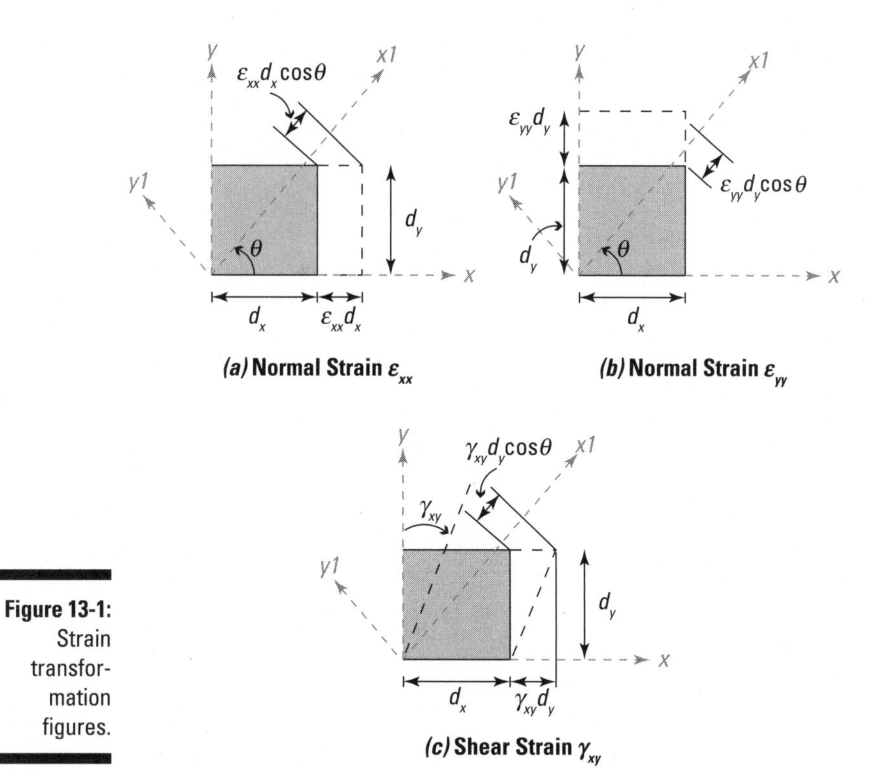

Figure 13-1: Strain transformation figures.

For example, if the current state of strain on an element is $\varepsilon_{xx} = 200\mu$, $\varepsilon_{yy} = -400\mu$, and the shear strain $\gamma_{xy} = 550\mu$, you can easily compute the state of strain at an orientation angle of 43 degrees by calculating the following:

$$\varepsilon_{x1} = 200\mu \cos^2(43°) + (-400\mu)\sin^2(43°) + 550\mu \sin(43°)\cos(43°) = +195\mu$$

And the transformed shear strain is then

$$\gamma_{x1y1} = -(200\mu - (-400\mu))\sin(2(43°)) + 550\mu \cos(2(43°)) = -560\mu$$

To determine the state of strain in the *y1*-direction, you can use the transformation equation and substitute ($\theta = 90° + 43° = 133°$) for the orientation angle:

$$\varepsilon_{y1} = 200\mu \cos^2(133°) + (-400\mu)\sin^2(133°) + 550\mu \sin(133°)\cos(133°) = -395\mu$$

With these transformed strains, you can then sketch the transformed strain element, using the procedure that I show in the following section.

Sketching a rotated strain element

The *rotated strain element* is a method for graphically displaying transformed strains. Rotated strain elements simultaneously show the states of strain on two mutually perpendicular planes, which is very useful when you start calculating maximum strain values (I cover those later in this chapter). Just follow these steps.

1. **Draw the original Cartesian axis with respect to the bottom edge of an unrotated strain element.**
2. **Draw the rotated *x1*-axis for the specified orientation angle θ.**
3. **Draw the rotated *y1*-axis at an angle of $\theta + 90°$ from the original *x*-axis.**

   You can also locate this axis by rotating an angle θ from the original *y*-axis as well.
4. **Draw the undeformed rotated element such that the *x1*-face is perpendicular to the *x1*-axis and the *y1*-face is perpendicular to the *y1*-axis.**

   Complete the basic square shape of an undeformed strain element at this new rotated orientation.
5. **Sketch a line parallel to the *x1*-face to illustrate the ε_{x1} strain condition.**

   For example, consider the transformed element I use in the preceding section where

   $\varepsilon_{x1} = +195\mu$, $\varepsilon_{y1} = -395\mu$, and the shear strain $\gamma_{x1y1} = -560\mu$.

   The fact that the transformed strain of this example is a positive value ($\varepsilon_{x1} = +195\mu$) indicates that the strain in the *x1*-direction causes the undeformed element of Step 4 to elongate in the *x1*-direction.
6. **Sketch a line parallel to the *y1*-face to illustrate the ε_{y1} strain condition.**

   Because this strain in this example is a negative strain ($\varepsilon_{y1} = -395\mu$), the undeformed element you drew in Step 4 shrinks (becomes smaller with respect to the *y1*-axis).
7. **Draw the basic strain element bounded by the *x1*- and *y1*-axes and the normal strain deformation lines from Steps 4 and 5.**

   Connecting these lines creates a basic rectangle that accurately displays the transformed normal strains.
8. **Apply the transformed shear strains by rotating the *x1* faces by an amount equal to γ_{x1y1}.**

   If the transformed shear strain is a negative value as in this example ($\gamma_{x1y1} = -560\mu$), these faces rotate in a counterclockwise direction from the *y1*-axis.

You can see the final transformed strain element for this example in Figure 13-2.

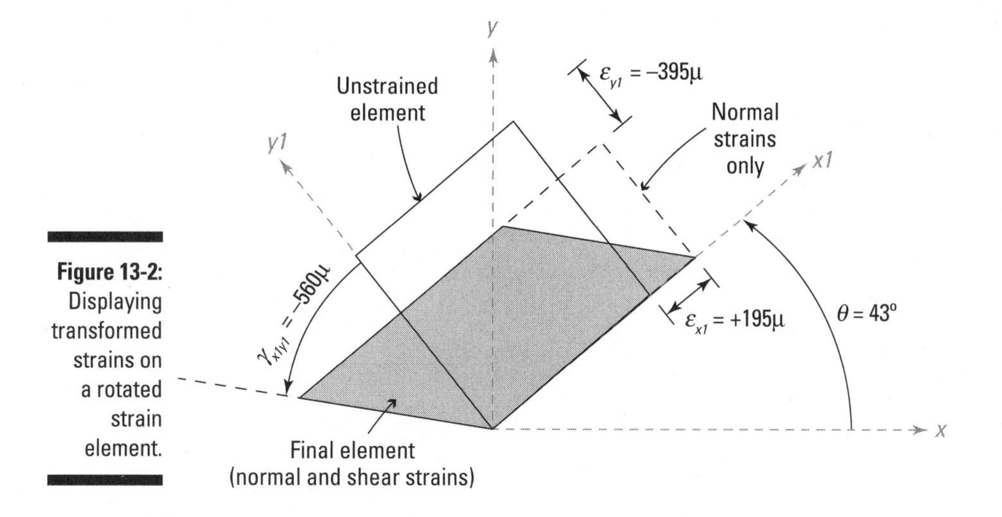

Figure 13-2: Displaying transformed strains on a rotated strain element.

Calculating and Locating Principal Strain Conditions

The strains at a point in an object vary depending on the orientation of the strain element (sometimes referred to as a *material element*). At one special orientation, you have a maximum normal strain, and at another you have a maximum shear strain for a given plane strain element.

The *principal strains* for a plane strain element are the maximum and minimum strains that occur in an object. You need to be aware of two types of principal strain values, each of which affects a strain element in different ways:

- **Principal normal strains:** *Principal normal strains* are the maximum and minimum strains that make an element elongate or contract.
- **Maximum shear strain:** *Principal shear strains* (or *maximum shear strains*) are a maximum shearing strain that makes one edge of an element deform relative to the opposite (and parallel) edge of the element. However, for a three-dimensional strain problem, the principal shear strain in a given plane may not actually be the maximum shear strain.

You need to be able to determine these values in order to form a basis for all your basic strain transformation calculations. In this section, I present the basic formulas for those calculations.

Defining the principal normal strains

Using the current state of strain (defined by ε_{xx}, ε_{yy}, and γ_{xy} in the XY plane), you can compute the basic principal normal strains for a plane strain condition by using the following equation:

$$\varepsilon_{P1,P2} = \frac{\varepsilon_{xx} + \varepsilon_{yy}}{2} \pm \sqrt{\left(\frac{\varepsilon_{xx} - \varepsilon_{yy}}{2}\right)^2 + \left(\frac{\gamma_{xy}}{2}\right)^2} \quad t$$

where ε_{xx} is the normal strain in the *x*-direction; ε_{yy} is the normal strain in the *y*-direction; and γ_{xy} is the shear strain in the plane of the normal strains.

When you evaluate this expression, you actually get two different values because of the ± in front of the second term. This operation is what determines the maximum and minimum value.

Consider an example like the one in the preceding section, where $\varepsilon_{xx} = 200\mu$, $\varepsilon_{yy} = -400\mu$, and the shear strain $\gamma_{xy} = 550\mu$. The principal normal strains are as follows:

$$\varepsilon_{P1,P2} = \frac{(200\mu + (-400\mu))}{2} \pm \sqrt{\left(\frac{200\mu - (-400\mu)}{2}\right)^2 + \left(\frac{550\mu}{2}\right)^2} = -100\mu \pm 407\mu$$

$$\varepsilon_{P1} = -100\mu + 407\mu = +307\mu$$

$$\varepsilon_{P2} = -100\mu - 407\mu = -507\mu$$

From these principal strains, you can conclude that the maximum normal strain $\varepsilon_{P,MAX}$ is the larger of these two values, or $+307\mu$, and the minimum normal strain $\varepsilon_{P,MIN}$ is the smaller of these two values, or -507μ.

Depending on the signs of the principal normal strains for the current state of strain, the maximum value can actually be either ε_{P1} or ε_{P2}, so make sure you calculate both.

Determining the angles for principal normal strains

After you have the principal normal strains determined (see the preceding section), the next step is to determine how they're oriented within the object. This task becomes especially important because it serves as a basis for your design calculations (which I show in Chapter 19) and for determining the state of strain at specific orientation angles within the section.

To determine the principal strain angles, you use a simple relationship that relates the shear strain γ_{xy} to the normal strains (ε_{xx} and ε_{yy}) to the original state of strain:

$$\tan 2\theta_P = \frac{\gamma_{xy}}{\varepsilon_{xx} - \varepsilon_{yy}}$$

So for the example in the preceding section, the principal strain angle θ_p is

$$\tan 2\theta_P = \frac{(550\mu)}{(200\mu - (-400\mu))} = +0.917$$
$$\Rightarrow 2\theta_P = 42.51^\circ$$
$$\Rightarrow \theta_P = +21.25^\circ \text{ (counterclockwise)}$$

which indicates that one of the principal stresses occurs at an orientation angle of 21.25 degrees positive (or counterclockwise) from the original x-axis. But just as with the principal angles for stress (see Chapter 7), you don't know which of the two principal values is related to this angle. To determine that, you need to substitute the current state of strain and this angle into the transformation equation.

$$\varepsilon_{x1} = 200\mu \cos^2(21.25^\circ) + (-400\ \mu)\sin^2(21.25^\circ) + 550\mu \sin(21.25^\circ)\cos(21.25^\circ)$$
$$= +307\mu$$

So based on this calculation, you know that the principal angle of 21.25° corresponds to $\varepsilon_{P1} = +307\mu$, or the maximum strain $\varepsilon_{P,MAX}$. You also know that the other principal angle occurs at 90 degrees counterclockwise from the first angle, or (90° + 21.25°) = 111.25° from the original x-axis, making ε_{P2} the minimum strain $\varepsilon_{P,MIN}$ with a value of –507μ.

The principal strains occur at an orientation that has zero shear strain acting on the element, which you can prove by substituting into the shear strain transformation equations I present in the earlier section "Transforming strains."

Computing the principal shear strain

Another significant state of strain that engineers want to investigate is the principal shear strain, which happens to be the other principal strain value you can compute for a plane-strain problem. Principal shear strains are important calculations for deformations of keyways in motor shaft connections or for elastomeric bearing pads (a type of rubber support) used in applications such as machinery supports or bridge girders.

To compute the principal shear strain, you use the relationship

$$\frac{\gamma_P}{2} = \sqrt{\left(\frac{\varepsilon_{xx} - \varepsilon_{yy}}{2}\right)^2 + \left(\frac{\gamma_{xy}}{2}\right)^2}$$

You may recognize this expression for the principal shear strain as the term after the ± sign in the equation for the principal normal strains I discuss in the earlier section "Defining the principal normal strains."

For the example I lay out in that earlier section, the principal shear strains are as follows:

$$\frac{\gamma_P}{2} = \pm\sqrt{\left(\frac{200\mu - (-400\mu)}{2}\right)^2 + \left(\frac{550\mu}{2}\right)^2} = \pm 407\mu$$

$$\Rightarrow \gamma_P = 2(\pm 407\mu) = \pm 814\mu$$

As with all principal calculations, you also need to determine the orientation angles that result in these maximum shear strain values. You can compute the angle of the principal shear strain orientation from

$$\tan(2\theta_S) = \frac{-(\varepsilon_{xx} - \varepsilon_{yy})}{\gamma_{xy}}$$

which means the orientation angle (θ_S) of one of the principal shear strains for this example is

$$\tan(2\theta_S) = \frac{-(200\mu - (-400\mu))}{550\mu} = -1.09$$

$$\Rightarrow 2\theta_S = -47.49^\circ$$

$$\Rightarrow \theta_S = -23.75^\circ(\text{clockwise})$$

So the principal shear strains occur at an angle of –23.75° and (90° – 23.75°) = 66.25°. Finally, you just need to check which of these two angles corresponds to the maximum in-plane shear strain $\gamma_{P,MAX}$ = +814µ. You can perform this check by using the basic transformation equation for shear strains:

$$\gamma_{x1y1} = -(200\mu - (-400\mu)) \sin(2(-23.75^\circ)) + 550\mu \cos(2(-23.75^\circ)) = +814\mu$$

Thus, you now know that the maximum positive shear strain occurs at an angle of –23.75° from the original *x*-axis.

An invariant rule for normal strains

In Chapter 7, I discuss a relationship known as the *stress invariant rule,* which states that the two normal stresses on perpendicular faces of any stress element are constant (or invariant) regardless of the element's orientation. A similar relationship known as the *strain invariant* relates normal strains (ε_{xx} and ε_{yy}) on one set of mutually perpendicular planes to the normal strains (ε_{x1} and ε_{y1}) on another set of mutually perpendicular planes:

$$\varepsilon_{xx} + \varepsilon_{yy} = \varepsilon_{x1} + \varepsilon_{y1}$$

If you look at the angle for the principal normal strain (which was +21.25°, or counterclockwise from the *x*-axis) and the principal shear strain (which was –23.75°, or clockwise from the *x*-axis), the angle between these two values is

$$\Delta\theta = \theta_P - \theta_S = +21.25° - (-23.75°) = +45.00°$$

It turns out the maximum shear strain is always oriented on a strain element that is rotated 45 degrees clockwise from the maximum principal normal strain element. So if you have already calculated the principal normal strain angles, you don't even need to perform the angle check I show in this section — just subtract 45 degrees from the principal normal strain angle, and you're done!

Exploring Mohr's Circle for Plane Strain

In Chapter 7, I show you *Mohr's circle for plane stress,* a graphical method that uses a state of stress (shown on a stress element) to find transformed stresses at any other orientation. You can apply the same basic methodology to strain transformation as well. Although the methods are very similar, Mohr's circle for plane strain contains a couple of slightly different steps. In this section, I outline the basic procedure for creating *Mohr's circle for plane strain* for a strain element in the XY plane.

If you make a couple of simple modifications to the variables, the technique for Mohr's circle for plane strain is identical to Mohr's circle for plane stress for all steps. Make the following substitutions:

- Normal strain ε_{xx} for any normal stress σ_{xx}
- Normal strain ε_{yy} for any normal stress σ_{yy}
- Half of the in-plane shear strain $\gamma_{xy}/2$ for the in-plane shear stress τ_{xy}

As with Mohr's circle for stress, a Mohr's circle for strain represents the state of strain for a given strain element by coordinates on the opposite ends of a diameter on the circle. As the element is transformed (or the diameter is rotated), the state of strain on the element changes because the coordinates at the end of the diameter of Mohr's circle also change. By using a bit of simple geometry, you can determine the state of strain at any orientation relative to the current state of strain.

Otherwise, the other basic assumptions remain similar, with the word "strain" substituting for "stress" in the procedure I outline in Chapter 7:

- **Normal strain plots on the horizontal axis.** Positive strains make the elements elongate, and compressive strains make the element shorten. Your strain elements may contain both positive and negative normal strain values (or even values of zero!) simultaneously.
- **All angles measured from Mohr's circle are twice their real value.** If you want to find the state of strain for an element that is rotated +10 degrees, the angle that you measure on Mohr's circle is 2(+10) = +20 degrees. You need the double angles so that the circle produces the same results as the transformation equations I mention earlier in this chapter.

Before you can apply Mohr's circle to plane strain, you need to know the current state of strain (which you can see on a properly configured plane strain element) with known shear strain (γ_{xy}) and normal strains (ε_{xx} and ε_{yy}). After you have the basic state of strain established, you're ready to construct the circle. At this point, the procedure for Mohr's circle for strain becomes identical to the Mohr's circle for stress.

1. **Establish the Cartesian axes.**

   On the horizontal axis, you plot the normal strain with positive normal strains (or tension) at the right end of the axis, and negative normal strains (or compression) at the left end. The upper end of the vertical axis is for negative shear strains and the lower end of the vertical axis is for positive shear strains.

   This first major difference between Mohr's circle for stress and Mohr's circle for strain is in dealing with the vertical axis (or the shear strain axis) of the circle. The vertical axis is reserved for one half of the shear strain value (or $\gamma_{xy}/2$), and it crosses the normal strain (or horizontal) axis at a normal strain value of zero. You must divide your strain values by two when plotting them.

2. **Determine the strain coordinates for the positive *x*- and *y*-axes of the current strain element.**

   The first point you plot is the V coordinate, which for an XY plane is of the form (ε_{xx}, $\gamma_{xy}/2$), where ε_{xx} is the normal strain on the *x*-face (or vertical face) and γ_{xy} is the shear strain for the element. The second point is the H coordinate, which for an XY plane is of the form (ε_{yy}, $-\gamma_{xy}/2$), where ε_{yy} is the normal strain on the *y*-face (or horizontal face) and γ_{xy} is the shear strain for the element.

   If the shear strain is negative for the element, the shear strain for the V coordinate is negative and the shear strain for the H coordinate is positive. The shear strain assigned to the H coordinate always has an opposite sign to the shear strain assigned to the V coordinate.

   Figure 13-3 shows an example of a strain element and the corresponding Mohr's circle for strain.

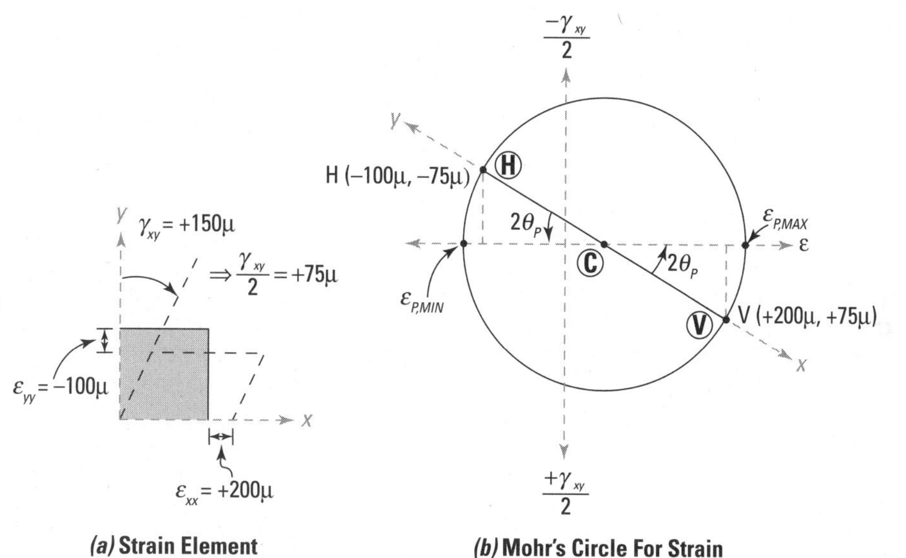

Figure 13-3: A strain element and the corresponding Mohr's circle.

3. **Draw a line connecting the two points of Step 2.**

   With these two points plotted, you can then draw the diameter of the circle by connecting Point V and Point H.

   Note: At this point, the steps and calculations become identical to the calculations for Mohr's circle for stress. You can find more on those calculations in Chapter 7.

4. **Determine the coordinates of the center of the circle.**

   In Figure 13-3, this point is labeled as Point C.

5. **Draw a circle (with the center you located in Step 4) that connects Points V and H.**

6. **Compute the radius *R* of the Mohr's circle.**

7. **Calculate the principal strains by adding the radius *R* to the *x*-coordinate of the center point.**

8. **Find the principal angle of the nearest principal strain.**

9. **Compute the principal angle to the other principal strain.**

Gauging Strain with Strain Rosettes

Because you can't measure stress directly, you have to focus on experimentally measuring the strain behavior of an object and then compute the corresponding stresses by using the principles of Hooke's law (which I discuss in Chapter 14). To help with these measurements, engineers and scientists utilize special sensors known as strain gauges.

A *strain gauge* is an electromechanical sensor that consists of a foil filament; the gauge is affixed to an object (usually by gluing it to the surface). As the object is stressed and consequently experiences strain, the strain gauge lengthens or shortens, which changes the electrical resistance characteristics of the wire/foil. A Wheatstone bridge circuit connected to the gauge by wires then calibrates the changes in electrical resistance to actual strains. Each strain gauge has a unique calibration factor that then relates these resistance changes to strains.

The strains read by a single strain gauge are normal strains in the direction of the gauge's axis, which means you can't actually measure a shear strain with an individual strain gauge by itself. However, if you orient multiple strain gauges in different orientations, you can use the strain transformation equations to take your experimental measurements of normal strains and compute the complete state of strain at a point. That's where strain rosettes come in.

A *strain rosette* is a grouping of three individual strain gauges at different orientation angles used to determine the state of strain at a point. Strain rosettes come in a wide variety of orientations. By far, the two most common arrangements (shown in Figure 13-4) are the 45- and 60-degree rosettes:

- **45-degree rosette:** The 45-degree strain rosette pattern features three gauges oriented at 0, 45, and 90 degrees so that the distance between each sensor is 45 degrees.

- **60-degree rosette:** The 60-degree strain rosette pattern includes three gauges oriented at 0, 60, and 120 degrees so that the distance between each sensor is 60 degrees.

  The 60-degree rosettes are frequently used in one of two variations. The first variation is similar to the 45-degree pattern, except that the angles between the gauges are 60 degrees. The second variation, known as a *delta rosette pattern,* forms a triangle configuration. The difference between the delta rosette and the 60-degree strain rosette pattern has to do with the implied location of the strain (shown as Point O in Figure 13-4) for the three patterns. For the 60-degree rosette pattern, the strains are assumed to be acting at the point of intersection of the axis of the three individual gauges. For the delta rosette pattern, the strains are assumed to be acting at a point inside the triangle made by the rosette patterns.

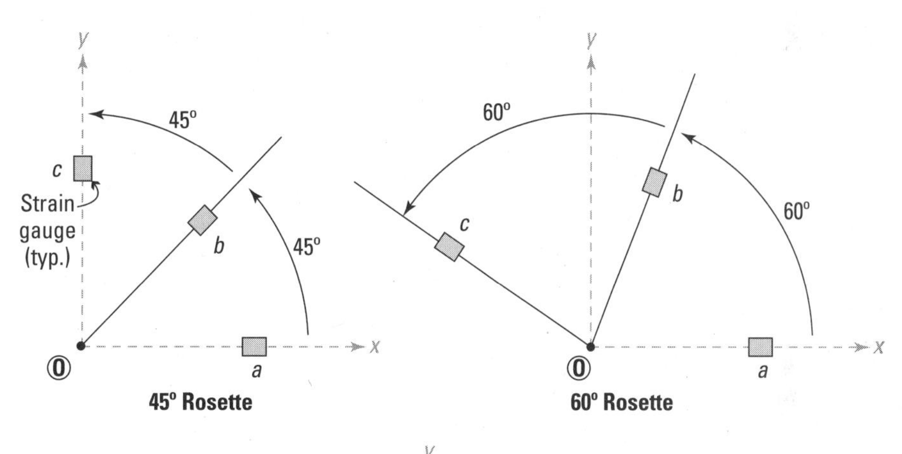

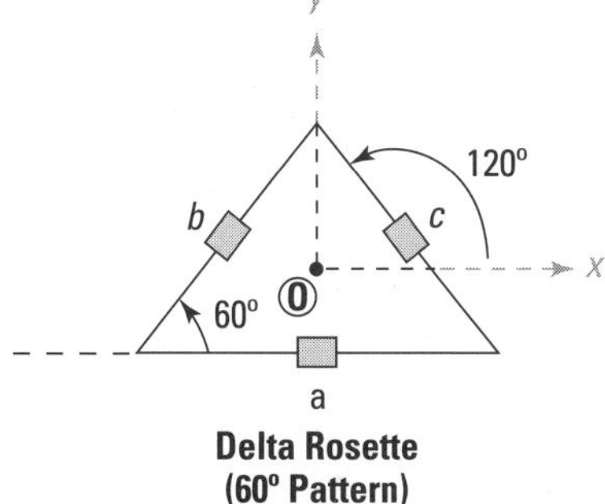

Figure 13-4: Strain rosette patterns.

With improvements in both computers and the manufacturing of sensors, the 45- and 60-degree rosette patterns are now nearly equally accurate. However, for many years, engineers and scientists preferred the 60-degree pattern because it maximized the orientation angle of three gauges within a 180-degree arc.

To illustrate how you can use three normal strain readings to determine a state of strain, suppose a 60-degree strain rosette reports $\varepsilon_a = 200\mu$ at 0°, $\varepsilon_b = -150\mu$ at 60°, and $\varepsilon_c = 400\mu$ at 120°. With these three readings, you can then write three separate transformation equations as follows based on the same current state of unknown strain (normal strains ε_{xx} and ε_{yy} and the shear strain γ_{xy}) — one for each of the three strain gauge readings.

You can then substitute the readings and their corresponding angles into these equations and determine the states of strain:

$$\varepsilon_a = \varepsilon_{xx}\cos^2\theta_a + \varepsilon_{yy}\sin^2\theta_a + \gamma_{xy}\sin\theta_a\cos\theta_a$$
$$\varepsilon_b = \varepsilon_{xx}\cos^2\theta_b + \varepsilon_{yy}\sin^2\theta_b + \gamma_{xy}\sin\theta_b\cos\theta_b$$
$$\varepsilon_c = \varepsilon_{xx}\cos^2\theta_c + \varepsilon_{yy}\sin^2\theta_c + \gamma_{xy}\sin\theta_c\cos\theta_c$$

$$+200\mu = \varepsilon_{xx}\cos^2(0°) + \varepsilon_{yy}\sin^2(0°) + \gamma_{xy}\sin(0°)\cos(0°)$$
$$-150\mu = \varepsilon_{xx}\cos^2(60°) + \varepsilon_{yy}\sin^2(60°) + \gamma_{xy}\sin(60°)\cos(60°)$$
$$+400\mu = \varepsilon_{xx}\cos^2(120°) + \varepsilon_{yy}\sin^2(120°) + \gamma_{xy}\sin(120°)\cos(120°)$$

Solving these three relationships simultaneously allows you to find the three unknown strains at any point:

$$\varepsilon_{xx} = +200\mu$$
$$\varepsilon_{yy} = +100\mu$$
$$\gamma_{xy} = -634\mu$$

From this point, you can then find principal strains and move on to all those other transformation calculations I describe throughout this chapter.

Chapter 14

Correlating Stresses and Strains to Understand Deformation

In This Chapter

- Understanding basic material behavior
- Examining important locations in stress-strain diagrams
- Putting Young's modulus of elasticity, Poisson's ratio, and Hooke's law to work
- Connecting stresses to strains

The deformation of an object under load is very important to engineers. Although you can easily determine the strength of a member by simply looking at stresses and computing required areas, strength isn't the only criteria that makes a structure successful. If you design the beams of your floor such that the vertical deflections are excessive, the floor will be awkward to walk on, and most people won't be happy with the design (though skaters may be able to do some cool tricks). For a design to be truly successful, you have to make sure that it doesn't deflect or vibrate too much (known as *serviceability conditions*) in addition to being strong enough.

Different applications have tighter tolerances with regards to deformation. Machinery may require deformations within a few thousandths of an inch, while a fishing pole may deflect several inches and still perform satisfactorily.

Understanding the relationship between stress and strain is your first step in relating applied forces to their deformation responses. To be able to relate stress and strain, you must have a solid understanding about the properties and behavior of materials. I start this chapter by introducing some basic terminology, and then I present two important constants, Poisson's ratio and Young's modulus of elasticity, that you need when relating stress to strain. Finally, I show you the Hooke's law relationship that provides the necessary equations that you actually use to relate stress to strain.

Describing Material Behavior

In reality, no object is truly rigid; all objects are deformable, so you need to take deformations and material behaviors into account by way of the strain calculations that I show you in Chapter 12. Fortunately, if you make a few simplifying assumptions, the equations of statics still remain valid.

You can experimentally determine many of the material properties by conducting a tension test on a sample of materials. In a *tension test,* you subject a material of a prescribed length (known as the *gauge length*) and cross-sectional area to an applied axial load. Using these values, you determine an average normal stress (see Chapter 6) and plot it versus the corresponding strain (see Chapter 12). These stress-versus-strain relationships provide the basis for the material properties I discuss in this chapter.

But before you can work with material properties, you need to get some basic terminology down first. In this section, I explain elastic and plastic behavior, describe the difference between a ductile material and a brittle material, and introduce the concept of material fatigue.

Elastic and plastic behavior: Getting back in shape?

When you design an object, you typically perform your calculations and select your member sizes with the anticipation that the object will return to its original position — after all, if you apply a load to a structure and the structure doesn't rebound, pretty soon it's unusable because deformations become permanent.

The ability of an object to return to its original shape when you remove a load is known as *elastic behavior;* all materials have basic elastic characteristics up to a point. You can see an example of a highly elastic behavior with the simple rubber band: It stretches when you pull on it but returns to its original dimensions or shape easily.

Design is almost always done to ensure that a material maintains its elastic behavior. So engineers need to know especially where elastic behavior ends (known as the *elastic limit*). I show you where to find the elastic limit in the later section "Defining the regions of a stress-strain curve."

After you reach a certain stress and its corresponding strain level in a material, the behavior changes. After a material is stressed beyond its elastic limit, it experiences permanent deformations (known as *plastic deformations*) that remain even after the object is unloaded.

Ductile and brittle materials: Stretching or breaking

Ductility (the ability of a material to undergo large plastic deformations prior to failure) is one of many very important characteristics that engineers consider during design. Ductility is an important factor in allowing a structure to survive extreme loads, such as those due to earthquakes and hurricanes, without experiencing a sudden failure or collapse. Materials that are very ductile include many types of metal (such as steel) and some types of plastic.

Ductility is often evaluated in one of two ways: by measuring the change in the length of a member or by measuring the change in cross-sectional area of a sample under load to failure. To calculate the percent elongation due to a change in length, use the following equation:

$$\text{percent elongation} = \left(\frac{L_f - L_O}{L_O}\right) \cdot 100$$

where L_f is the length of the specimen when it finally *ruptures* (or breaks) and L_o is the initial length of the specimen.

If you recognize this formula, it's the same calculation as the basic normal strain equation in Chapter 12, but now it's expressed as a percentage. The larger the strain that a material can sustain before rupture, the more ductile the material is said to be.

Another representation of ductility, particularly for tension applications, is measured by a percent reduction in area. It measures the amount of *necking* (or change in cross-sectional area) that occurs prior to the ultimate failure as follows:

$$\text{percent reduction in area} = \left(\frac{A_i - A_f}{A_i}\right) \cdot 100$$

where A_i is the initial area of the test specimen and A_f is the final area of the test specimen when it ruptures.

A material that behaves with very little ductility is said to be *brittle*. An errant baseball effectively demonstrates the brittle behavior of your living room window (not to mention the impact it has on your wallet). A brittle material displays very little visible deformation before it ruptures, and it usually fails without advance warning. Examples of materials that are typically brittle are cast iron, stone, and glass.

Unfortunately, the distinction between brittleness and ductility isn't readily apparent, especially because both ductility and brittle behavior are dependent not only on the material in question but also on the nature and type of stress, the temperature, and the rate of loading.

In your high-school chemistry class, your teacher may have illustrated the effect of liquid nitrogen on a bouncy rubber ball. By subjecting the normally ductile rubber ball to extreme cold, your teacher could cause it to shatter (a brittle behavior) simply by dropping it on the ground.

Fatigue: Weakening with repeated loads

Fatigue is caused by the repeated loading and unloading of a material, which results in damage to the material on a microscopic level. Consider twisting a simple paper clip. As you unfold the paper clip and bend it back and forth slightly, you may not actually reach a yield point of the base metal. However, if you repeat that simple bending process enough times (sometimes measured in the hundreds of thousands of cycles), you can actually break the paper clip without ever reaching the ultimate strength of the material. (You can read more about yield point and ultimate strength in the later section "Site-seeing at points of interest on a stress-strain diagram.")

Fatigue is a very serious concern for objects and structures subjected to repeated and cyclical loadings, such as vibrations from machinery, or even very old buildings subjected to repeated wind loadings. Member connections are also very susceptible to fatigue effects. A failure due to fatigue is always a brittle failure — and brittle failures are bad news because they're sudden and often unexpected. (See the preceding section for more on brittle behavior.)

Strength of a material under fatigue is dependent on the number of load cycles and the basic intensity of the repeated loading. The larger the load (and hence the stress), the fewer cycles you need to break an object due to fatigue. You determine the *fatigue performance* of a material by plotting the number of cycles to failure under a specified applied stress.

A brief history of common material tests

In addition to a simple tension test to determine Young's modulus of elasticity, designers have used a number of other tests for many years to determine useful material properties:

- **The Mohs hardness test:** Around 1812, Friedrich Mohs, a German mineralogist, created a relative scale of material hardness by studying which materials were capable of scratching other materials. The Mohs hardness test conceptually dates back to ancient Greece.
- **The Brinell hardness test:** Johan Brinell developed this test in 1900 as a basic method for determining the hardness of a material by measuring the amount a standard round object could be pushed into a test specimen under a specific load. The test provided numerical results to quantify the hardness of a material.
- **The Rockwell hardness test:** Hugh Rockwell and Stanley Rockwell (no relation) created this hardness measurement for their work at a ball bearing manufacturer around 1915 to help quantify the hardness of a material. Results of this test have been correlated to the tensile strength of the base material.
- **The Charpy *V*-notch test:** This method came from French scientist Georges Charpy around 1905; it helps quantify the amount of impact energy an object can absorb before it ruptures (or breaks). The Charpy *V*-notch test can also qualitatively describe the amount of ductility a material possesses.

Creating the Great Equalizer: Stress-Strain Diagrams

In experimental analysis, you can easily measure the load and corresponding deformation from your basic tests (such as the tension test I describe earlier in the chapter). But the real challenge is to describe the behavior of materials in a way that's independent of the size or shape of whatever you're trying to design. Using stress compensates for any variations in loads and cross-sections, and using strain compensates for differences in deformation. In other words, comparing stress and strain lets you focus on the intrinsic properties of a material.

Justifying stress-strain relationships

Imagine that you tie a small cable to a weight lying on the floor. If your cable is exactly the right size, you can't pick up any additional weight without failing the cable. The only way to pick up a heavier weight (assuming you use the same cable material) is to change the size of the cable, which means that you change the cross-sectional area.

However, when working with load-deformation relationships, you soon realize that getting a feel for the material's true behavior is difficult. You can create plots of load versus deformation — which are useful when you're trying to determine whether the floor beam you're sitting on will deflect too much and cause your furniture to slide to the middle of the room — but these diagrams don't directly take into account the size of the actual member doing the work. For this reason, you're better off working with stresses.

Figure 14-1a shows three different objects made of the same material that can have vastly different load versus deflection curves.

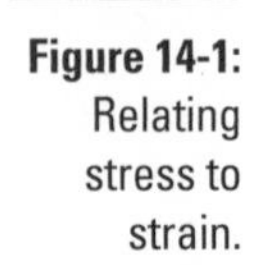

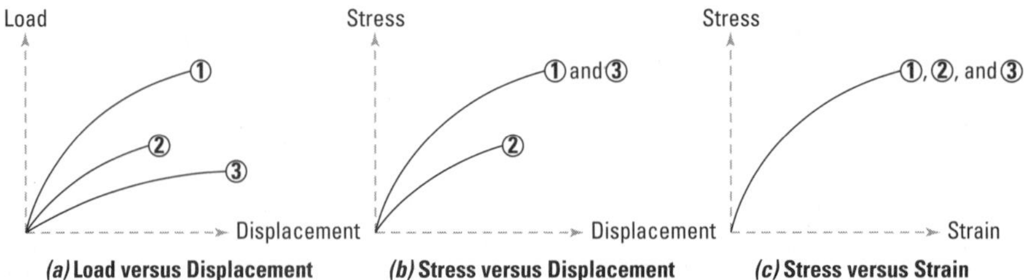

Figure 14-1: Relating stress to strain.

Plotting stress versus deformation isn't a much better option. As you can see in Figure 14-1b, two objects can have the same deformation while experiencing significantly different stresses. This discrepancy commonly occurs in flexural members. For beams, stresses (see Chapter 9) and deformations (see Chapter 16) are both based on the cross section's moment of inertia. However, stress is also a function of the depth of the beam, whereas deflection is also a function of the length. Thus, two different beams can experience the same stress but have uniquely different deformations and moments of inertia. That's why even the stress versus deformation plots aren't completely sufficient.

Describing materials with stress versus strain

Analyzing the behavior of a member based on its characteristic stresses and strains (as shown in Figure 14-1c) is often a better method than plotting load or stress versus deformation (see the preceding section). With this information, you can

- Normalize different loads and parameters, such as cross-sectional properties and member length, into the stress calculations
- Normalize the deformation and member length into the strain calculations

Doing so allows you to investigate the behavior of the material itself without having to worry about the loads or geometry of a given application. With these parameters aside (because you've already incorporated them into the stresses and strains), you can easily select an appropriate material type because the stress-strain diagrams for a given material are always identical.

If you plot stress versus strain rather than load versus deflection, you automatically take into account the deformation of the object with respect to the intensity of the load on a given cross-sectional property.

Exploring Stress-Strain Curves for Materials

After you've constructed the stress-strain diagram from experimental testing, you may be ready to ask the question "Which points on a materials stress-strain plot do I need to consider?" Figure 14-2 shows a typical stress-strain curve. The following sections explore the different regions of the stress-strain curve and the importance of several specific locations.

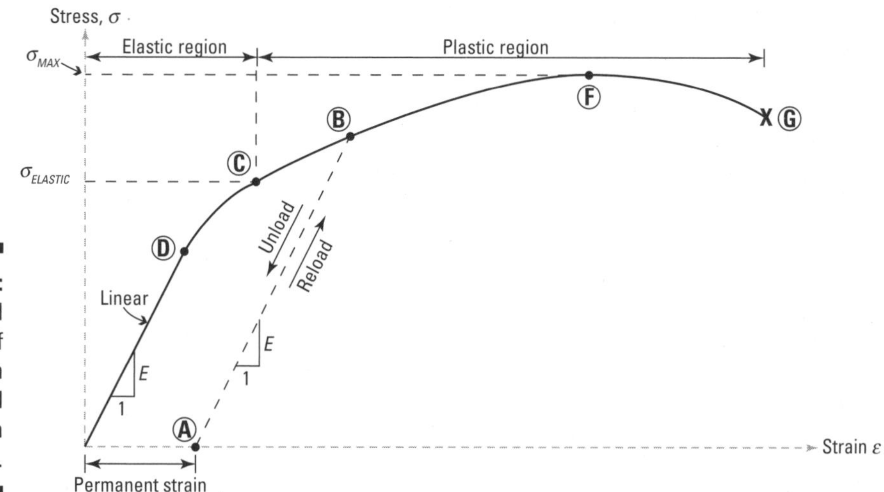

Figure 14-2: Points and regions of interest on a typical stress-strain curve.

Defining the regions of a stress-strain curve

The stress-strain relationship shown in Figure 14-2 divides into two primary regions:

- **Elastic region:** The *elastic region* is the region of the stress-strain curve where removing a load from an object results in the object returning to its original or unloaded shape. When an object's current stress level is within the elastic region, it rebounds to its original state by retracing the original stress-strain curve without retaining permanent deformation whenever the applied load is removed.
- **Plastic region:** The *plastic region* is the region of the curve that extends beyond the elastic region of the curve. When a material is stressed to a point on the stress-strain curve within the plastic region (such as Point B), the curve recedes with a straight-line segment that's parallel to the straight-line portion within the elastic region when the load is removed. As a result, even when the entire load is removed, a permanent strain remains even when stress is no longer applied (Point A). When load is reapplied, the stress-strain plot typically retraces the unloading portion of the curve back up to the original stress-strain curve (at Point B) and then continues to follow the original material curve as the strain increases.

The point on the stress-strain curve that separates the elastic and plastic regions is known as the *elastic limit* (at Point C) and is found by observing the plot of the data from experimental tests (such as the tension test). Exceeding the elastic limit results in a *plastic deformation.* See the following section for more on this location.

If a load is repeatedly applied and released, the end of the straight-line portion of the reloading actually increases slightly with each reloading. This phenomenon is known as *strain hardening* (or *work hardening*). If you bend a paper clip back and forth repeatedly, you may observe that it becomes increasingly difficult to deform after several bends (known as *cyclic strain hardening*) or as you increase the angle of bend (even if it's only bent one time and not repeatedly).

Some types of deformations actually increase without actually increasing the applied stress. Plastic deformation that continues to increase under a constant or sustained load is known as *creep.*

Site-seeing at points of interest on a stress-strain diagram

After you understand the basic regions on a stress-strain diagram (see the preceding section), you can turn your attention to several specific and very important points, some of which I include in Figure 14-2:

- **Proportional limit:** The *proportional limit* (at Point D) corresponds to the location of stress at the end of the *linear region* (or the straight-line portion within the elastic region), where the stress and strain values remain linearly (or proportionally) related.
- **Elastic limit:** The *elastic limit* (at Point C) is where the material stops behaving elastically, and you can measure permanent (irreversible) deformation. As I note earlier in the chapter, it's the transition point between the elastic region and the plastic region of the stress-strain diagram.
- **Yield point:** The *yield point* is a point on the curve where strain increases significantly more for an incremental increase in stress and is often located somewhere between the proportional limit (Point C) and the elastic limit (Point D), though for some materials it may occur beyond the elastic limit as well.

  The yield point for many materials can be difficult to locate because of a gradual transition between elastic and plastic behavior, which is why I don't note it in Figure 14-2. However, you can often define it as the point beyond the elastic limit where the strain increase becomes bigger for a given increment of applied stress.

  One common method for locating the yield point uses a strain of 0.2 percent at zero stress as a starting point. You trace a line parallel to the linear region of the stress-strain diagram from this point until the straight line crosses the stress-strain curve. The point where this line crosses the curve is an often accepted measure of the yield point of a material.

  In some materials, such as certain types of metal, a well-defined yield point is visible on a stress-strain diagram. For these materials, the elastic limit, proportional limit, and the yield point often occur very close to each other.

- **Ultimate strength:** The *ultimate strength* (or the *ultimate tensile stress*) is the absolute maximum stress a material feels (which occurs at Point F) before it actually ruptures. Often, this value is significantly more than the yield stress (as much as 50 to 60 percent more than the yield for some types of metals). When a ductile material reaches its ultimate strength, it experiences necking where the cross-sectional area reduces locally. The stress-strain curve contains no higher stress than the ultimate strength. Even though deformations can continue to increase, the stress usually decreases after the ultimate strength has been achieved.
- **Rupture point:** The *rupture point* is the point of strain where the material physically separates (Point G). At this point, the strain reaches its maximum value and the material actually ruptures (or fractures), even though the corresponding stress may be less than the ultimate strength at this point.

In design, the yield point and the ultimate strength point become locations of interest to an engineer. The stresses that occur at these points are two stresses that structural designers frequently use.

Knowing Who's Who among Material Properties

When you start calculating the relationship between stress and strain, two material constants quickly rise to the top of the heap: Young's modulus of elasticity and Poisson's ratio. These constants quantify a material's stiffness and deformation under load, and I cover them in the following sections.

Finding stiffness under load: Young's modulus of elasticity

Young's modulus of elasticity was named for British scientist Thomas Young (1773–1829), who helped quantify the stiffness of a material under load. Though this property bears Young's name, several other scientists and mathematicians (such as Leonhard Euler and Giordano Ricatti) actually established this relationship earlier.

Young's modulus of elasticity is a material property that defines the relationship between an elastic uniaxial strain in one direction to the elastic uniaxial stress (in the same direction) that's causing it. (See Chapter 8 for more on uniaxial stresses and Chapter 13 for more on uniaxial strains.) In a material test, the Young's modulus of elasticity is actually the slope of the linear region of the stress-strain diagram. A quick way to determine this value is to take a ratio of the stress and strain at the proportional limit:

$$E = \frac{\sigma_{PROP}}{\varepsilon_{PROP}}$$

where σ_{PROP} is the uniaxial stress at the proportional limit and ε_{PROP} is the corresponding uniaxial strain at the same point (at Point D on Figure 14-2 earlier in the chapter). Young's modulus of elasticity is typically characterized by the variable E and has units that match the units of the applied stress at the proportional limit. (Flip to "Site-seeing at points of interest on a stress-strain diagram" earlier in the chapter for more on the proportional limit.) Typically, Young's modulus of elasticity has SI units of giga-Newton per square meter (GN/m^2 or GPa) and U.S. customary units of kip per square inch (ksi).

For a certain mild steel, the normal stress at the proportional limit may be around σ_{PROP} = 32 ksi, and the corresponding strain at this point is ε_{PROP} = 1,100μ. The Young's modulus of elasticity in this example is then E = (32 ksi) ÷ (1,100μ) = 29,000 ksi. Table 14-1 shows examples of several values for Young's modulus of elasticity for different materials. Fortunately, many design guides have values for Young's modulus of elasticity for various materials tabulated for you already.

Table 14-1 Values of Young's Modulus of Elasticity

Material	*SI (GPa)*	*U.S. customary (ksi)*
Aluminum	69	10,000
Concrete	30	4,350
Steel	200	29,000

Young's modulus of elasticity is actually one of several modulii that you can use to relate stress to the strain in a material.

- **Secant modulus of elasticity**: This modulus is the relationship of stress to strain for any point on the stress-strain curve. You represent it as a line from the origin (0,0) to the point on the curve.
- **Tangent modulus of elasticity**: This relationship is the measure of the slope of the line at a particular point on the stress-strain curve and is very useful in nonlinear or plastic analysis of mechanics of materials.

Young's modulus of elasticity is both a secant modulus and a tangent modulus in that it represents the slope of the linear region with respect to the origin (0,0) of the stress-strain curve.

Another modulus of elasticity that you may see from time to time is the *shear modulus of elasticity* (or the *modulus of rigidity*), which relates shear stress to shear strain. I explain more about modulus of rigidity in "Relating Stress to Strain" later in this chapter.

Getting longer and thinner (or shorter and fatter) with Poisson's ratio

Poisson's ratio was named for French mathematician Siméon Denis Poisson (1781–1840) who quantified the relationship for strains in multiple directions of objects under load. He explained a phenomenon, now referred to as *Poisson's effect.* An object that's subjected to tension experiences an elongation (which is a normal longitudinal strain) in the direction of the applied stress. At the same time, it also experiences a reduction in the dimensions that are transverse (perpendicular) to the direction of stress causing the deformation — that is, a lateral strain also occurs in a direction due to a longitudinal stress. You can demonstrate Poisson's effect with a simple ball of clay: When you mash the ball between your hands, it gets smaller in one dimension (between your hands) while getting larger in the perpendicular directions. Figure 14-3 illustrates such an object.

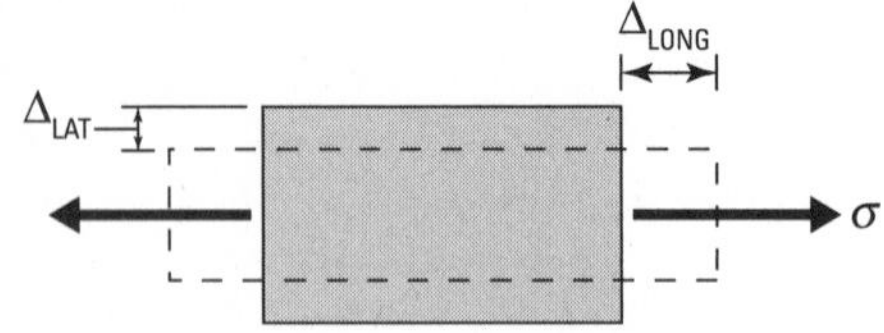

Figure 14-3: Poisson's effect.

The lateral strain has the opposite sign of the longitudinal strain (which is the strain in the direction of the stress). If an object gets longer due to a uniaxial stress, the lateral strain is negative; if an object gets shorter, the lateral strain is positive.

Depending on the material, the relationship between these longitudinal strains and the lateral strains varies; however, these strains are related to each other by a constant ratio for a given material (as long as the strains you're dealing with remain very small). This constant of variation is known as *Poisson's ratio,* and it relates the normal strain in the lateral direction to the normal strain in the longitudinal direction of an applied stress as follows:

$$\nu = -\frac{\varepsilon_{LAT}}{\varepsilon_{LONG}}$$

where ε_{LONG} is the longitudinal normal strain in the direction of the applied stress and ε_{LAT} is the lateral normal strain in the transverse (or lateral) direction. Poisson's ratio is usually assigned to the Greek letter nu (ν) and is a unitless parameter, so it doesn't require conversion for SI or U.S. customary units.

Poisson's ratio is a single numeric value that quantifies the magnitude of Poisson's effect. The bigger the Poisson's ratio value is, the more lateral (or transverse) strain in relation to a longitudinal strain that an object experiences under a given stress.

One use of Poisson's ratio is to help identify a material based on its strain relationships. For example, perhaps you experimentally measure the strains at a point in a tension rod to have a longitudinal strain of ε_{LONG} = 1,000μ and ε_{LAT} = –300μ. Using Poisson's ratio, $\nu = -(-300\mu) \div (1{,}000\mu) = +0.3$, which corresponds to most types of steel.

The directions of the lateral normal strains don't necessarily have to align with one of the Cartesian *x*-, *y*-, or *z*-axes — the lateral strains just have to act in any perpendicular direction to a longitudinal strain. A round bar subjected to a tensile load experiences an elongation along the axis of the bar, but decreases in the radial directions by equal amounts.

Poisson's ratio is often a positive value and is typically in the range of 0.25 to 0.35 for many common materials, including metals. Materials such as concrete and some members of the wood family have Poisson's ratios closer to 0.15. An incompressible material at low strains can have a Poisson's ratio as high as 0.5. Some types of materials (known as *auxetic materials*) such as foams actually can have a negative value for Poisson's ratio. Other materials that are not isotropic can sometimes have different values of Poisson's ratio in different directions, and in some cases their Poisson's ratio can exceed even 0.5.

Relating Stress to Strain

Relating stress to strain is one of the key concepts you use in mechanics of materials. In this section, I show several basic assumptions to keep in mind as well as the actual equations that make it work. In Parts IV and V of this text, I illustrate various examples that utilize these relationships.

Making assumptions in stress versus strain relationships

In order for the stress-strain relationships I present in the coming section to work, you must make sure your problem satisfies the following assumptions:

- **Linear elastic behavior:** The analysis of stress and strain (using the methods in this chapter) assumes that the current state of stress versus strain is still within the elastic region (see "Defining the regions of a stress-strain curve" earlier in the chapter).
- **Homogeneous and isotropic materials**: A material is homogeneous if the material properties don't vary with position within an object, and it's isotropic if the material properties don't vary with direction.
- **Validity of principle of superposition:** The principle of superposition allows you to separate multiple stress and strain behaviors, analyze them independently, and then recombine their effects for a combined (or net) effect. Don't worry about the technical aspects here; I describe this concept in more detail in Chapter 15.

Hooke springs eternal! Using Hooke's law for one dimension

Robert Hooke (1625–1703) was an English scientist who investigated the elastic properties of materials and made an important revelation about the behavior of a spring object (shown in Figure 14-4).

He concluded that the force F in any spring is proportional to the *extension* (the deformation from the relaxed state) Δ as follows:

$$F = k \cdot \Delta$$

where the term k is the stiffness of the spring.

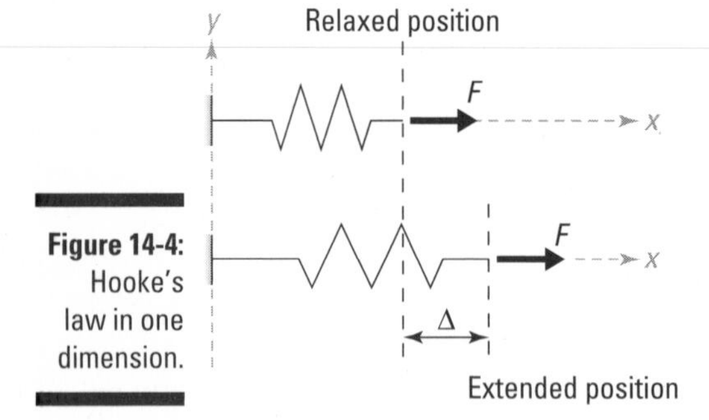

Figure 14-4: Hooke's law in one dimension.

All objects deform under load, so in a sense, all objects behave as springs. The stiffness of an object is directly related to the section properties of the cross section, the material from which the object is made, and the length or span of the member. The stiffness of the member varies depending on the type of loading; the member may have one stiffness when loaded axially, and another when the member is subjected to bending.

If you recognize that this formula is an axial application, you may notice that you can manipulate this expression to reflect terms for stress (*F/A*) and strain (Δ/L) and their relationship within the elastic region. Based on the material's proportional limit (which I introduce earlier in the chapter), you can show that $\sigma_{xx} = E\varepsilon_{xx}$ for a uniaxial stress state, where *E* is Young's modulus of elasticity (which I also describe earlier in this chapter), ε_{xx} is the axial strain, and σ_{xx} is the axial stress. See Chapter 8 for more on axial stress applications and Chapter 12 for more on axial strains.

This relationship is the same as the one in the straight-line portion of the stress-strain graph in Figure 14-2 earlier in the chapter, meaning that as the stresses increase in a material, the strains increase by a proportional amount. This correlation is true up to the proportional limit. After the proportional limit, a relationship still exists, but Young's modulus of elasticity is no longer valid.

Some approaches for analyzing stress-strain behavior beyond the proportional limit actually use a pseudo-Young's modulus of elasticity as the value for *E* in this basic stress-strain relationship equation. The value of this modified modulus is actually the slope of a tangent line to the stress-strain curve at the location of interest (or the tangent modulus I describe earlier in this chapter).

So if you can define the equation of the curve and then find the tangent (which you can do by taking the first derivative of the stress-strain curve at the point of interest), this relationship still holds. Remember that the tangent modulus of the linear region of the stress-strain curve is the same as Young's modulus of elasticity. Outside the linear region, you need to use the appropriate tangent modulus value at the point of interest on the curve.

You can extend the same idea of relating stress to strain to shear applications in the linear region, relating shear stress to shear strain to create Hooke's law for shear stress: $\tau_{xy} = G\gamma_{xy}$, where τ_{xy} is the shear stress, γ_{xy} is the corresponding shear strain, and *G* is the shear modulus of elasticity (or the modulus of rigidity).

For materials within the elastic region, you can relate Poisson's ratio (ν), Young's modulus of elasticity (*E*), and the shear modulus of elasticity (*G*):

$$E = 2(1 + \nu)G$$

So if you happen to know two of these three material properties, you can easily find the third value.

Developing a generalized relationship for Hooke's law in two or three dimensions

You can extend the ideas behind Hooke's law for one dimension to help you analyze stress and strain in multiple dimensions. Consider the stress element shown in Figure 14-5a, which is subjected to a normal stress in the *x*-direction and a second normal stress in the *y*-direction. The material is aluminum, which has a Young's modulus of elasticity E of 10,000 ksi and a Poisson's ratio ν of 0.333.

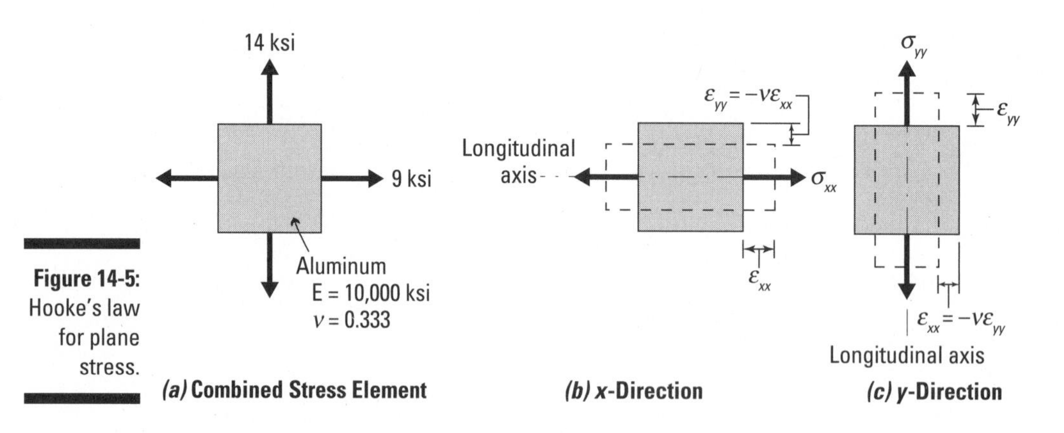

Figure 14-5: Hooke's law for plane stress.

Hooke's law for plane stress

Because the element remains elastic and the deformations remain very small, the principle of superposition lets you actually analyze this element as two separate one-dimensional problems as shown in Figures 14-5b and 14-5c. When you look at the element of 14-5b, you may notice that it's the same uniaxial case I describe in the previous section. The tensile stress of σ_{xx} = 9 ksi results in a longitudinal strain ε_{xx} that's positive and a corresponding lateral strain $\varepsilon_{yy} = -\nu\varepsilon_{xx}$ due to Poisson's effect. Similarly, from Figure 14-5c, the uniaxial tension σ_{yy} = 14 ksi results in a longitudinal strain ε_{yy} and a lateral strain $\varepsilon_{xx} = -\nu\varepsilon_{yy}$.

The principle of superposition says that you can then add the combined effects for each of these strains to compute a single net strain in a given direction:

$$\varepsilon_{xx,TOT} = \varepsilon_{xx,B} + \varepsilon_{xx,C} = \varepsilon_{xx} + (-\nu\varepsilon_{yy})$$

where $\varepsilon_{xx,B}$ is the strain for the uniaxial case shown in Figure 14-5b and $\varepsilon_{xx,C}$ is the uniaxial strain from the case shown in Figure 14-5c. Substituting into Hooke's law, you can rewrite the expression as follows:

$$\varepsilon_{xx,TOT} = \frac{\sigma_{xx}}{E} - \nu\frac{\sigma_{yy}}{E} = \frac{1}{E}\left(\sigma_{xx} - \nu\sigma_{yy}\right)$$

So for this example, if you substitute in the appropriate normal stress values in each of the three Cartesian directions, you can compute the net strain in a given direction. For the strain in the *x*-direction,

$$\varepsilon_{xx,TOT} = \frac{1}{E}\left(\sigma_{xx} - \nu\sigma_{yy}\right) = \frac{1}{(10{,}000\ \text{ksi})}\left((9\ \text{ksi}) - (0.333)\cdot(14\ \text{ksi})\right) =$$
$$+0.000433 = +433\mu$$

From this calculation, you can see that the combined effect of this loading results in a net positive strain in the *x*-direction. You can then compute the net strain in the *y*-direction through a similar derived relationship:

$$\varepsilon_{yy,TOT} = \frac{1}{E}\left(-\nu\sigma_{xx} + \sigma_{yy}\right) = \frac{1}{(10{,}000\ \text{ksi})}\left(-(0.333)\cdot(9\ \text{ksi}) + (14\ \text{ksi})\right) =$$
$$+0.001100 = +1{,}100\mu$$

which indicates that under this loading, the element experiences a positive net strain in the *y*-direction as well.

Hooke's law in three dimensions

You can apply the same logic you use to create the two-dimensional generalized equations (see the preceding section) to three dimensions as well; you just need to add a third term to represent the *z*-Cartesian axis as follows:

$$\varepsilon_{xx} = \frac{1}{E}\left(\sigma_{xx} - \nu\sigma_{yy} - \nu\sigma_{zz}\right)$$
$$\varepsilon_{yy} = \frac{1}{E}\left(-\nu\sigma_{xx} + \sigma_{yy} - \nu\sigma_{zz}\right)$$
$$\varepsilon_{zz} = \frac{1}{E}\left(-\nu\sigma_{xx} - \nu\sigma_{yy} + \sigma_{zz}\right)$$

From Hooke's law, the strain in any direction is a result of the stress in that direction plus the Poisson effects from the stresses in the other two Cartesian directions.

The generalized relationship for three dimensions works for the uniaxial case as well. If the uniaxial stress (σ_{yy}) is in the *y*-direction, the other two stresses are both zero: $\sigma_{xx} = \sigma_{zz} = 0$ ksi. (See Chapter 8 for more on uniaxial stress.) Substituting into the generalized expressions, you get the original formula for one-dimensional Hooke's law. Just remember, even with stress in a single direction, you still get strains in all three directions.

Similarly, you can extend Hooke's law for shear to relate the shear stress in a given direction to the shear strain in the same direction:

$$\gamma_{xy} = \frac{2(1+v)}{E}\tau_{xy} = \frac{\tau_{xy}}{G}$$
$$\gamma_{yz} = \frac{2(1+v)}{E}\tau_{yz} = \frac{\tau_{yz}}{G}$$
$$\gamma_{xz} = \frac{2(1+v)}{E}\tau_{xz} = \frac{\tau_{xz}}{G}$$

Calculating stress from known strain values

You can also rework the strain relationship equations in reverse. Applying a bit of algebra to the generalized relationships I show in the preceding section, you can solve for each of the three stresses:

$$\sigma_{xx} = \frac{E}{(1+v)(1-2v)}\left[(1-v)\varepsilon_{xx} + v\left(\varepsilon_{yy} + \varepsilon_{zz}\right)\right]$$
$$\sigma_{yy} = \frac{E}{(1+v)(1-2v)}\left[(1-v)\varepsilon_{yy} + v\left(\varepsilon_{xx} + \varepsilon_{zz}\right)\right]$$
$$\sigma_{zz} = \frac{E}{(1+v)(1-2v)}\left[(1-v)\varepsilon_{zz} + v\left(\varepsilon_{xx} + \varepsilon_{yy}\right)\right]$$
$$\tau_{xy} = \frac{E}{2(1+v)}\gamma_{xy} = G\gamma_{xy}$$
$$\tau_{yz} = \frac{E}{2(1+v)}\gamma_{yz} = G\gamma_{yz}$$
$$\tau_{xz} = \frac{E}{2(1+v)}\gamma_{xz} = G\gamma_{xz}$$

If you happen to know the strains in each of the three Cartesian directions, you can determine a corresponding stress in each of those directions.

Part IV
Applying Stress and Strain

In this part . . .

This part shows you how to apply the principles of stress and strain to different types of problems you encounter in mechanics of materials. I start by showing how you can combine different types of stresses from Part II into one combined net effect. I then show you how you can predict deformation based on loads by utilizing stress-strain relationship. You discover how to use these deformation relationships to analyze statically indeterminate structures, and I explain how compression members can experience failure due to geometric considerations as opposed to material stress failure.

You also find out how to use all these basic applications to perform basic designs for different types of members. Finally, the part explores analysis techniques that allow you to handle advanced situations, such as impact, that use physics-based work-energy methods and their relationships to stress and strain.

Chapter 15

Calculating Combined Stresses

In This Chapter

- Applying the principle of superposition
- Combing stresses from axial, shear, and moment loads
- Incorporating torsion into shear applications

Stresses come in a wide array of sources and directions. You may be analyzing the axial loads in the column of a building due to vertical (or gravity) loads when a wind force suddenly hits your building and causes extra bending moments to appear. If you designed only for the vertical axial stresses, your design may not be sufficient for the added stress effects from the bending moments.

For scenarios such as this (and for many more different scenarios in general), you need to be able to compute the combined effects of multiple stresses on an object. In this chapter, I show you the basic methodology for combining stresses through a process known as the principle of superposition. I start by explaining several of the key assumptions of this method, and then I show you how to calculate combined stresses for several different categories of stresses.

After you have this step completed, you can start calculating all those maximum and minimum stress values that I show you in Part II for transforming stresses.

Understanding the Principle of Superposition: A Simple Case of Addition

As you learned in statics, the *principle of superposition* states that multiple actions on an object are equivalent to the sum of each of the effects when applied separately. This technique is very handy for dealing with complex load cases when you're finding support reactions and internal forces. But you can also apply this method in the area of mechanics of materials, and it's especially useful when working with stresses and strains as long as you recall the basic assumptions behind the principle — most of which I cover in Chapter 14:

- **Small displacement theory:** *Small displacement theory* implies that when a structural system has a load applied to it, the system *deflects* (or deforms), but only a small or negligible amount.
- **Linear system behavior:** *Linear system behavior* indicates that a given load results in a given deformation. If the system is linear, it experiences twice the deformation when you apply twice the load. I talk more about this requirement in Chapter 14.

  Although many common structures behave linearly under small loads, this response may no longer be valid when a structure begins to deform plastically or fail due to larger and larger loads.

- **Elastic material behavior:** Elastic material behavior is what occurs when a deformable body completely returns to its original and undeformed shape after an applied load is removed. This behavior implies a direct relationship between the applied stress on an object and the object's deformation (or strain) response. If this assumption is valid for your situation, you can use Young's modulus of elasticity to move quickly and easily between these two values. For more information on this topic, turn to Chapter 14.

Statics problems naturally satisfy the major assumptions in this list because in statics you assume that all objects are rigid bodies. However, that isn't the case in mechanics of materials, so you cheat a little bit by making the major assumption that although objects do deform under load, their deformations remain small enough that they still don't cause a problem.

In mechanics of materials, you can use the principle of superposition in two different but important applications: combining stress (which I cover in this chapter) and deformation of structures (which I cover in Chapter 16).

The principle of superposition allows you to break up an otherwise-complex problem into more manageable pieces. After you have the actions separated, you can compute the stress magnitudes individually and then recombine them based on the signs of their values as I show in Figure 15-1 for a simple stress element.

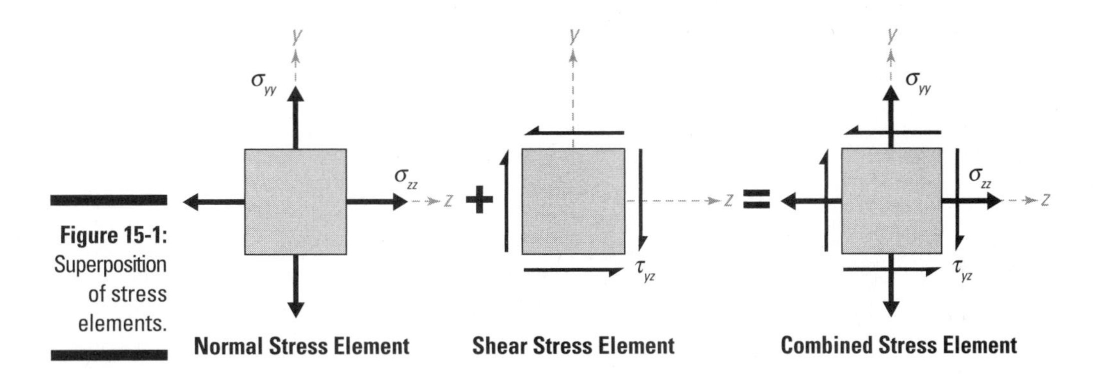

Figure 15-1: Superposition of stress elements.

If you determine the individual state of stress at your point of interest for each individual action, you can then combine their effects by using the guidelines I explain in the following section.

Setting the Stage for Combining Stresses

Combining stresses is fundamental to your ability to be able to accurately apply mechanics of materials to the objects in the world around you. Although the concept itself is fairly simple, you need to make sure to remember a couple of rules and conventions.

Following some simple rules

The basic idea for combining stresses boils down to this basic idea: Normal stresses only combine with other normal stresses, and shear stresses only combine with other shear stresses, as long as you follow a few simple rules:

- **Normal stresses combine only if they're acting along the same direction.** For example, you can only combine a normal stress in the x-direction with another normal stress in the x-direction, and the other axes behave similarly. You can, however, combine stresses that have a different sense as long as their lines of action are all in the same direction.
- **Shear stresses combine only if they're in the same plane and same direction.** You can only add a shear stress in a given plane (such as the XY plane) to other shear stresses in the same plane and to shear stresses that are acting in the same direction.

  You can think of this idea as adding shear stresses if the lines of action of the individual shear stress arrows are in the same direction.

If a normal stress and a shear stress act at the same time, you must consider both as you apply the principle of superposition even though a normal stress contributes a value of zero to the superposition involving the shear stresses. Likewise, a shear stress contributes a value of zero to the superposition involving the normal stresses.

Establishing a few handy conventions

For the purposes of the examples in this chapter, I use the axis convention shown in Figure 15-2, where the cross section of the member lies in the XY plane and the longitudinal axis is parallel to the z-axis. This orientation helps illustrate how you can determine whether you should add or subtract stresses in the same direction or plane.

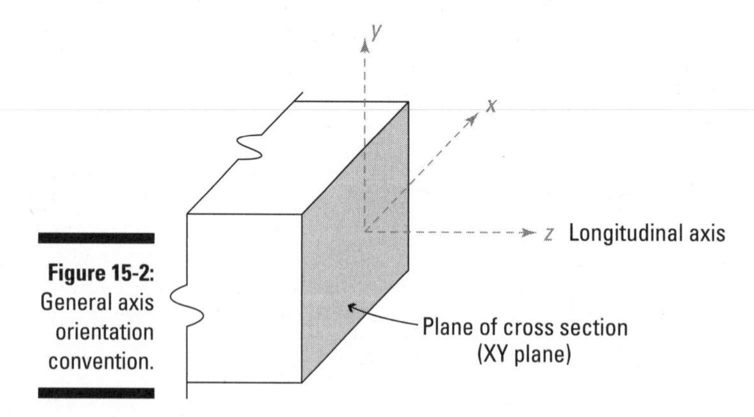

Figure 15-2: General axis orientation convention.

For normal stresses from axial loads, if the arrows are pulling on the object (which is a tensile normal stress), they're a positive value. If the arrows are pushing on the object (or a compressive normal stress), the stresses associated with that arrow have a negative value. The direction of these normal stresses is always in the direction of the axial force that causes them. Remember, for many objects (including the pressure vessels, which I discuss in Chapter 8), you can actually have multiple normal stresses acting simultaneously in multiple directions.

For normal stresses from bending moments, the sign convention follows the convention I establish in Chapter 9: positive for a tensile effect, and negative for compression. However, the directions of these stresses vary within the cross section based on the direction of the moment. To illustrate these directions, consider a cross section that lies in the XY plane of the Cartesian coordinate system shown in Figure 15-2.

- A bending moment about an x-axis gives you normal stresses in the z-direction; the maximum positive and minimum negative occur on opposite sides of the neutral axis at the extreme fibers that are parallel to the x-axis.
- A bending moment about a y-axis gives you a normal stress in the z-direction; the maximum positive and minimum negative occur on opposite sides of the neutral axis at the extreme fibers that are parallel to the y-axis.

Shear stresses, whether they come from torsion or flexure are handled according to the sign convention I establish in Chapter 6: positive if the shear effects on opposite vertical edges of the stress element tend to rotate the element in a counterclockwise direction and negative if the rotation is clockwise. However, in the case of torsion (see Chapter 11), you also have to pay special attention to which edge of the cross section you are working with because the direction of these stresses is dependent on the direction of the internal torque on the object. On one side of the exposed cross section, you get a positive shear stress, and on the opposite side, you get a negative shear stress. Just watch the direction of the torque to give you a hint on the direction; shear stress is always in the same direction as the direction of the torque on a given face.

Handling Multiple Axial Effects

After you've got your conventions squared away (see the earlier section "Setting the Stage for Combining Stresses"), you're ready to explore the different load and stress combinations you may encounter. The simplest of these cases involve only normal stresses. For these, you simply need to examine each of the cases separately and then add their effects together (assuming they meet the requirements of the principle of superposition — flip to "Understanding the Principle of Superposition: A Simple Case of Addition" earlier in the chapter). Consider the simple cylindrical pressure vessel (shown in Figure 15-3a), which is also subjected to a compressive axial force at the same time. I discuss both of these scenarios separately in Chapter 8.

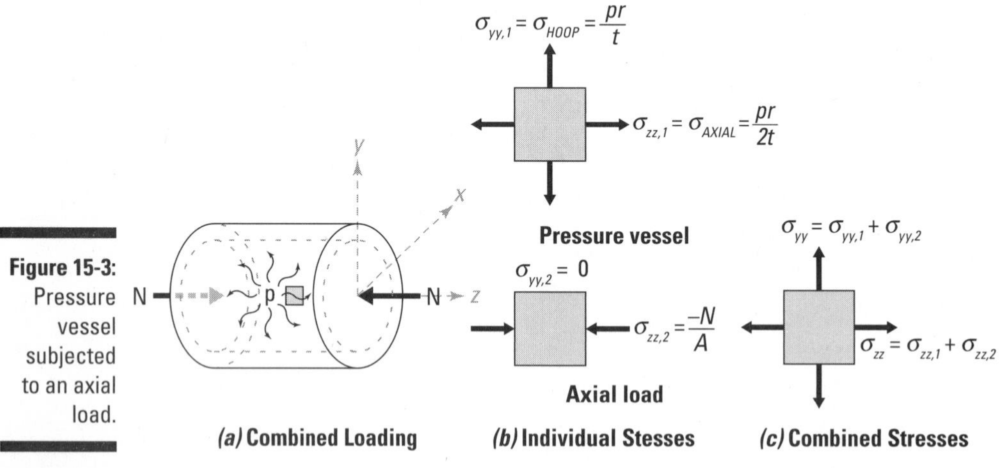

Figure 15-3: Pressure vessel subjected to an axial load.

If the applied loads don't produce a stress in a given direction, you can always display them as a zero value.

Assume that for the presssure vessel loads, you computed the hoop stress to be 16 MPa (T) and the axial stress to be 8 MPa (T). If the normal stress from only the axial load is 14 MPa (C), you have all the information you need to determine the combined normal stress element by using the method of superposition. You can analyze each of the cases separately and display one stress element for the pressure vessel and a second for the axial load (for this example both of these elements are shown in Figure 15-3b).

Using the Cartesian coordinate system I establish in Figure 15-2, you know that the hoop stress of 16MPa is a tension stress in the *y*-direction. Because the axial load is acting only in the z-direction in this problem, this load creates no stress in the y-direction. You compute the net normal stress acting in the y-direction, or σ_{yy}, as follows:

$$\sigma_{yy} = \sigma_{yy,1} + \sigma_{yy,2} = (+16\text{ MPa}) + (0\text{ MPa}) = +16\text{ MPa} = 16\text{ MPa (T)}$$

The axial (or longitudinal) stress from the pressure vessel is +8 MPa in the z-direction, and the stress from the applied axial load is a compressive stress of –14 MPa in the z-direction. Thus, the net normal stress in the z-direction, or σ_{zz}, is

$$\sigma_{zz} = \sigma_{zz,1} + \sigma_{zz,2} = (8\text{ MPa}) + (-14\text{ MPa}) = -6\text{ MPa} = 6\text{ MPa (C)}$$

You also need to examine the shear stresses, but because all the stresses of this example are normal stresses, you automatically know that $\tau_{xy} = 0$ MPa. You can see the final combined stress element in Figure 15-3c.

Including Bending in Combined Stresses

A large number of combined stress problems are the result of bending effects in conjunction with other effects (such as axial loads, torsion, and so on). A shear force on the end of a cantilever beam produces a bending moment at the fixed support (as well as other locations along the length) at the same time as a flexural shear force. An axial force that doesn't pass through the centroid of the cross section (or along the longitudinal axis) can even create multiple bending moments about multiple axes simultaneously. Sounds like a great opportunity for combined stress techniques, doesn't it? Good news: That's what I cover in the following sections!

Bending biaxially from inclined point loads

Sometimes you encounter a beam where the load doesn't act parallel to Cartesian axes but rather acts at some angle such that its line of action is through the centroid (see Chapter 4) of the cross section. To solve this, you must find the component of the moment parallel to the each of the Cartesian axes.

Consider the rectangular cross section shown in Figure 15-4a, which is 60 millimeters high and 20 millimeters wide and is loaded at the end of the cantilever with a load of 1,000 Newton at an inclined angle of 30 degrees. In this example, you want to compute the stresses at the support (or Point 2).

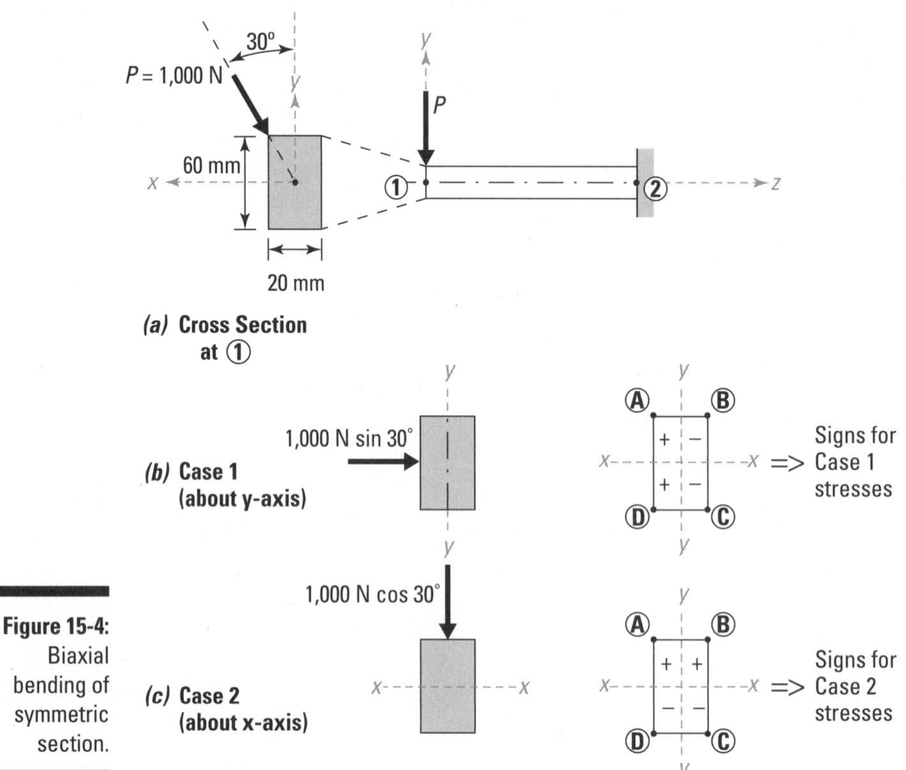

Figure 15-4: Biaxial bending of symmetric section.

For this example, you can establish the Cartesian coordinate system such that the cross section of the beam lies in the XY plane and the origin is located at the centroid of the cross section — making the *z*-centroidal axis the longitudinal (or axial direction) axis for the member.

To keep the signs of the stresses straight for biaxial bending problems, I like to compute the magnitude of the bending stresses (see Chapter 9), assuming that the moments are all positive (as indicated by absolute value brackets) and apply the signs by hand at the end of the calculation based on the direction of bending.

For Case 1 (shown in Figure 15-4b), which contains the horizontal component of the load, you calculate the applied moment at Point 2 about the *y*-axis M_y as 1,000 N(sin 30°)(3 m) = 1,500 N-m. To compute the stress at a particular location (such as the extreme fibers) within the cross section, you also need the moment of inertia about the y-centroidal axis, (see Chapter 4) as $I_{yy} = 4.0 \times 10^{-8}$ m^4. You can then compute the stress due to Case 1 as

$$\left|\sigma_{zz,1}\right| = \left|-\frac{M_y x}{I_{yy}}\right| = \left|-\frac{(1{,}500\text{ N-m})(0.01\text{ m})}{\left(4.0\times10^{-8}\text{ m}^4\right)}\right| = \left|-375.0\text{ MPa}\right| = 375.0\text{ MPa}$$

Similarly, Case 2 (shown in Figure 15-4c) contains the vertical component of the load. You find the applied moment about the x-axis at Point 2 as $M_x = -1{,}000$ N (cos 30°)(3 m) = –2,598 N-m. To compute the stress at a particular location, you need to first compute the moment of inertia about the x-centroidal axis as $I_{xx} = 3.6 \times 10^{-7}$ m^4. The stress due to Case 2 is then

$$\left|\sigma_{zz,2}\right| = \left|-\frac{M_x y}{I_{xx}}\right| = \left|-\frac{(-2{,}598\text{ N-m})(0.03\text{ m})}{\left(3.6\times10^{-7}\text{ m}^4\right)}\right| = \left|+216.5\text{ MPa}\right| = 216.5\text{ MPa}$$

To combine these stresses, you need to look at the signs of the stress relative to the direction of bending. That is, you know that on either side of the neutral axis, a member is either in tension or compression, and it can't be in both on the same side of the neutral axis. If the member is in tension, the normal stress is positive, and if it's in compression, the normal stress is negative.

Using the magnitude of the normal stresses ($\sigma_{zz,1}$ for Case 1 and $\sigma_{zz,2}$ for Case 2), you're now ready to apply the signs manually for specific points and finally compute the combined stresses. For Case 1, Point A and Point D experience a tension stress, so the signs of the normal stress at these locations are positive, or $+\sigma_{zz,1}$, such that $\sigma_{zz,A1} = \sigma_{zz,D1} = +375.0$ MPa. Points B and C experience a compression stress of $-\sigma_{zz,1}$. Therefore, $\sigma_{zz,B1} = \sigma_{zz,C1} = -375.0$ MPa.

You can repeat this process for the moment of Case 2. The moment about the x-axis in this example results in a tension (or positive) stress above the neutral axis and a compression (or negative) stress below the neutral axis. That means that $\sigma_{zz,A2} = \sigma_{zz,B2} = +216.5$ MPa and $\sigma_{zz,C2} = \sigma_{zz,D2} = -216.5$ MPa.

Finally, to determine the combined stress from both of these two cases together, you simply add (or superimpose) these stress quantities together:

$$\sigma_{zz,A} = \sigma_{zz,A1} + \sigma_{zz,A2} = (+375.0 + 216.5\text{ MPa}) = +591.5\text{ MPa} = 591.5\text{ MPa (T)}$$

$$\sigma_{zz,B} = \sigma_{zz,B1} + \sigma_{zz,B2} = (-375.0 + 216.5\text{ MPa}) = -158.5\text{ MPa} = 158.5\text{ MPa (C)}$$

$$\sigma_{zz,C} = \sigma_{zz,C1} + \sigma_{zz,C2} = (-375.0 - 216.5\text{ MPa}) = -591.5\text{ MPa} = 591.5\text{ MPa (C)}$$

$$\sigma_{zz,D} = \sigma_{zz,D1} + \sigma_{zz,D2} = (+375.0 - 216.5\text{ MPa}) = +158.5\text{ MPa} = 158.5\text{ MPa (T)}$$

Figure 15-5 shows a plot of the normal stresses at their respective locations in the cross section. In this figure, I've rounded the normal stresses to the nearest MPa.

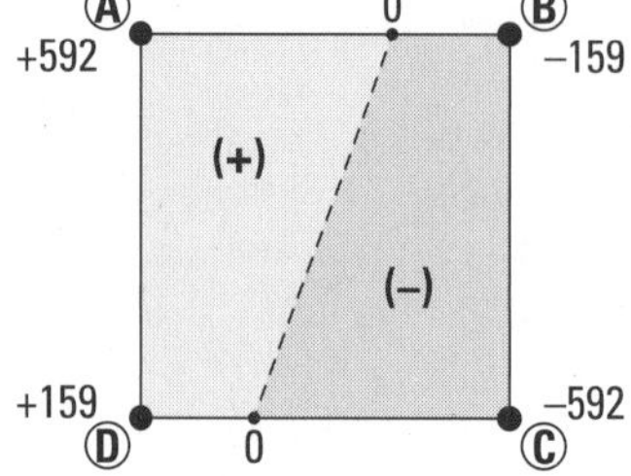

Figure 15-5: Results of biaxial bending example.

Plotting the four stresses, you can see that at Point A, the combined stress in this example, is a maximum positive value, and the maximum negative stress appears in the opposite corner at Point C. The dotted line along the cross section represents a line that connects the location of zero stress in the member. To the left of this line, all normal stresses are tension, and to the right, all normal stresses are compression. Regardless of whether the stress is in tension or compression, each of these stresses still acts in the longitudinal or *z*-direction.

Incidentally, you also have multiple shear effects in this example, but this example deals only with the normal stresses from the bending effects. I show you how to include shear effects from flexural stresses into your combined stress elements in the following section.

Combining flexural shear and bending stresses

One of the most common combined stress cases you encounter appears when you start working with *flexural stresses* (stresses resulting from transverse loads that produce bending and shear simultaneously). The vast majority of flexural load cases result in both normal stresses from bending (see Chapter 9) as well as shear stresses from shear (see Chapter 10).

Just as with the other cases of combined stresses, you can use the principle of superposition to separate the effects of normal stresses due to flexural bending and the shear stresses due to flexural shear.

When you work with flexural stresses, remember that both the shear and the moment vary with location within the cross section. To compute these basic stresses for vertical forces and moments about the *x*-centroidal axis, you use the following relationships:

$$\sigma_{zz} = -\frac{M_x y}{I_{xx}} \quad \text{and} \quad \tau_{xy} = \frac{V_y Q_x}{I_{xx} t}$$

Although the moment M_x, the shear V_y, and the moment of inertia I_{xx} are constant values for a given cross section at a given location on the flexural member, the values for y, first moment of area Q_x, and thickness t are all dependent on the exact point within the cross section at which you want to compute the stress.

You can have shear stresses in multiple directions from flexural loads when you have an inclined load, such as the one I show in Figure 15-4.

Consider the simply supported beam shown in Figure 15-6 that has a cross section of width b and height h lying in the XY plane. The beam is subjected to a vertical uniform load along the full length of the beam. Typically, you should analyze the combined stress state for this problem at multiple points within the cross section, but you definitely want to include at least the following points:

- **Point of minimum shear stress:** Point A and Point E are important because they're the locations of minimum shear stress; actually, the value of shear stress at both of these points is exactly zero (you can read more about why in Chapter 10).
- **Point of minimum normal stress:** Point C is a vital analysis point because it's located at the horizontal neutral axis. It also happens to be the location of zero bending stress as well as the location of maximum shearing stress.

You also want to include additional points within the cross section because sometimes the largest combined effects (such as principal stresses, which I discuss in Chapter 7) don't occur at the locations of maximum normal or minimum shear stresses. Other possible locations you may want to consider include

- **Geometric changes:** In members such as I-sections and T-sections, the section changes width at different locations within the cross section. You usually want to include a point on either side of these sudden geometry changes in your analysis, particularly because shear stress can change dramatically at these points.
- **Material property changes:** For members composed of multiple materials or multiple pieces of similar materials that are attached by some mechanical means (such as nails, screws, glues or welds), the interfaces of these materials are typically points worth investigating.

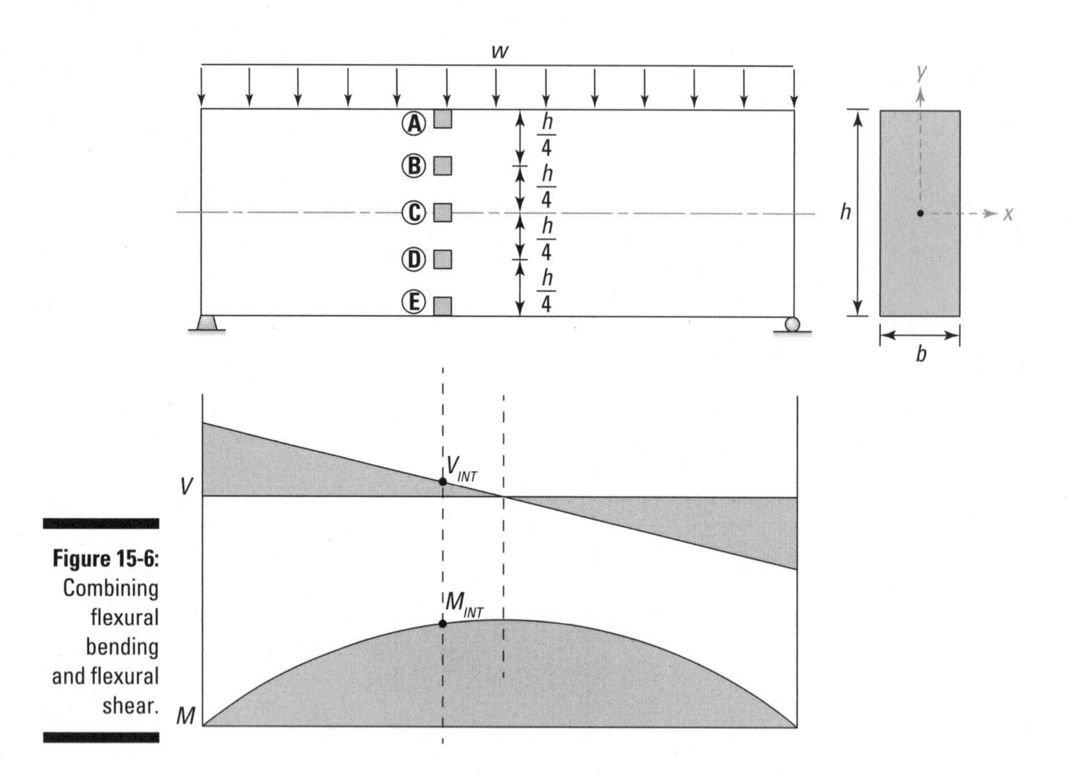

Figure 15-6: Combining flexural bending and flexural shear.

In this example, I don't actually have either of these cases, so I arbitrarily choose a point between the maximum and zero bending stress locations and the maximum and zero shear stress locations. I choose Point B to be a distance of $h/4$ from the top and Point D to be a distance of $h/4$ from the bottom.

Because the loading in this example causes a positive moment, the normal stresses at Point A and Point B are all negative (compressive) values. Point A has the largest negative normal stress because it's at a larger distance (+*y*-value) than Point B. The normal stress at Point C is zero because it's the location of the neutral axis. Point D and Point E are both positive normal stress values because they both have a negative *y* value (because they're below the neutral axis). Point E has the largest tensile stress because it's the largest negative distance (in the –*y*-direction) to Point E. Figure 15-7 shows the normal stresses for these locations.

You then determine the shear stresses from the flexural loads on a point-by-point basis as well. At Points A and E, you know that the shear stress is zero, and at the neutral axis (Point C), the shear stress is maximum. The shear stresses at Points B and D have a value between zero and the maximum.

You can see the shear stresses for these locations in Figure 15-7. You can then combine these normal and shear stresses by using the principle of superposition at each point to create the combined stress elements (which I also show in Figure 15-7).

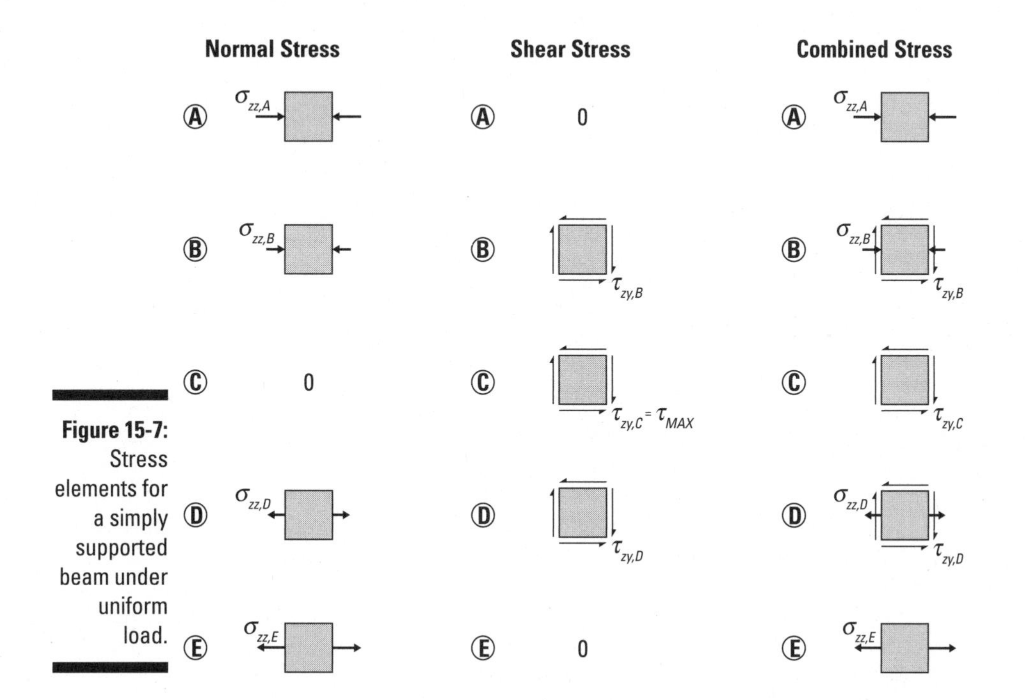

Figure 15-7: Stress elements for a simply supported beam under uniform load.

Acting eccentrically about axial loads

When working with axial loads, you typically assume that they act at the centroid of the cross section (or *concentrically*) because the longitudinal axis perpendicular to the plane of the cross section acts through that point. When the axial loads act through the centroid of the cross section, you can usually assume that these normal stresses are actually average normal stresses from axial loads (as I discuss in Chapter 8) and that they have the same numeric value at all locations within the cross section. However, in some structures, the axial load may well be applied *eccentrically* (or not concentrically) to the centroid of the object.

Eccentric load behavior can happen in structures when a beam sitting on a column or foundation experiences a rotation that moves the center of the reaction away from an intended position.

Fortunately, when you handle these types of situations, you can use the same basic superposition techniques that I present in the preceding section; you just have a couple of extra bending stresses to consider.

To illustrate this process, consider the member shown in Figure 15-8a that is subjected to an axial tension load applied at Point O.

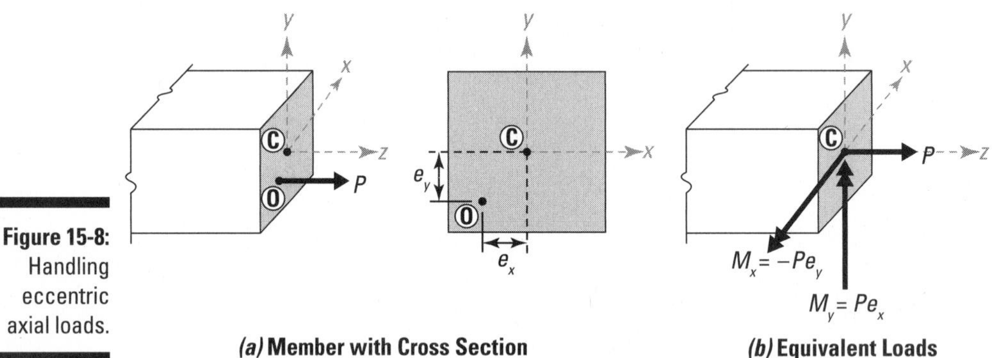

Figure 15-8: Handling eccentric axial loads.

To handle eccentric axial loads, you simply need to relocate the axial load to the centroid of the cross section. Remember from statics that when you relocate a force on a rigid body, you get an equivalent axial force of the same magnitude acting at the centroid (producing an axial normal stress), and you create additional bending moments about the *x*- and *y*-axes, which both produce normal stresses in the *z*-direction.

For this example, the location of the axial load is in the lower-left quadrant of the cross section as shown in Figure 15-8a, with eccentric dimensions e_x and e_y, which are both negative when measured with respect to the centroid at Point C. That means that for a positive axial force *P*, the corresponding moment of this force about the *x*-axis is negative and the moment of the force about the *y*-axis is positive.

To compute the magnitude of each these moments, you simply take the magnitude of the force and multiply it by its perpendicular distance from the centroid: $M_x = -Pe_y$ and $M_y = Pe_x$. I show these moments on the cross section of Figure 15-8b.

In these equations, you want to be sure to input both the magnitudes of the forces and their respective eccentricities as positive values. You can then apply the signs (or the sense) to these moments by using a bit of logic.

After you have the moments computed, you can then determine the bending stresses in the z-direction (as I outline in Chapter 9) separately about each axis in the plane of the cross sections. Finally, you can then combine each of these normal stresses with the axial stress (which is also in the z-direction) to get a single normal stress acting in the z-direction.

Eccentric axial loads can actually create moments about both Cartesian axes within the cross section. To handle these cases, you compute the effective moment about each axis and then treat the problem as a type of biaxial bending problem (which I discuss in "Bending biaxially from inclined point loads" earlier in this chapter). Just remember that you also have an equivalent axial load effect that produces an extra normal stress that you must include in the combined stress element.

Putting a Twist on Combined Stresses of Torsion and Shear

Another stress that can play a role in your combined stress analysis is the shear stress as a result of torsion. As I describe in Chapter 11, *torsion* is a twisting phenomenon that causes rotation about the longitudinal axis of rotation, which is assumed to pass through the centroid. Because it's a twisting effect, torsion applies to the principle of superposition a bit differently than other bending moments do: It can be positive in one location on a cross section and negative in another. For direct shear, shear stresses in a cross section usually have the same sign at every location. Although the methodology for superposition is still sound, the signs associated with a torsional stress are worth investigating. (Head to "Understanding the Principle of Superposition: A Simple Case of Addition" earlier in the chapter for more on superposition.)

As long as your situation meets all the criteria for the principle of superposition, you can still use the basic idea to include shear stresses from torsion. But here's the warning up front: Beware of the signs and direction of the torsional stresses.

For circular shafts, where warping isn't an issue, the stresses from the applied torque are all shear stresses. Consider the round shaft of Figure 15-9a, which is subjected to both direct shear V and an applied torque T. (Remember that whenever you have a direct shear, you need to be very mindful about the flexural bending stresses when you perform your superposition. However, in this example, I assume the bending stresses are zero.)

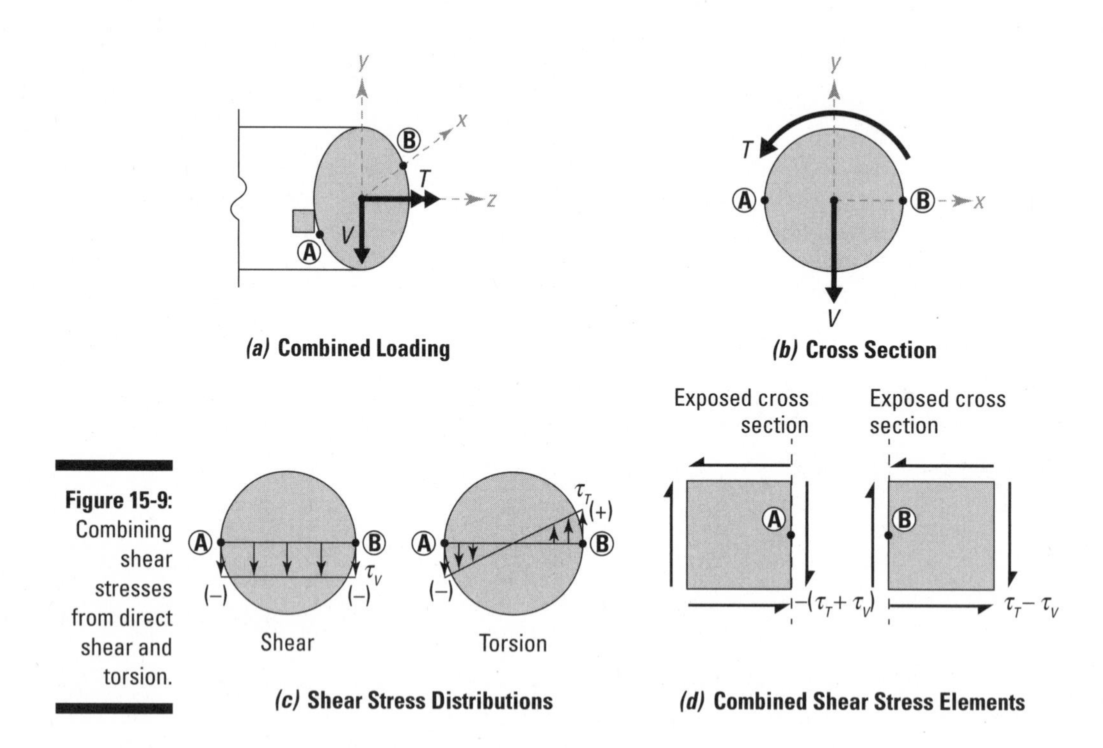

Figure 15-9: Combining shear stresses from direct shear and torsion.

In Figure 15-9b, the applied torque is acting counterclockwise on the cross section and produces a negative shear stress at Point A in the cross section and a positive shear stress at Point B. However, for the direct shear load V, the resulting shear stress is a negative value (because the direction of the applied shear is negative). (I show both of these shear stress distributions in Figure 15-9c.) So at Point A, the shear stresses from torsion and direct shear are both in the same direction and become additive. But at Point B, the two shear stresses are opposing each other, and the net shear stress is reduced. Figure 15-9d shows the combined stress elements that result for both of these points.

After you determine these combined stress elements, you're free to continue your stress analysis by computing principal stresses and angles using the techniques of Chapter 7 and then relating those stresses to strains using the concepts of Chapter 14.

Chapter 16

When Push Comes to Shove: Dealing with Deformations

In This Chapter

- Squaring away conventions
- Using integrals to compute axial and flexural deflections in beams
- Determining the angle of twist due to torsion

Aside from computing the actual stresses in a member due to applied load (which I cover in Part II), another computation that engineers find extremely important is the calculation of deformations (or displacements and rotations), which come in many shapes and varieties. In this chapter, I show you how objects subjected to loads actually deform.

Displacements from bending moments are what occur while you're standing on a diving board pondering a back flip into the deep end of a pool. Axial displacements are what cause pressure vessels to expand or small columns to shorten. Torsional deformations can be used in the study of rotation in power-transmission shafts.

You express the basic relationship for determining deformations in structures as differential equations that relate the deformation to an internal force. However, before you panic at the thought of spending your time solving differential equations, I keep it fairly simple here. In this chapter, I show you that a little integration goes a long way. And in some cases, you don't even have to do that.

Covering Deformation Calculation Basics

Before you can begin to calculate deformations, you need to understand how applied loads and deformations are actually interrelated; this interrelation is known as the *stiffness* of a member. In this section, I define the parameters of stiffness and discuss the key assumptions behind the deformation calculations that I introduce later in this chapter.

Defining stiffness

You can define the basic relationship between load and deformation as follows:

load = stiffness × deformation

where the *stiffness* term refers to the properties of an object to resist deformations under applied loads. The stiffness of an object is related to the following factors:

- **Cross-sectional properties:** Properties such as the cross-sectional area (see Chapter 4) and the second moment of area (or the area moment of inertia) and the torsion constant (both in Chapter 5) play a significant role in helping you determine the stiffness of an object.
- **Structural dimensions:** Dimensions of the object within a structure often affect the stiffness of the object. An object's length is an important factor in its deformation response under load.
- **Material properties:** Material properties such as Young's modulus of elasticity, Poisson's ratio, and the modulus of rigidity are specific to the material of the object and can affect the deformation of the object.

For each load type, calculating stiffness helps determine the response of the structure. You can calculate some structural stiffnesses (for members subjected to axial loads and torsion) with a simple formula, but others (such as those due to bending) require more complex calculations involving calculus or differential equations.

Depending on the type of applied load, the corresponding deformation may be a displacement or rotation. An object can also have multiple load effects and consequently experience multiple types of deformation simultaneously. This situation occurs frequently in objects such as columns of buildings that can be subjected to both axial loads and bending moments.

Making some key assumptions

Displacements are usually measured in inches for U.S. customary units and in meters for SI units (though for very small displacements, you can always convert to millimeters). Rotations in the calculations of this chapter have no units, but they're generally expressed as radians in both U.S. customary and SI systems.

For the deformation calculations I show in this chapter, a few key assumptions are constant throughout, many of which I mention in Chapter 14:

- **The structural object behaves linearly with respect to the load.** A member deflects under a load. It deflects twice as much under twice the load.

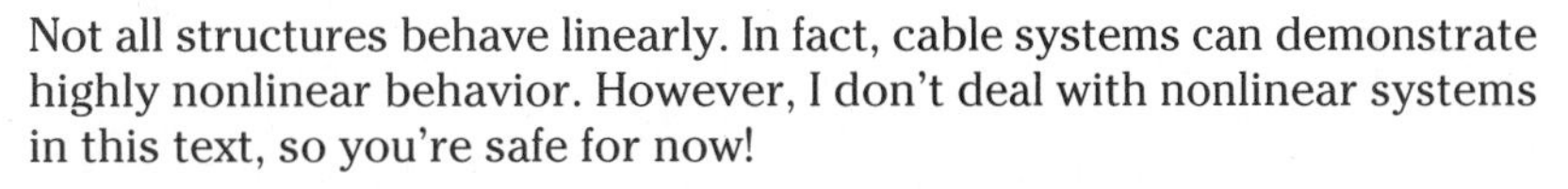

  Not all structures behave linearly. In fact, cable systems can demonstrate highly nonlinear behavior. However, I don't deal with nonlinear systems in this text, so you're safe for now!

- **The material of the object is elastic, isotropic, and homogenous.** Remember that *elastic* refers to the region of material behavior where any deformation due to applied load totally disappears when the load is removed. A material is *homogeneous* if its properties don't vary with position within the object and *isotropic* if its properties don't vary with direction.
- **The member is prismatic.** A member that is prismatic has a cross-sectional area that is constant (or uniform) along the length (or longitudinal axis) of the member. However, for some calculations in this chapter, I show you how you can work around this requirement and actually work with non-prismatic sections.
- **All displacements remain small.** The calculations observe small displacement theory, meaning that the displacements remain very small. I discuss this theory more in Chapter 15.

Addressing Displacement of Axial Members

Displacements are always relative calculations, meaning you measure them with respect to some reference location. For axial members, you usually use one end of the member as the reference and determine axial displacements with respect to that end. Even though a bar subjected to tension grows at both ends, the calculations I present here assign the combined displacement to one end.

For some problems, you often conveniently take the reference location for displacement calculations as a support or wall location (if one is available) because those locations don't usually move.

If you're calculating a displacement, you must include a reference location, such as a support location or some other arbitrary point, from which you can base your calculations. Without this reference clearly defined, the techniques I present in this chapter only provide deformation over a given length, and you can't necessarily always use them to describe a system's total displacement.

Computing axial deformations

Determining axial deformations is fairly straightforward, and fortunately, it doesn't usually require solving a differential equation. For simple cases, you calculate axial deformation by taking Hooke's law for stress and strain and substituting the equation for axial stress (from Chapter 8) and the equation for strain (from Chapter 12):

$$\sigma = E\varepsilon \Rightarrow \frac{P_{INT}}{A} = E\left(\frac{\Delta}{L}\right)$$

If you perform a bit of algebra, you can relate the axial deformation in terms of the load:

$$\Delta = \frac{P_{INT}L}{AE}$$

where P_{INT} is the internal force in the member; L is the length over which the internal force is constant; A is the cross-sectional area of the member; and E is Young's modulus of elasticity for the material.

Imagine an aluminum bar with a cross-sectional area A of 2 square inches, a length L of 12 feet, a Young's modulus of elasticity E of 10,000 ksi, and a single tensile force of 40,000 pounds at the end — which makes the internal force P_{INT} = 40,000 pounds at every point along the length. You can compute the deformation (or axial elongation) to the load as follows:

$$\Delta = \frac{(40{,}000 \text{ lbs})\left(\dfrac{1 \text{ kip}}{1{,}000 \text{ lbs}}\right)(12 \text{ ft})\left(\dfrac{12 \text{ in}}{1 \text{ ft}}\right)}{\left(2 \text{ in}^2\right)(10{,}000 \text{ ksi})} = 0.288 \text{ in}$$

Even though the calculations are usually fairly simple, you must be careful about the units. In this example, the load is in pounds, and the Young's modulus of elasticity is in ksi, which actually has a *kip* in it. If you don't make the units in your calculations agree, you'll be off by a factor of 1,000 (because 1 kip = 1,000 lbs)! Also, another potential unit hazard comes in the length of structural members. For larger members, you may see the length expressed in feet (or meters), where the cross-sectional area is in square inches (or square millimeters).

Determining relative displacements

For members with different cross-sectional areas along the length, or subjected to different internal loads throughout, you have to use a slight variation on the axial displacement formula:

$$\Delta_i = \sum\left(\frac{P_i L_i}{A_i E_i}\right)$$

where the terms in the equation are the same as the original equation, but the *i* subscript indicates that you need to do this calculation for different subregions. The subregions that you must analyze occur when you split an axial member because of changes in geometry, in material properties, or in internal force.

You actually compute the elongation of each individual region separately and then combine them to determine total deformation. To illustrate the process, consider the aluminum assembly (where E = 10,000 ksi) shown in Figure 16-1a that has three different cross sections lying in XY planes, lengths, and multiple axial loads applied at the locations shown.

To compute the total displacement at the end of the axial member (as well as intermediate displacements) with respect to a reference location, you can use the following simple procedure:

1. **Divide the axial member into subregions for your analysis.**

   You can divide the member in Figure 16-1a into distinct subregions. The figure has three different cross-sectional areas, so you know that you need at least three regions. However, the 30-kip load applied at Point C causes different internal forces within Section BCD, so you need to break this region into separate subregions as well, each of which has a cross-sectional area of $A_{BC} = A_{CD} = 1.2 \text{ in}^2$. For this reason, you need a total of four subregions: AB, BC, CD, and DE.

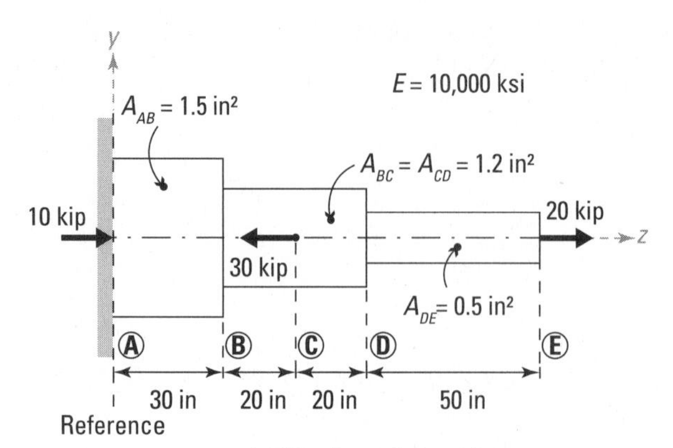

(a) **Aluminum Assembly**

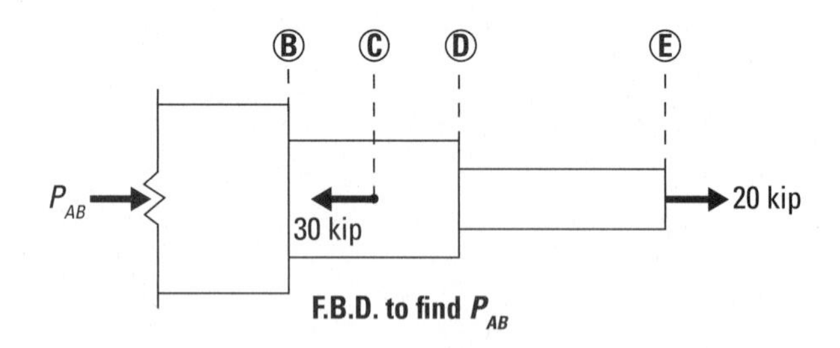

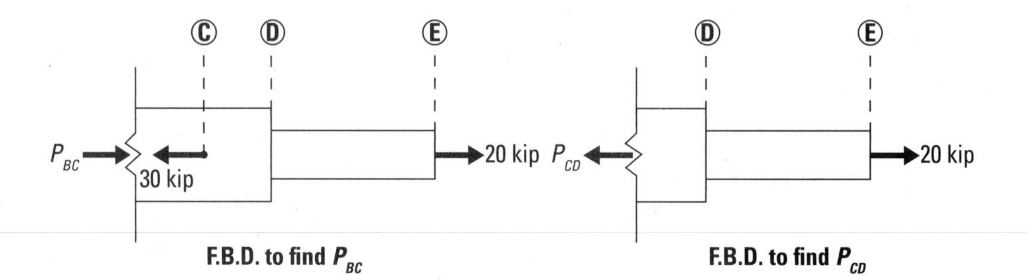

E

P_{DE} 20 kip

F.B.D. to find P_{DE}

(b) **Free-Body Diagrams**

Figure 16-1: Computing relative axial deformations.

2. **Determine the internal force in each subregion from Step 1.**

   For each of the subregions, use statics to determine the internal forces in each subregion. You can use the free-body diagrams shown in Figure 16-1b to help you with your equilibrium calculations, or you can draw an axial load diagram for the entire system. (Notice that I've omitted the shear and moment from these internal forces because all loads are axial and acting through the centroid of the member.)

   P_{AB} = 10 kip (C) P_{BC} = 10 kip (C) P_{CD} = 20 kip (T) P_{DE} = 20 kip (T)

3. **Calculate the relative deformation within each subregion.**

   Next, you compute the relative axial deformation within each subregion by substituting the length, cross-sectional area, Young's modulus of elasticity, and the internal load for each subregion. For example,

   $$\Delta_{AB} = \frac{P_{AB}L_{AB}}{A_{AB}E_{AB}} = \frac{(-10\text{ kip})(30\text{ in})}{(1.5\text{ in}^2)(10{,}000\text{ ksi})} = -0.02\text{ in}$$

   The negative sign in this value indicates that the length of the member gets shorter (or decreases), which should make sense because the internal axial force in this section is also negative (indicating compression). You can perform a similar calculation for the remaining three subregions and find that

   $\Delta_{BC} = -0.017$ in $\Delta_{CD} = +0.033$ in $\Delta_{DE} = +0.20$ in

4. **Determine displacement of points of interest, starting at the reference location.**

   If you choose the wall at Point A as the reference, you know that Point B experiences a movement in the amount of $\Delta_{AB} = -0.02$ in (or 0.02 in to the left) from where it started.

   Relative to Point B, you know that Point C moves an amount equal to $\Delta_{BC} = -0.017$ in. The total displacement with respect to the wall is then $\Delta_{AB} + \Delta_{BC}$ = (–0.02 in + (–0.017 in)) = –0.037 in of total displacement from its original position.

   Similarly, relative to Point C, Point D moves +0.033 in (or to the right), and the total displacement with respect to the wall is $\Delta_{AB} + \Delta_{BC} + \Delta_{CD}$ = (–0.02 in + (–0.017 in) + 0.033 in) = –0.004 in (which is still slightly to the left of its original position).

   Finally, the tip of the member moves a total amount equal to the sum of all of the displacements, $\Delta_{AB} + \Delta_{BC} + \Delta_{CD} + \Delta_{DE}$ = (–0.02 in + (–0.017 in) + 0.033 in + 0.20 in) = +0.204 in, which is to the right of the original position of Point E.

   Despite the fact that Points A, B, C, and D all end up moving to the left (as indicated by their negative values), the total deformation at the end of the object ends up being a net elongation of +0.204 in.

Handling non-prismatic sections under axial load

Another type of axial problem that you may encounter deals with computing deformations of a tapered section (or a section that doesn't have a uniform cross section along the length), such as the one shown in Figure 16-2a. An example of this type of problem involves determining the behavior of a tapered chimney smokestack.

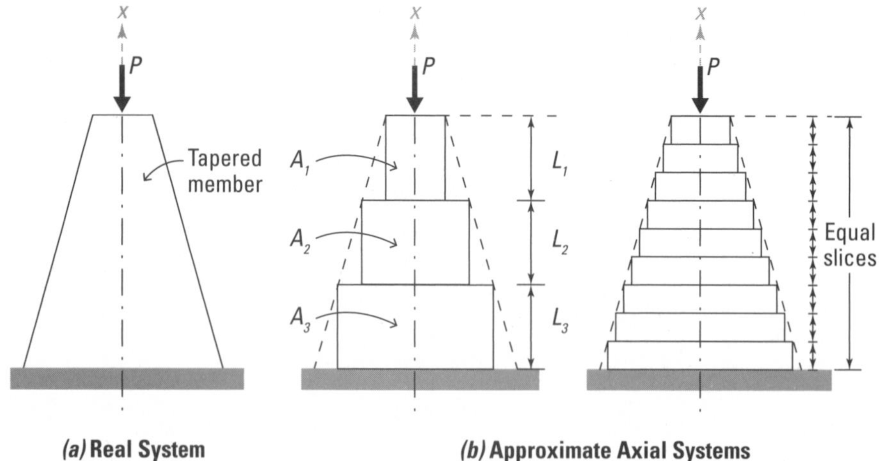

Figure 16-2: Computing axial deformations on tapered sections.

Although you can determine an exact solution to this problem if you know the profile of the taper and if all cross sections are concentric with a longitudinal axis, you have to revert to a slightly different variation of Hooke's law (see Chapter 14) where the axial strain, $\varepsilon_{xx} = du/dx$ and the axial stress is $\sigma_{xx} = P/A$. In this variation, both the stress and the strain of the object become dependent on the location of the cross section. You can determine the displacement of a member by integrating both sides of this equation:

$$\frac{du}{dx} = \frac{P}{AE} \Rightarrow \int du = \int \frac{P}{AE} \cdot dx$$
$$\Rightarrow \Delta = \int \frac{P}{AE} \cdot dx$$

In this equation, the variable A is no longer constant but becomes a function of its position (x) within the member, so you need to include that term in your integral. The benefit of this equation is that you can also handle problems where the load varies along the length (such as from the self weight or from problems involving surface friction or drag on objects) or where material properties vary with position.

An alternative way to analyze this system is to slice it into pieces that have a constant cross section of a small subregion within the member, such as either of the figures shown in Figure 16-2b. You then analyze this system as I show in the preceding section. The problem is that if you don't use enough subregions to define the taper, your calculations may not be very accurate. The more pieces you slice an object into, the closer your answer usually becomes to the theoretical (or exact) values; you just need to perform a lot more work in your calculations. The idea of slicing an object into a lot of different pieces is actually the basis for mathematical solution techniques such as differential methods and finite element methods.

Discovering Deflections of Flexural Members

The classical form of flexural analysis utilizes the *Euler-Bernoulli beam theory* (sometimes known as the *engineer's beam theory*), which was developed around 1750. The Euler-Bernoulli beam equation relates a beam's deflection to the moments applied to the member and produces the following second-order differential equation:

$$EI\frac{d^2v}{dx^2} = M$$

where v is the displacement at a location x on the beam; E is Young's modulus of elasticity (which I discuss in Chapter 14); and I is the second moment of area (as I describe in Chapter 5). The M value always turns out to be those generalized moment equations I mention in Chapter 3.

Setting up flexural assumptions

To use the classic beam equations from the preceding section, the beam must meet a couple of requirements:

- **All loads act laterally to the member.** The loads are applied perpendicular to the beam, meaning no axial force is present. But the beam can have shear and moments.
- **The beam neglects deformations due to shear.** The classic beam relationship neglects the effects of axial and shear forces on the overall displacements of the beam due to flexural loads.

  In reality, shear deformations should also be included, but it becomes a much more involved process (see the nearby sidebar for more on this topic).

In this text, I work only with basic Euler-Bernoulli beam theory, which neglects shear deformations. This assumption is typically okay because in most applications, transverse deformations due to shear are usually less than 5 percent of the total deformation for very long members — although this amount varies depending on the length of the beam and the depth of the cross section.

Defining the elastic curve for displacements

The equation represented by $v(x)$ is the solution of the classic beam equation (or the differential equation) I show at the beginning of this section. This equation, which is known as the *elastic curve* for the member, is an equation that defines the displacements of a beam at all locations as a function of position x along the length. If you plotted this function, you could actually show the entire deflected shape of the beam.

You can actually write the equations for elastic curves over any incremental length as long as you write enough equations to account for every point along the length of the member. For an elastic curve, you have to keep a couple of requirements in mind:

- **Elastic curve is a smooth function.** A *smooth function* has no kinks (or instant changes in slope). These kinks can actually occur whenever you have an internal hinge in a structure — where one side of a point rotates more than the other side (see Figure 16-3a).
- **Elastic curve is a continuous function.** The function for classic bending-type problems must be *continuous,* meaning that it doesn't have any sudden jumps or breaks in the elastic curve (see Figure 16-3b).

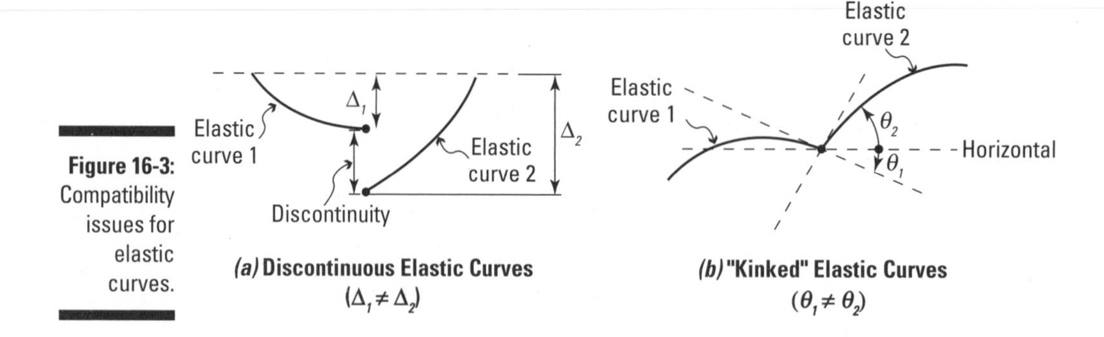

Figure 16-3: Compatibility issues for elastic curves.

Shear displacement and Timoshenko's beam theory

Stephen Timoshenko was an instrumental figure in the formulation of modern engineering mechanics. His work in flexural theory expanded upon the classic Euler-Bernoulli beam equation by also incorporating the effects of shear displacement into the flexural calculations as follows:

$$EI\frac{d^4v}{dx^4}=w(x)-\frac{EI}{\kappa AG}\cdot\left(\frac{d^2w}{dx^2}\right)$$

where κ is a shear deformation coefficient based on the shape of the cross section, and w is the load function acting on the member. If you neglect shear deformations, you usually take this value as zero.

The Euler-Bernoulli classic beam equation is actually just a specialized case of the Timoshenko beam theory. The solution of this fourth-order differential equation yields a higher-order function, which is typically more accurate in dealing with transient (or time-dependent) loads and is a better predictor of total beam displacements. For more on this topic, I leave you to consult an advanced mechanics textbook or your friendly neighborhood graduate professor.

A *compatibility condition* (or *continuity requirement*) is a relationship that relates displacements or rotations on one side of a cut line to their similar values on the other. These compatibility conditions prevent your elastic curve from becoming disjointed or kinked.

Closing in on a twice-integrated solution

To actually find the equation for the elastic curve, for statically determinate structures, you simply need to apply a little bit of calculus. To illustrate the process, consider the simply supported beam with span L, shown in Figure 16-4, which is subjected to a triangular load of magnitude, w.

To find the equation of the elastic curve, you just follow a few basic steps.

1. **Determine the generalized moment equation.**

   To start the process, you must generate an expression for the generalized moments as a function of the position x on the beam as I discuss in Chapter 3. The generalized moment $M(x)$ for the general section shown in Figure 16-4b is

   $$M(x)=-\frac{wLx}{6}+\frac{wx^3}{6L}$$

   where x is the position along the beam measured from the left end (or Point A). You can then substitute this expression into the classic beam equation and start the integration process.

Figure 16-4: Double integration technique for beam displacements

(a) **Complete System F.B.D.**

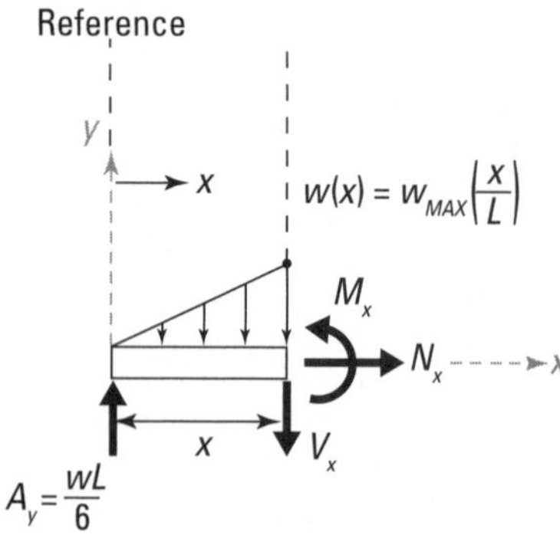

(b) **Generalized F.B.D. at *x***

2. **Integrate the beam equation to find the equation for the slope of the elastic curve.**

$$\frac{dv}{dx} = \theta(x) = \int \frac{d^2v}{dx^2} = \int \frac{M}{EI} \cdot dx = \int \frac{1}{EI}\left[-\frac{wLx}{6} + \frac{wx^3}{6L}\right] \cdot dx$$

$$\Rightarrow \frac{dv}{dx} = \theta(x) = -\frac{wLx^2}{12EI} + \frac{wx^4}{24EIL} + C_1$$

where C_1 is a *constant of integration* that appears as a result of the integration process. I show you how to solve for this constant in the following section.

Although this expression isn't actually the equation of the elastic curve that you're after, it does have another important significance: The first derivative of the displacement, or *dv/dx*, is actually the slope of the elastic curve. With this equation, you can determine how much a beam has rotated from its original positions at any location on the beam. Engineers use both displacements and slopes in their design and analysis of beams and bending members.

3. **Integrate the equation of Step 2 to determine the equation for the elastic curve.**

To get $v(x)$, the equation of the elastic curve, you simply need to integrate the slope equation from Step 2 once more:

$$v = \int \frac{dv}{dx} \cdot dx = \int \left[-\frac{wLx^2}{12EI} + \frac{wx^4}{24EIL} + C_1\right] \cdot dx$$

$$\Rightarrow v = -\frac{wLx^3}{36EI} + \frac{wx^5}{120EIL} + C_1x + C_2$$

where C_2 is a second constant of integration.

A situation where the double integration technique may not be sufficient is if the loads applied to the beam are fairly complex or the beam itself is statically indeterminate. In these cases, determining the support reactions and generalized moment equation can be a bit tricky. For indeterminate problems, you may prefer to use the method I show in the later section "Integrating the load distribution to solve for beam displacements."

Using boundary conditions to find constants of integration

As you may recall from your calculus or differential equations class, anytime you integrate a function (such as the solution to the beam equation I show in Step 3 in the preceding section), you end up with one constant of integration for each time you integrate; in that example, you actually end up with two constants, C_1 and C_2. These constants represent a unique solution to the equation that takes into account known values of the function at specific locations known as *boundary conditions.*

Boundary conditions come in a wide variety of forms, and they help make a function for an elastic curve meet a specific set of criteria, such as support reactions for displacements and rotations, or even for internal moments. I show several common boundary conditions for different types of supports and internal forces in Figure 16-5.

Here's how you calculate constants of integration:

1. **Determine the boundary conditions for the problem.**

   For Figure 16-4, the support at Point A (located at $x = 0$) is a pinned support, which has no vertical displacement, and the support at Point B (located at $x = L$) is a roller support, which also has no vertical displacement. From the diagrams shown in Figure 16-5, the boundary conditions for this problem are

   $$x = 0 \rightarrow v(0) = 0$$
   $$x = L \rightarrow v(L) = 0$$

2. **Solve for the constants of integration.**

   From the first boundary condition, you can substitute a value of $x = 0$ into the equation for the elastic curve and set it equal to the value of $v = 0$. Doing so gives you a value for the constant of integration $C_2 = 0$. Repeating the process for the second boundary condition requires a bit more algebra, but substituting in the values of $x = L$ and $v = 0$ into the equation gives you the value for the constant of integration C_1:

   $$C_1 = +\frac{7wL^3}{360EI}$$

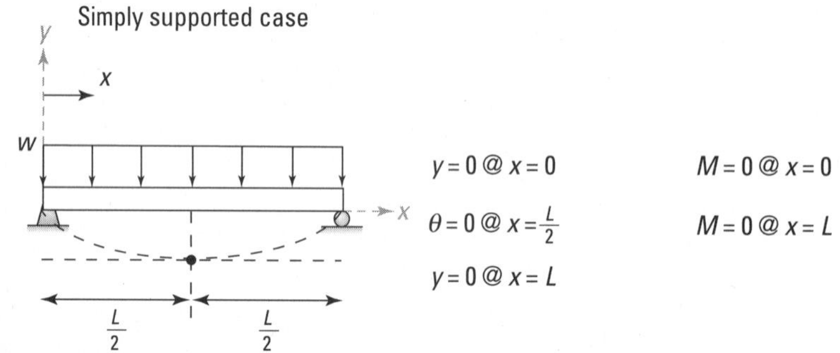

Figure 16-5: Boundary conditions.

3. **Substitute the constants of integration into the equations for the elastic curve and slope.**

   With the constants determined from the boundary conditions, you can then write the complete equations for the elastic curve and the slope for any position x for this example:

   $$v = -\frac{wLx^3}{36EI} + \frac{wx^5}{120EIL} - \frac{7wL^3x}{360EI} = \frac{w}{360EIL}\left[3x^5 - 10L^2x^3 + 7L^4x\right]$$

   The slope of the elastic curve is

   $$\frac{dv}{dx} = \theta(x) = -\frac{wLx^2}{12EI} + \frac{wx^4}{24EIL} + \frac{7wL^3x}{360EI} = \frac{w}{360EIL}\left[15x^4 - 30L^2x^2 + 7L^3x\right]$$

Defining differential equations for deformation

You can apply the basic concepts of this section to other types of deformation and load as well; you just need to know the differential equation that relates the internal loads to the deformation and determine the appropriate boundary conditions. For torsion, the following second-order differential equation relates the angle of twist to the torque applied to the member:

$$GJ\frac{d^2\phi}{dx^2} = T$$

where ϕ is the angle of twist (a measure of rotation) at a location x on the beam or shaft, G is the shear modulus of elasticity (see Chapter 14); J is the torsion constant for the cross section; and T is the applied internal torque with respect to x. To solve this equation, you simply integrate this equation twice with respect to x and apply boundary conditions to find the two unknown constants of integration that arise.

For axially loaded members, the following equation relates the axial deformation to the internal axial loads on the member:

$$AE\frac{du}{dx} = P$$

where u is the axial displacement at a location x on the member; E is Young's modulus of elasticity; A is the cross-sectional area; and P is the internal axial force with respect to x. To solve this equation, you need to integrate once only, and apply only a single boundary condition to obtain the constant of integration.

Integrating the load distribution to solve for beam displacements

Unfortunately, for *statically indeterminate* beams (or beams with more unknown support reactions than available equilibrium equations), you typically have one or more of the unknown reaction supports appearing on the free-body diagram when you go to write the generalized moment equations. Without more information, the equations of static equilibrium may be insufficient to determine all of the support reactions and internal forces of the member.

You can actually modify the methodology I present in the preceding section and come up with a couple of new expressions that allow you solve certain indeterminate problems (though it also works on statically determinate problems as well). Starting with the classic beam equation, you know the relationship between the moment and the displacement function:

$$EI\frac{d^2v}{dx^2} = M(x)$$

As I note in Chapter 3, if you differentiate this equation, you get an expression for shear because $dM/dx = V$:

$$EI\frac{d^3v}{dx^3} = V(x)$$

Finally, differentiating this equation once more gets you an expression for the load, because $dV/dx = w$. The final equation is a fourth-order differential equation and relates the displacement of the elastic curve (see the earlier section "Defining the elastic curve for displacements") to a function related to the load:

$$EI\frac{d^4v}{dx^4} = \left(\frac{dV}{dx}\right) = w(x)$$

This equation is actually a special case of Timoshenko's beam theory without the contribution from shear deformations, which you often neglect because they tend to be very small. Flip to the sidebar "Shear displacement and Timoshenko's beam theory" in this chapter for more on this theory.

As with the double integration technique I show earlier in the chapter, working with the fourth-order equation in this section requires you to find boundary conditions. However, the advantage of using the fourth-order equation is that you can use additional information about the beam aside from the known displacements and rotations at the support reactions; now you can also use internal moments and shears. The disadvantage of the fourth-order equation is that you now have to perform a series of four integration steps, and each step creates an extra constant of integration — that is, your equation for the elastic curve has a C_1, C_2, C_3, and C_4 in the final expression. ("Defining the elastic curve for displacements" earlier in the chapter gives you the lowdown on these boundary conditions and how to find them.)

Four consecutive integrations

Consider the indeterminate beam shown in Figure 16-6, which is subjected to a uniform load w as shown.

The fourth-order beam equation from the preceding section tells you that

$$EI\frac{d^4v}{dx^4} = \left(\frac{dV}{dx}\right) = -w$$

where $w = w(x)$ is constant. So if you start integrating, you can develop the expressions for the lower-order derivatives — just don't forget the constants of integration at each step:

$$EI\frac{d^3v}{dx^3} = V(x) = -wx + C_1$$

$$EI\frac{d^2v}{dx^2} = M(x) = \int V(x)\cdot dx = \frac{-wx^2}{2} + C_1x + C_2$$

$$EI\frac{dv}{dx} = \theta(x) = \int M(x)\cdot dx = \frac{-wx^3}{6} + C_1\frac{x^2}{2} + C_2x + C_3$$

$$EIv = \int \theta(x)\cdot dx = \frac{-wx^4}{24} + C_1\frac{x^3}{6} + C_2\frac{x^2}{2} + C_3x + C_4$$

which generates the equation for the elastic curve, $v(x)$.

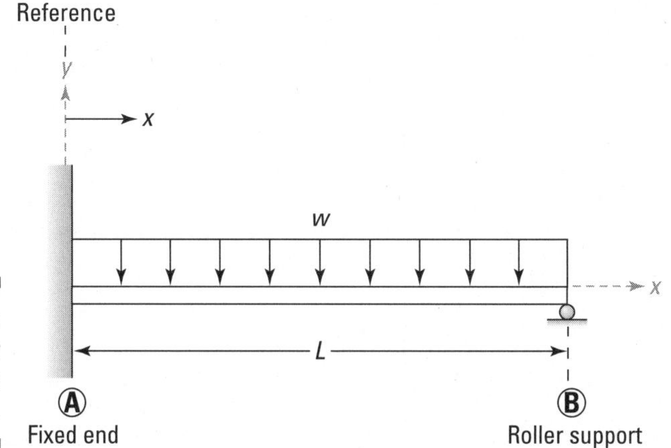

Figure 16-6: Fourth-order solution method.

Finding boundary conditions to help solve for constants of integration

After you find the elastic curve equation, you're ready to apply the boundary conditions. In addition to the displacement and rotation boundary conditions (or support boundary conditions) I show in Figure 16-5a earlier in the chapter, you can also use the value of shears and moments at specific locations if you know them. For these items, you usually want to look for places where the moments and shear are equal to zero, such as the situations I show in Figure 16-5b.

For the beam of Figure 16-6, start by looking at locations of known displacement. At the fixed support of Point A, you know that both the displacement and rotation must be zero because it's a fixed support. That is, at $x = 0 \rightarrow v(0) = 0$ and $\theta(0) = 0$. At the support at Point B, the displacement is 0, giving a boundary condition of $x = L \rightarrow v(0) = 0$. Finally, at the roller support at Point B, the value of the moment is zero, which gives you a boundary condition of $x = L \rightarrow M(L) = 0$.

If you have an internal moment boundary condition, you can't have a rotational (or slope) related boundary condition. Likewise, if you have a shear force boundary condition, you can't have a displacement boundary condition, and vice versa.

From this information, you can substitute these boundary conditions into the appropriate differential equations and determine the constants of integration:

$$C_1 = \frac{5wL}{8} \quad C_2 = \frac{-wL^2}{8} \quad C_3 = 0 \quad C_4 = 0$$

These calculations then allow you to write the equation of the elastic curve.

$$v = \frac{1}{EI}\left[\frac{-wx^4}{24} + \frac{5wLx^3}{48} - \frac{wL^2x^2}{16}\right] = \frac{w}{48EI}\left(-2x^4 + 5Lx^3 - 3L^2x^2\right)$$

Fortunately, many textbooks and design aids provide you with some of the more-common basic equations for internal loads and elastic curves already computed for you. Just remember to pay special attention to where they measure their reference position x from. Most measure it from the left end of the diagram, but some resources may do it differently, so keep your eyes open!

Finding maximum slope and deflection from elastic curve equations

After you have the equations of the elastic curve defined as a function, you can determine the maximum and minimum values of the curve — the maximum and minimum displacements. Using some basic calculus, which I refresh in Chapter 2, you can take the derivative of the elastic curve (or the slope equation) and set that equation to equal zero. Solving for the location x gives you the location of the maximum and minimum values. With these locations, you can then substitute into the equation for the elastic curve $v(x)$ and actually compute these maximum or minimum values.

Angling for a Twist Angle

When a load is applied in objects such as beams and axial members, the deformation (or displacement) is usually a linear displacement. However, objects under torsion (such as the ones I discuss in Chapter 11) behave a bit differently. The following sections examine how torsion affects deformation.

Measuring the angle of twist in prismatic shafts

Under torsion loads, as long as the object is a circular shape (either hollow or solid), the object twists within the plane of the cross section when a torque is applied. Figure 16-7a shows a circular shaft subjected to a torque at each end.

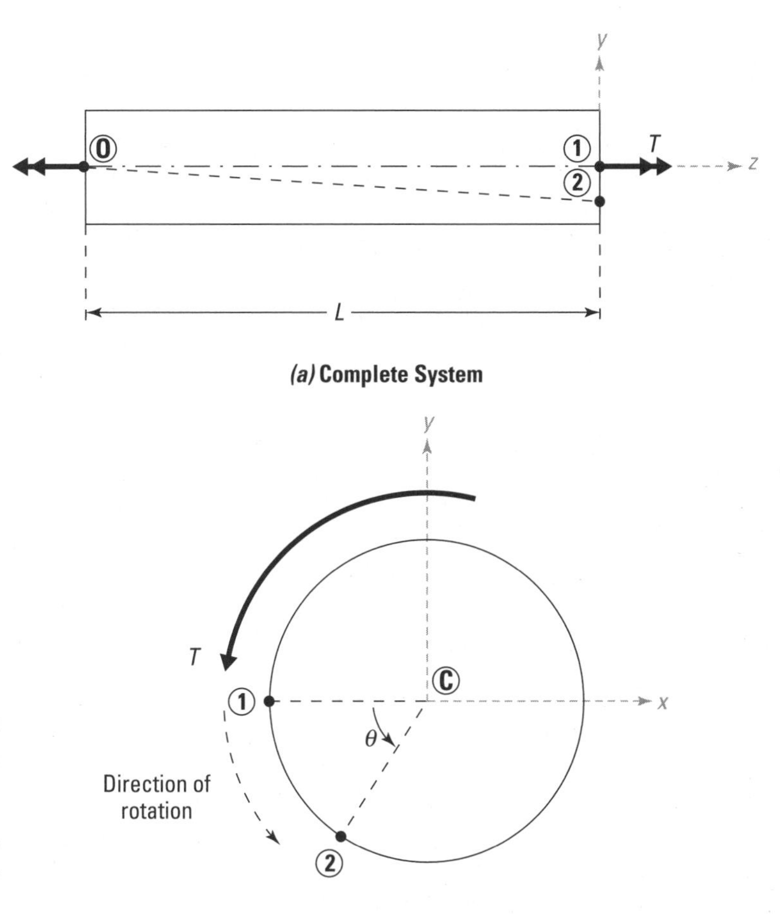

Figure 16-7: Measuring the angle of twist for circular shafts.

Within circular shapes, such as the one in Figure 16-7b, Point 1 on the outer surface rotates to its final position at Point 2. Points within the cross section remain in the same plane of the cross section, and they simply orbit around the center of the cross section. Although you can measure the distance that this point moves around the circumference of the cross section, a better representation of this displacement is to measure the change in angle between these two points with respect to the axis of rotation (or the longitudinal axis) of the shaft.

This change in angle is known as the *angle of twist* and can be calculated as

$$\phi = \frac{TL}{GJ}$$

where T is the internal torque at the cross section of interest; L is the length of the shaft (or distance from a reference location) on which the torque T is measured; G is the shear modulus of elasticity (or the modulus of rigidity, which I explain in Chapter 14); and J is the torsion constant I describe in Chapter 11.

For example, consider a 2-meter-long solid circular steel shaft (G = 79 GPa) that has a radius r of 10 millimeters and is subjected to an applied torque of 100 kN-m. Computing the torsion constant for this shaft, $J = 0.5\pi(r^4)$ = 15,700 mm^4. To get the angle of twist, use the following equation:

$$\phi = \frac{TL}{GJ} = \frac{(100\ \text{kN}\cdot\text{m})(2\ \text{m})}{(79\ \text{GPa})\left(\frac{10^9\ \text{Pa}}{1\ \text{GPa}}\right)(15{,}700\ \text{mm}^4)\left(\frac{1\ \text{m}}{1{,}000\ \text{mm}}\right)^4} = 0.161\ \text{rad}$$

When calculated correctly, the units on the angle of twist should all cancel out, which means the units are typically measured in radians (or rad) for both U.S. customary and SI units.

You must be careful with the units of this calculation. For many objects, particularly those measured in the metric (or SI) system, the shear modulus G is usually reported in GPa (or 1×10^9 Pa); at the same time, shaft dimensions are commonly reported in millimeters (which means the torsion constant J is calculated in mm^4). You can see the necessary conversions I use to avoid this problem in the preceding calculation.

Measuring the angle of twist in compound torsion problems

Shafts are frequently used to transmit power through torque and are often subjected to multiple pulley loads or other torque connections such as gears and cogs. To determine the angle of twist for shafts subjected to multiple applied torques or consisting of multiple diameter shafts, you follow a procedure that is very similar to the procedure I outline for axial loads in the earlier section "Determining relative displacements." Instead of calculating relative displacements, you now calculate relative angles of twist.

To illustrate the process, consider the brass shaft (where G = 5,800 ksi) shown in Figure 16-8a that consists of two different shaft diameters: a 100-inch-long shaft *ABC* with a diameter of 0.75 inches and a 60-inch-long shaft *CD* with a 0.25-inch diameter. Two torques are applied at the locations indicated; their magnitudes are T_1 = 50 kip-ft at the end of the shaft and T_2 = 75 kip-ft applied at 80 inches from the wall.

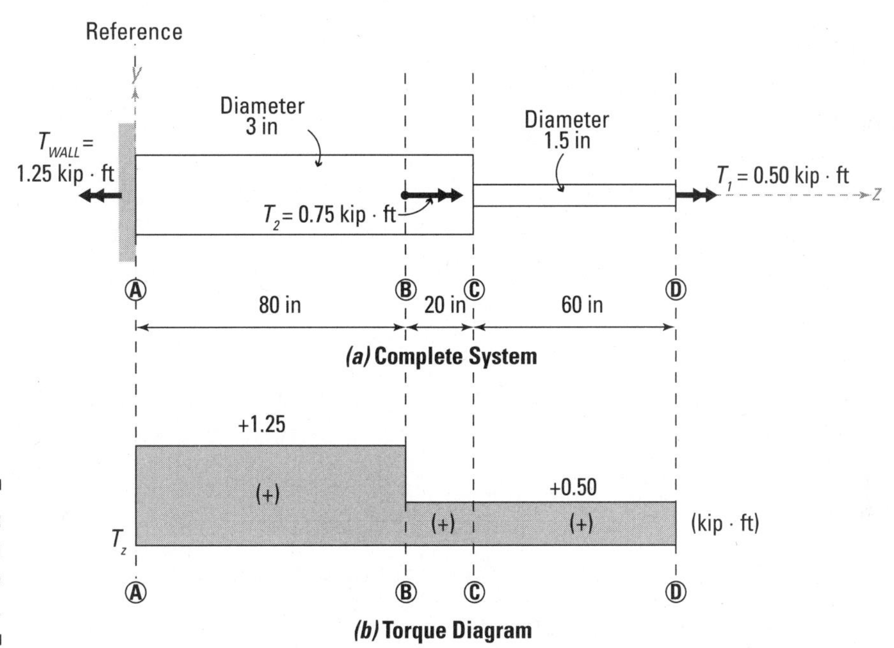

Figure 16-8: Computing relative twist angles.

Just as with axial members, you must divide a shaft into subregions and determine the internal torsion applied within each region.

To solve angle of twist problems, follow these steps:

1. **Divide the shaft into subregions for your analysis.**

   You can divide the member in Figure 16-8a into three distinct subregions. The figure has two different cross-sectional areas, so you know that you need at least two subregions. However, the 0.75-kip-ft torque applied at Point B causes different internal torques within shaft *ABC*. Thus, for this problem, you need three subregions: AB, BC, and CD.

2. **Determine the internal torque in each subregion from Step 1.**

   Use statics to determine the internal loads in each subregion. You can use the free-body diagrams to draw the torque diagram, which I show in Figure 16-8b. (Turn to Chapter 3 for more guidance.) As the torque diagram shows

   T_{AB} = +1.25 kip-ft T_{BC} = +0.50 kip-ft T_{CD} = +0.50 kip-ft

3. **Calculate the torsion constant for each subregion.**

   For the angle of twist calculation, you need to compute the torsion constant *J* for each subregion as I outline in Chapter 11. For subregions *AB* and *BC*, the torsion constant is the same:

$$J_{AB} = J_{BC} = \frac{\pi}{2}\left(\frac{3.00\text{ in}}{2}\right)^4 = 7.95\text{ in}^4$$

$$J_{CD} = \frac{\pi}{2}\left(\frac{1.50\text{ in}}{2}\right)^4 = 0.497\text{ in}^4$$

4. **Calculate the relative angle of twist within each subregion.**

   Next, you compute the relative angle of twist by substituting values for each subregion into the basic angle of twist equation. For example,

$$\phi_{AB} = \frac{T_{AB}L_{AB}}{G_{AB}J_{AB}} = \frac{(+1.25\text{ kip-ft})\left(\frac{12\text{ in}}{1\text{ ft}}\right)(80\text{ in})}{(7.95\text{ in}^4)(5{,}800\text{ ksi})} = +0.0260\text{ rad}$$

   The positive sign in this value indicates that the cross section rotates counterclockwise under the internal torque. You can perform a similar calculation for the remaining three subregions and find that

   ϕ_{BC} = +0.00260 rad ϕ_{CD} = +0.1249 rad

5. **Determine the total angle of twist at points of interest, starting at the reference location.**

   If you choose the wall at Point A as the reference, you now know that the cross section at Point B experiences a positive rotation relative to Point A in the amount of ϕ_{AB} = +0.0260 rad.

   Relative to Point B, you know that the cross section at Point C rotates an amount equal to ϕ_{BC} = +0.00260 rad. The total rotation with respect to the wall at Point A is then the sum of $\phi_{AB} + \phi_{BC}$ = (+0.0260 rad + (+0.00260 rad)) = +0.0286 rad of total rotation.

   Similarly, relative to Point C, the cross section at Point D moves +0.0286 rad and the total angle of twist with respect to the wall at Point A is $\phi_{AB} + \phi_{BC} + \phi_{CD}$ = (+0.0260 rad + (+0.00260 rad) + (+0.1249 rad)) = +0.1535 rad.

Chapter 17

Showing Determination When Dealing with Indeterminate Structures

In This Chapter

- Understanding indeterminate structures
- Computing behaviors by using redundant methods
- Working with systems of multiple materials
- Analyzing systems with rigid components

If a structure is statically determinate (or solvable using just the equations of equilibrium), you can use the principles of equilibrium from Chapter 3 to compute the internal loads and then use those loads to calculate stresses and strains. However, in the real world, many structures aren't statically determinate. Some of these systems have more supports than static equilibrium equations, while others may be built from multiple materials that resist applied loads by carrying different portions of their loads.

In this chapter, I show you several different methods for solving statically indeterminate systems. I introduce you to several different types of compatibility conditions and how you can apply them to problems of axial forces, torsional moments, and flexural effects. This chapter's techniques should add significantly to your arsenal of tricks for solving mechanics of materials problems.

Tackling Indeterminate Structures

In a *statically determinate* system, the number of support reactions and internal forces that you're trying to find are never more than the available equilibrium equations — you have a maximum of three unknowns for two-dimensional problems and no more than six unknowns for three-dimensional problems.

Unfortunately, most of the structural problems you actually encounter often have more unknowns than they do available equilibrium equations. Such structures are said to be *statically indeterminate,* meaning that equilibrium equations alone aren't sufficient to solve for all the unknown support reactions or internal forces.

Depending on the type of system and the number of degrees of indeterminacy, you actually have a few tricks up your sleeve for solving these problems. For example, if the structural system you're dealing with constitutes a statically indeterminate frame or machine that contains internal hinges, you can separate the members at internal hinge locations and give yourself additional equilibrium equations to work with; these extra equations can often give you enough extra information to solve for all the unknown support reactions. However, in some situations, even these statics tricks may not be sufficient for you to solve a particular problem.

For the majority of indeterminate applications, you must look at the behavior of the material itself to determine the characteristics of the deformation, and then you can develop equations that relate those deformations to each other. These equations are known as *compatibility conditions* (which I introduce in Chapter 16), and they become the basis for unlocking indeterminate problems.

In this section, I introduce several types of indeterminate structures, all of which require working with different types of compatibility equations. I also note some of the assumptions you make when solving indeterminate-structure problems.

Categorizing indeterminate structures

You can classify many indeterminate structural systems into one of the three following categories based on characteristics of the system:

- **Indeterminate supports:** Structures with *indeterminate supports* are structures that are indeterminate because they have an insufficient number of equilibrium equations (which I describe in Chapter 3) to solve for all the external support reactions. Solving these types of systems usually requires techniques known as *redundant methods.* Flip to the later section "Withdrawing Support: Creating Multiple Redundant Systems" for more on these methods.
- **Multiple materials:** Problems involving objects constructed from multiple materials require compatibility equations to relate the strains (or forces) in one material to the strains (or forces) in another.
- **Rigid bar problems:** *Rigid bar problems* are typically the easiest to identify because they contain a clearly defined rigid structural member. With these problems, you can define the compatibility conditions for deformations based on the behavior of the rigid member itself.

After you classify the type of indeterminate system, you're then ready to utilize an appropriate method to solve the problem. Each of these methods requires you to establish different compatibility conditions, which I explain later in the chapter.

Clarifying assumptions for indeterminate methods

To apply the techniques of this chapter, you must ensure that the structure or object you're dealing with satisfies the following criteria (the same criteria that I outline in Chapter 14 for materials and stress-strain relationship):

- **Linear system behavior:** The response of the structure is directly proportional to the load applied. If you double the load, you double the deformation.
- **Elastic material behavior:** When a material deforms under load, it returns to its original position when the load is removed.
- **Isotropic and homogenous material properties:** *Isotropic* means that the materials are constant in a given direction, and *homogenous* means the material properties are uniform throughout.

Withdrawing Support: Creating Multiple Redundant Systems

A *redundant system* is a statically indeterminate system that you make statically determinate by removing extra unknown internal forces or support reactions. You then reapply these removed forces and reactions as separate behaviors, compute the corresponding deformation, and finally recombine all this information by using compatibility conditions.

After you remove the extra degrees of indeterminacy from the original system, the new resulting statically determinant system is known as the *primary system.* You use basic statics to analyze the primary system under the effects of the original applied loads.

Regardless of whether the system is an axial, flexural, or torsional system, you follow a few basic steps to solve a problem with multiple redundant support systems:

1. **Determine the number of degrees of indeterminacy for the problem.**

   Degrees of indeterminacy are the number of extra unknown reactions or internal forces in your system in excess of equilibrium equations.

2. **Create a primary system by removing one support degree of freedom for every degree of indeterminacy.**
3. **Compute the deformation(s) of the statically determinate primary system at each removed degree of freedom in Step 2.**
4. **One at a time, apply unit loads to each degree of freedom from Step 2 and calculate the corresponding deformations at all removed degrees of freedom throughout the structure.**
5. **Write the compatibility equations and solve for the unknown redundant support reactions.**
6. **Apply the equilibrium equations for the unknown support reactions (if necessary).**

The following sections discuss the types of compatibility equations required for axial, flexural, and torsional problems and illustrate how you can use redundant systems to solve them.

Axial bars with indeterminate supports

Computing deformations with axial members is generally pretty straightforward with the help of Hooke's law (see Chapter 14). If the member is subjected to internal axial loads, you can compute the stress in the member and then use Hooke's law to compute the strain, which lets you estimate the deformation. *Thermal effects* (or strains induced by changes in temperature — see Chapter 12) are another cause of deformations, and they can also cause support reactions in indeterminate structures. However, the procedure for analysis is the same whether you're dealing with a thermal deformation or a deformation from applied loads.

Consider the axial bar in Figure 17-1a, which is restrained at both ends and experiences a temperature change of 60 degrees Celsius. The bar is aluminum and has a cross-sectional area A of 0.1 m^2; a length L_{AB} of 6 m; a coefficient of thermal expansion α of 23×10^{-6}/°C; and a Young's modulus of elasticity E of 69 GPa.

This example is an indeterminate problem because it has two unknown forces in the x-direction. Summing forces in the x-direction for the top figure of Figure 17-1 produces the following equation:

$$\overset{+}{\rightarrow}\sum F_x = 0 \Rightarrow R_A + R_B = 0$$

Because this one equation contains two unknowns (R_A and R_B), this problem is statically indeterminate to the first degree. You know this because the other equations of equilibrium provide no additional useful information about these axial support reactions.

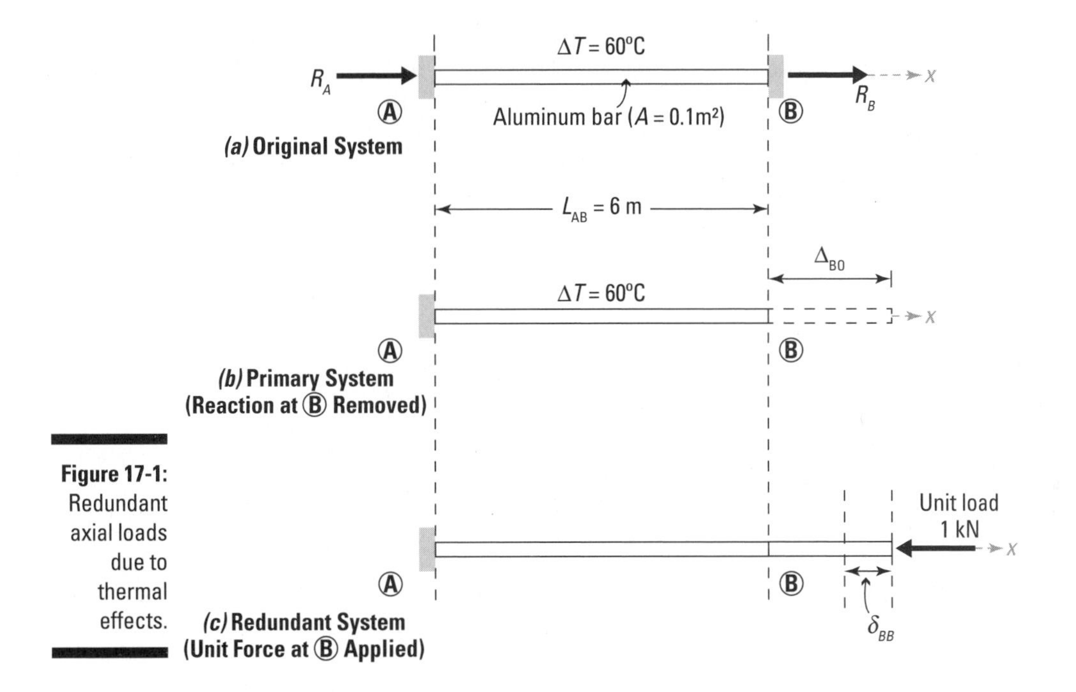

Figure 17-1: Redundant axial loads due to thermal effects.

Suppose you're interested in determining the reactions at Points A and B if the temperature increases by ΔT = 60 degrees Celsius. Applying the basic steps I outline earlier,

1. This problem has one degree of axial indeterminacy.
2. You can choose either support; it really makes no difference for axial systems. Here, I create the primary system by removing the support at Point B (as shown in Figure 17-1b).
3. You can find the deformation due to the increase in temperature at the free end of the primary system by using the equation for thermal deformation (which I cover in Chapter 12). Use the following equation:

   $$\Delta_{BO} = \alpha(\Delta T)(L_{AB}) = (23 \times 10^{-6}/°C)(50°C)(6\text{ m}) = 0.0069\text{ m} = 6.9\text{ mm}$$

4. The applied unit load is a 1-kN load applied at Point B (as shown in Figure 17-1c). Under this unit load, the axial bar deforms as follows:

   $$\delta_{BB} = \frac{(-1)L_{AB}}{A_{AB}E_{AB}} = \frac{(-1)(6\text{ m})}{\left(0.1\text{ m}^2\right)\left(69\times10^6\text{kPa}\right)} = -8.7\times10^{-7}\frac{\text{m}}{\text{kN}} = -8.7\times10^{-4}\frac{\text{mm}}{\text{kN}}$$

   This calculation shows that for every 1 kN of applied load, the bar experiences a deformation at Point B in the amount of 8.7×10^{-4} mm. For this problem, the direction of the applied load was to the left at Point B, so the deformation is negative, indicating that the bar has shortened.

5. The compatibility condition for this problem is that for every 1 kN, the bar experiences a deformation of 8.7×10^{-4} mm. However, you also know from Step 3 that for the primary system, the end of the bar at Point B experiences a total displacement of 6.9 mm (with respect to the support at Point A) due to the thermal changes. You can then express the compatibility relationship as

$$0 = \Delta_{BO} + R_B\delta_{BB} = (6.90 \text{ mm}) + R_B\left(-8.7\times10^{-4}\frac{\text{mm}}{\text{kN}}\right) = 0 \Rightarrow R_B = +7{,}931 \text{ kN}$$

The positive sign on this reaction force at Point B indicates that it is acting in the same direction to the direction of the unit load in the redundant system. This means that the redundant reaction is actually resisting the displacement.

From this calculation, you can see that if a thermal expansion is restrained, very significant restraint forces can occur.

6. From here, you can determine any of the remaining unknown support reactions and internal forces using the basic equations of equilibrium as I describe in Chapter 3.

Systems of axial members

The key to handling a system of axial rods is basically no different than working with a single axial rod. You can relate the deformation in any of the members to the internal force of the member through the basic relationship I describe in Chapter 16:

$$\Delta = \frac{PL}{AE}$$

where P is the internal force; L is the original length; A is the cross-sectional area; and E is the Young's modulus of elasticity for the member. Consider the system of axial members shown in Figure 17-2a, which consists of three axial bars with the dimensions shown and is subjected to a vertical load of 10 kip. To analyze a system of axial rods, you must relate the deformations in each bar to the deformation of the combined system by using compatibility:

1. **Draw the free-body diagram of the system and write the equations of equilibrium at the node that is experiencing displacement.**

   To start the analysis, you must develop a free-body diagram that relates all the internal forces in each rod to the applied load (as shown in Figure 17-2b) at Point D. Next you apply the equations of equilibrium to the free-body diagram.

$$+\uparrow \sum F_y = 0 \Rightarrow F_{AD}\sin(45°) + F_{BD} - (10\text{ kip}) = 0$$
$$\overset{+}{\rightarrow} \sum F_x = 0 \Rightarrow -F_{AD}\cos(45°) + F_{CD} = 0$$

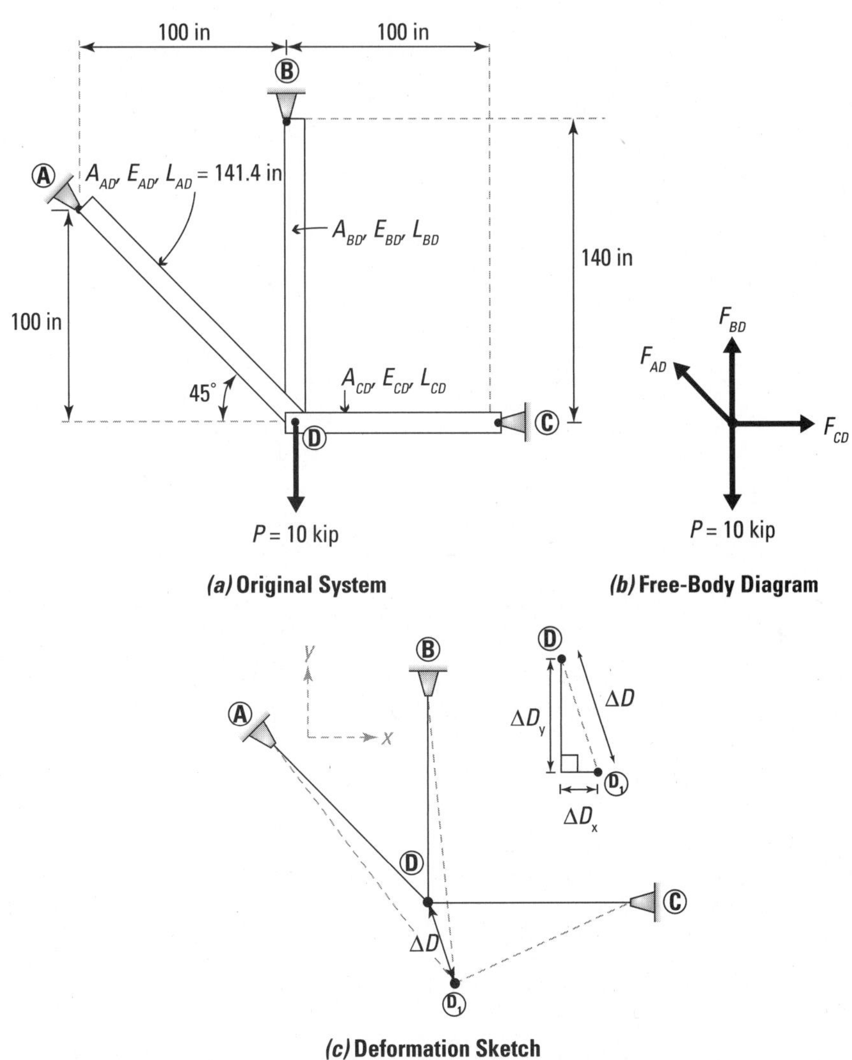

Figure 17-2: System of axial members.

2. **Sketch the deformed shape of the system under the load.**

   Due to the load, Point D moves to its new position at Point D_1 vertically downward an amount ΔD_y, as well as horizontally an amount ΔD_x as shown in Figure 17-2c. From this sketch, you can use the Pythagorean theorem to show the total displacement ΔD in terms of the horizontal and vertical displacements:

$$\Delta D = \sqrt{(\Delta D_x)^2 + (\Delta D_y)^2}$$

3. **Express the elongation of each axial bar in terms of its internal axial force.**

   For example, you can express the elongation in bar AD by rearranging the basic axial deformation equation from Chapter 16:

$$\Delta_{AD} = \frac{P_{AD}L_{AD}}{A_{AD}E_{AD}} \Rightarrow P_{AD} = \frac{\Delta_{AD}A_{AD}E_{AD}}{L_{AD}}$$

   You can similarly create expressions for the other bars in the system.

4. **Compute the final deformed length for each bar in terms of the unknown displacement.**

   For example, the final length of bar AD is given as the length of the line between Point A and Point D_1. Use the Pythagorean theorem:

$$L_{AD,FINAL} = \sqrt{(100 \text{ in} + \Delta D_x)^2 + (100 \text{ in} + \Delta D_y)^2}$$

5. **Calculate the axial deformation in each bar as the difference between the final length and the original length.**

$$\Delta_{AD} = L_{AD,FINAL} - L_{AD} = \sqrt{(100 \text{ in} + \Delta D_x)^2 + (100 \text{ in} + \Delta D_y)^2} - (141.4 \text{ in})$$

   Many engineers make a major simplifying assumption that because ΔD_x and ΔD_y are assumed to be very small, $(\Delta D_x)^2$ and $(\Delta D_y)^2$ are both practically zero — or at least very, very small. Mathematically, you can simplify the preceding equation as follows:

$$\Delta_{AD} = \sqrt{(10{,}000 \text{ in}^2) + (200 \text{ in})(\Delta D_x) + (\Delta D_x)^2 + (10{,}000 \text{ in}^2) + \ldots}$$
$$\overline{\ldots (200 \text{ in})(\Delta D_y) + (\Delta D_y)^2} - (141.4 \text{ in})$$
$$\Rightarrow \Delta_{AD} \approx \sqrt{20{,}000 \text{ in}^2 + (200 \text{ in}) \cdot (\Delta D_x + \Delta D_y)} - (141.4 \text{ in})$$

   You can then repeat this process for each of the remaining bars to express their deformations in terms of the unknown displacement components, ΔD_x and ΔD_y.

6. **For each bar in the system, express the bar's axial deformation in terms of the bar's internal force.**

   From the relationship between the bar's axial deformation and the vertical displacement of the system (as shown in Step 2), you can conclude that the corresponding internal force in the bar can be expressed as follows:

$$\Delta_{AD} = \frac{P_{AD}L_{AD}}{A_{AD}E_{AD}}$$
$$\Rightarrow P_{AD} = \frac{A_{AD}E_{AD}}{L_{AD}} \cdot \left(\sqrt{20{,}000 \text{ in}^2 + (200 \text{ in}) \cdot (\Delta D_x + \Delta D_y)} - (141.4 \text{ in}) \right)$$

   You can then repeat the process for each of the other bars of the system.

7. **Substitute the internal force equations from Step 6 into the equilibrium equations of Step 1 and solve for the unknown deformations for each of the bars in the system.**

   You now have two equilibrium equations with different expressions containing ΔD_x and ΔD_y. You can use basic algebra to simultaneously solve these equations for the displacements at Point D.

8. **Use the displacements of Step 7 to determine the internal forces in each bar by substituting into the equations of Step 6.**

After you have the internal forces in each bar determined, you're free to calculate stresses and strains within the member using the basic principles in Chapter 16 for axial deformations.

Flexural members of multiple supports

Redundant systems for flexural loads follow a similar procedure to the axial method I show in the preceding sections. To solve flexural problems with indeterminate supports, you remove extra supports from the model and compute the displacements at the locations of the removed supports. You then apply a unit load at the location of each removed support and determine the corresponding displacements. However, you also have to consider whether the system is a single indeterminate support system or a multiple indeterminate support system. The following sections explain the calculations for each.

When dealing with indeterminate flexural problems, you typically want to remove degrees of determinacy that result in either a simply supported beam or a cantilever system, because the corresponding calculations are usually much simpler. In fact, you can often find the necessary displacement calculations directly in design aids or tables for these support scenarios for a wide array of loading patterns.

Single indeterminate support systems

Single indeterminate support systems are systems where the beam has only one degree of static indeterminacy. Consider the simply supported continuous beam in Figure 17-3a, which has an extra support applied at Point B.

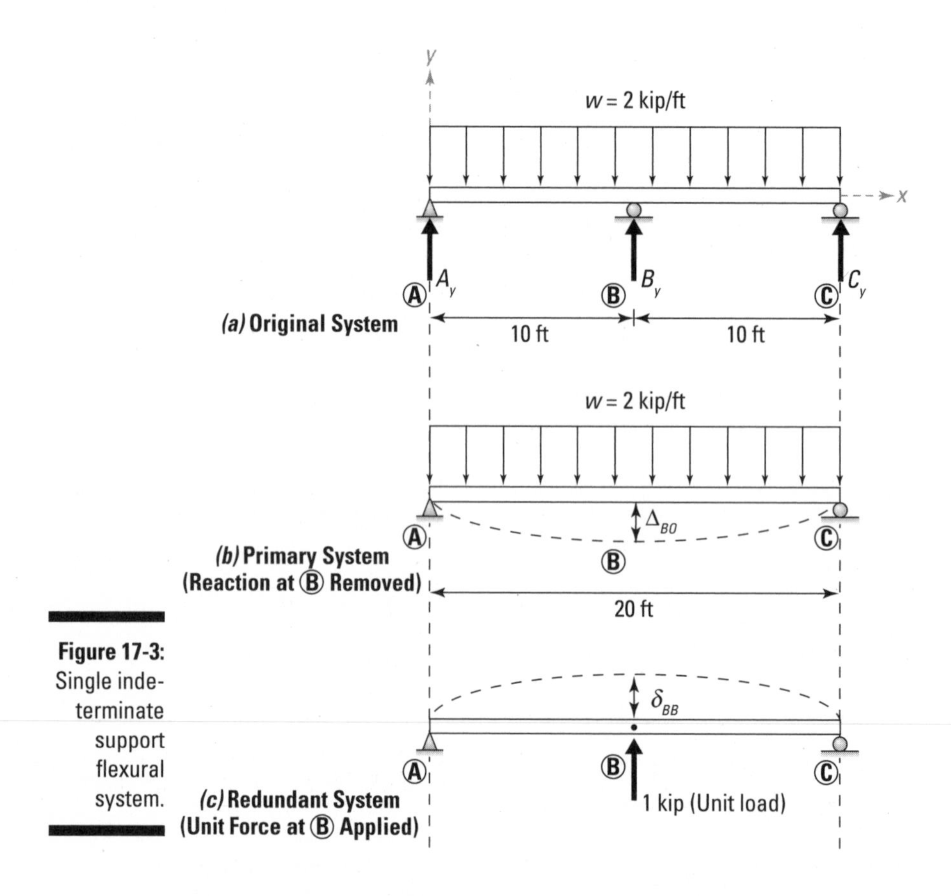

Figure 17-3: Single indeterminate support flexural system.

The beam is steel (with a Young's modulus of elasticity E of 29,000 ksi), and the moment of inertia I is 150 in^4. (***Note:*** For prismatic and uniform beams, these values actually don't matter because they end up canceling out in the final step, but including them anyway is a good habit.) If each span is $L_{AB} = L_{BC}$ = 10 ft and the distributed load is $w = -2$ kip/ft, you can apply the same basic redundant methodology I describe in the earlier section "Withdrawing Support: Creating Multiple Redundant Systems." You first must start by determining how many degrees of freedom to remove.

To determine the number of degrees of freedom in the system, you simply count up the total number of unknown support reactions — A_x and A_y at the pinned support at Point A; B_y at the roller at Point B; and C_y at the roller at Point C — which results in a total of four unknown reactions. Each free-body diagram gives you three equilibrium equations, so for this example, you have one degree of indeterminacy (four unknown support reactions minus the three equations). Now you know that you need to remove one of the supports to actually work this method.

As long as you're careful in your calculations, which degree of freedom you remove to make the primary system really doesn't matter. In the example of Figure 17-3a, I choose to remove the support at Point B (as I show in Figure 17-3b) because with this support removed, the primary system is actually a simply supported beam.

For this example, the displacement you need to calculate is $\Delta_{BO,}$ which you compute at Point B, the midspan of the primary system. Using the principles in Chapter 16, you can find Δ_{BO} as follows:

$$\Delta_{BO} = \frac{5wL^4}{384EI} = \frac{5(-2\text{ kip/ft})(20\text{ ft})^4\left(\frac{12\text{ in}}{1\text{ ft}}\right)^3}{384(29{,}000\text{ ksi})(150\text{ in}^4)} = -1.66\text{ in}$$

When computing these displacements, remember to be careful about the units. Distributed loads and beam spans are often measured in feet or meters, while section properties are often measured in inches or millimeters. Make sure to convert these units so that they cancel out and you can end up with the units of displacement that you want.

You then apply a unit load at Point B (as I show in Figure 17-3c) and determine the corresponding displacement (δ_{BB}) due to the applied unit load. For this example, you can compute this displacement as

$$\delta_{BB} = \frac{PL^3}{48EI} = \frac{(1)(20\text{ ft})^3\left(\frac{12\text{ in}}{1\text{ ft}}\right)^3}{48(29{,}000\text{ ksi})(150\text{ in}^4)} = +0.066\ \frac{\text{in}}{\text{kip}}$$

If the following compatibility equation for this system looks very familiar, you may be recognizing it as the expression for the axial problems in "Axial bars with indeterminate supports" earlier in the chapter:

$$0 = \Delta_{BO} + R_B\delta_{BB} = (-1.66\text{ in}) + R_B\left(0.066\frac{\text{in}}{\text{kip}}\right) = 0 \Rightarrow R_B = B_y = +25.2\text{ kip}$$

A positive numerical value once again tells you that the direction of R_B is in the same direction as the applied unit load. For this example, because it's positive, R_B is acting upward at the support at Point B, which is in the opposite direction of the displacement of the primary system.

Multiple indeterminate support systems

Flexural members (such as beams) commonly have multiple degrees of indeterminacy, often as a result of continuous spans and support. The redundant method you use for the single support systems in the preceding section still works (as long as you meet the assumptions I state in "Clarifying assumptions for indeterminate methods" earlier in the chapter); it just requires a bit of modification to the single support system compatibility expressions.

Consider Figure 17-4a, which shows a continuous beam with two degrees of indeterminacy. For this example, you need to remove two degrees of freedom to create the primary system. For this example, I choose to remove the supports at Points B and C simply because doing so results in a simply supported structure (as shown in Figure 17-4b).

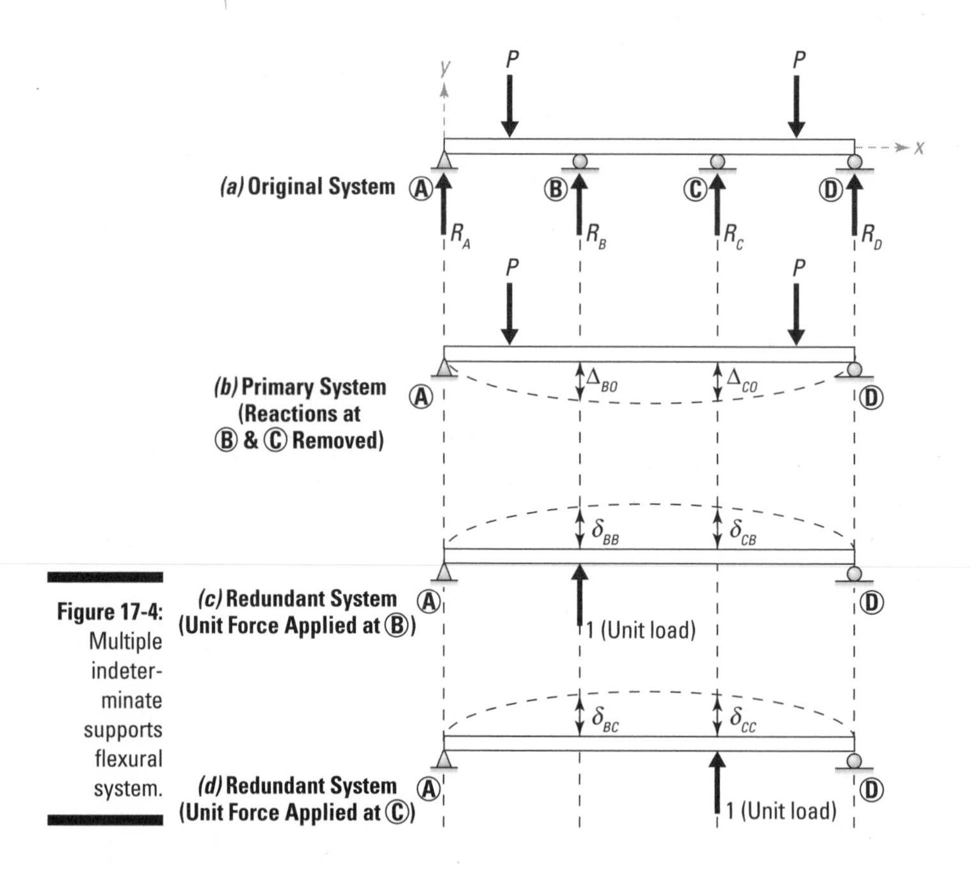

Figure 17-4: Multiple indeterminate supports flexural system.

The difference is that you need to create two redundant systems for this example — one system for each degree of freedom that you removed. For the unit load at Point B, you then calculate the displacements per unit load at both Point B and Point C and determine the values of δ_{BB} and δ_{CB} (as shown in Figure 17-4c). Similarly, you compute the values of δ_{BC} and δ_{CC} due to a unit load at Point C (as I show in Figure 17-4d). Finally, you establish compatibility relationships for each degree of freedom that you originally removed (which are similar to those for the earlier single support systems):

- At Point B: $0 = \Delta_{BO} + \delta_{BB}R_B + \delta_{BC}R_C$
- At Point C: $0 = \Delta_{BO} + \delta_{CB}R_B + \delta_{CC}R_C$

To find the support reactions R_B (which is B_y) and R_C (which is C_y), you simply need to simultaneously solve both of these compatibility equations (with the displacements substituted in, of course).

For problems with more than two supports, you need to write additional equations for each degree of indeterminacy and then create one redundant system for each degree of indeterminacy.

For one or two degrees of indeterminacy, these simultaneous equations may not be a challenge, but if you have a hundred degrees of indeterminacy, you have to solve a hundred simultaneous equations with a hundred different unknowns in each. Although this complication is a major drawback to this method, the method still works. You probably want to use a computer to solve all these simultaneous equations (and matrices), though.

Torsion of shafts with indeterminate supports

Indeterminate circular shafts are an important part of machine design and often have multiple moments or support reactions acting on them — especially if resistance or friction is present. You handle these problems in a method very similar to the method I cover earlier in the chapter for axial bars with indeterminate supports.

Consider the circular shaft shown in Figure 17-5a that is constrained from twisting at each end — the end supports are fixed — and is one degree indeterminate (which means you need to remove one degree of freedom when you start making your redundant systems). If an applied torque of $T = -3$ kN-m is applied at $L_1 = 0.67$ m from the left end of a shaft that has length $L = 3$ m, you're now ready to solve this indeterminate torsion problem. For this example, I assume that the diameter of the shaft is 40 millimeters and the shear modulus of the shaft material is $G = 80$ GPa. (Head to Chapter 14 for more on shear modulus.)

As with axially loaded members, the first step is to release one of the redundant supports. In this case, I arbitrarily choose Point B to create the primary system shown in Figure 17-5b. With the redundant support at Point B removed, you can then compute the deformation at Point B due to the applied loads (or torque in this case).

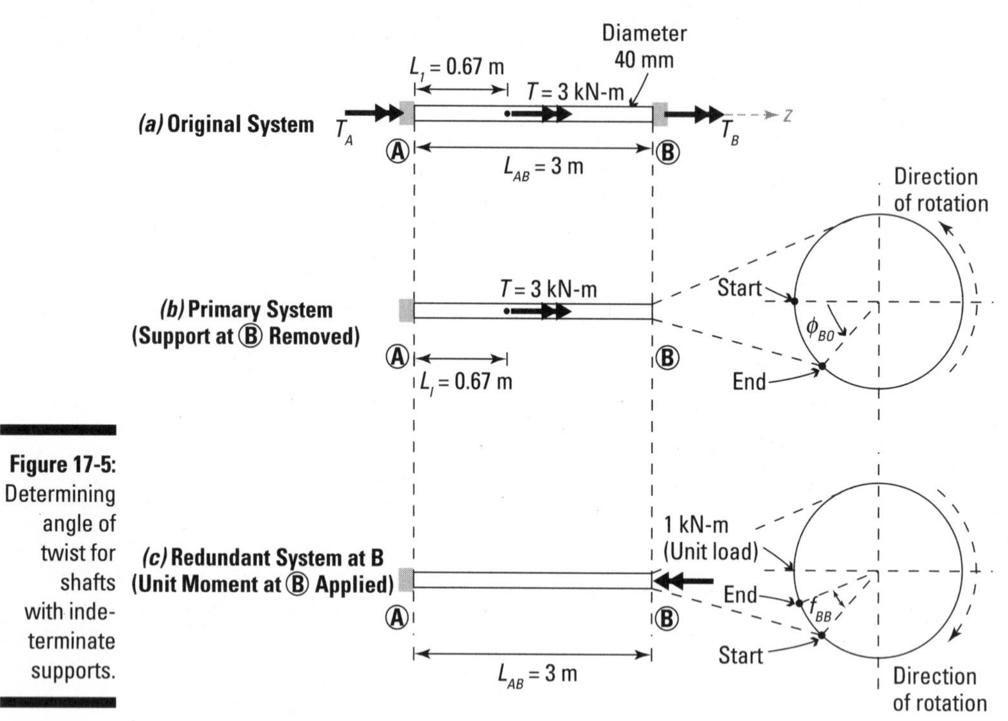

Figure 17-5: Determining angle of twist for shafts with indeterminate supports.

The deformation for torsion of circular shafts is always the angle of twist of the circular cross section. You can then calculate the angle of twist of the system with the released support at Point B using the basic angle of twist equation (see Chapter 16):

$$\phi_{BO} = \frac{TL_I}{G_{AB}J_I} = \frac{(+3\ \text{kN}\cdot\text{m})(0.67\ \text{m})}{\left(80\times10^6\ \text{kPa}\right)\left(\frac{\pi}{2}(0.04\ \text{m})^4\right)} = +0.00625\ \text{rad}$$

Next, you can apply a unit load (which in this case is a moment of 1 kN-m) to the end at Point B to determine the deformation per unit load (see Figure 17-5c):

$$f_{BB} = \frac{(-1)(L_{AB})}{G_{AB}J_{AB}} = \frac{(-1)(3.0\ \text{m})}{\left(80\times10^6\ \text{kPa}\right)\left(\frac{\pi}{2}(0.04\ \text{m})^4\right)} = -0.00933\ \frac{\text{rad}}{\text{kN}\cdot\text{m}}$$

After you have determined this angle of twist per unit torque, you can then create a compatibility expression that relates the three systems:

$$0 = \phi_{BO} + T_B\delta_{BB} = (+0.00625 \text{ rad}) + T_B\left(-0.00933\frac{\text{rad}}{\text{kN}\cdot\text{m}}\right) = 0 \qquad T_B = +0.667 \text{ kN}\cdot\text{m}$$

As with the other redundant methods, a positive value for T_B indicates that this support reaction is in the same direction as the unit torque from the redundant system. After you have this support reaction for T_B, you can then use the principles of equilibrium in Chapter 3 to solve for internal forces or the other support reactions at Point A.

Dealing with Multiple Materials

Members composed of multiple materials are also known as *composite members.* The trick to dealing with multiple materials is recognizing that all materials in a cross section contribute to the resistance of the total applied load. Structural members made from reinforced concrete (where a concrete cross section contains steel reinforcing bars) are a common use of multiple materials in engineering. Unfortunately, prior to solving the problem, you don't actually know how much each material contributes toward the total resistance. But with the right free-body diagrams, you can quickly create expressions that relate these forces to the total and then apply a bit of logic to get your compatibility condition. This section shows you how.

Axial bars of multiple materials

The simplest of the multiple material problems you encounter appears when you're working with the axial loads of composite members. Materials in axial members must be *concentric* (or symmetric) about the longitudinal axis (otherwise, you need to include flexural effects, which I show in the following section).

To analyze multiple materials subjected to the axial load P = 18 kip, consider Figure 17-6a, which shows an axial member of length $L = L_1 = L_2$ = 48 in. It's composed of two materials having known areas A_1 and A_2 and material properties E_1 and E_2. To illustrate the calculations, assume that Material 1 has a cross-sectional area A_1 of 2 in^2 and a Young's modulus of elasticity E_1 of 29,000 ksi. Material 2 has a cross-sectional area A_2 of 4 in^2 and a Young's modulus of elasticity E_2 of 10,000 ksi.

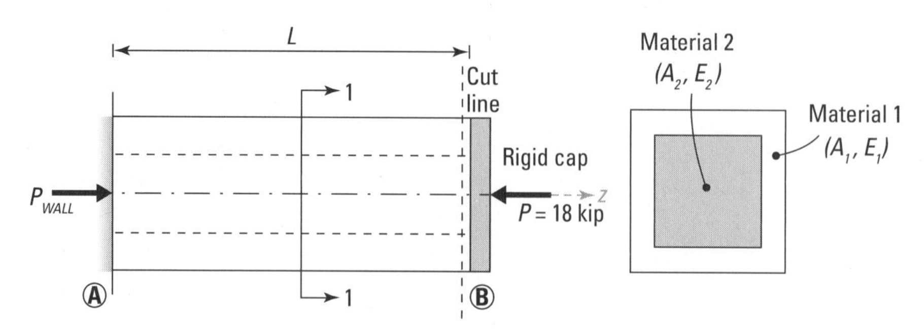

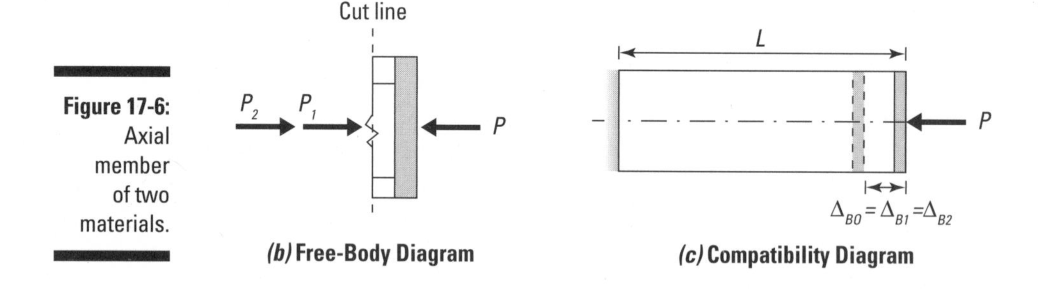

Figure 17-6: Axial member of two materials.

To analyze this system (or any axial system with two concentric materials), you follow several simple steps:

1. **Cut a section through both materials, including any known loads, and write the axial equilibrium equation for that piece.**

   Figure 17-6b shows you the section cut through the end of the member (including the load) for this example in terms of the internal axial forces P_1 and P_2. The following is that section's axial equilibrium equation:

   $$\sum F_z = 0 \Rightarrow P_1 + P_2 = P$$

2. **Determine the compatibility requirements for both materials.**

   For systems of multiple materials, the compatibility equation usually comes from the idea that for the object to deform, all the materials in the object must deform equally (as you can see in Figure 17-6c). (Of course, this rule assumes that one material can't slip relative to another — that is, they're perfectly fused or joined together.) For this example, the compatibility equation is $\Delta_1 = \Delta_2 = \Delta$.

3. **Write expressions for Δ_1 and Δ_2 of each material in terms of their respective loads (P_1 and P_2) and section properties.**

   Because both materials in this example are being axially loaded, the deformation of each material is based on the internal load in the material and that material's section properties. Here are the expressions:

$$\phi_1 = \frac{T_1 L_1}{J_1 G_1} = \frac{T_1(2\text{ m})}{\frac{\pi}{2}(0.04\text{ m})^4(40\times10^9\text{ Pa})} = (1.243\times10^{-5})T_1$$

$$\phi_2 = \frac{T_2 L_2}{J_2 G_2} = \frac{T_2(2\text{ m})}{\frac{\pi}{2}\left[(0.08\text{ m})^4-(0.04\text{ m})^4\right](80\times10^9\text{ Pa})} = (4.144\times10^{-7})T_2$$

4. **Substitute the expressions from Step 3 into the compatibility equation of Step 2 and solve for one of the two internal forces in terms of the other.**

Whether you solve for the load in Material 1 or Material 2 first doesn't matter. For this example, I choose to solve for the force in Material 1 (or P_1) in terms of the force in material 2 (or P_2):

$$\Delta_1 = \Delta_2 = \frac{P_1 L_1}{A_1 E_1} = \frac{P_2 L_2}{A_2 E_2} \Rightarrow 0.00083P_1 = 0.0012P_2 \Rightarrow P_1 = 1.45P_2$$

5. **Substitute the equation of Step 4 into the equilibrium equation of Step 1 and solve for the unknown reaction (P_2).**

For this example,

$$P_1 + P_2 = 18\text{ kip} \Rightarrow 1.45P_2 + P_2 = 18\text{ kip} \Rightarrow P_2 = +7.35\text{ kip} = 7.35\text{ kip (C)}$$

After you know P_2, you can then substitute into the expression from Step 4 and solve for P_1:

$$P_1 + 7.35\text{ kip} = 18\text{ kip} \Rightarrow P_1 = +10.65\text{ kip} = 10.65\text{ kip (C)}$$

Be careful when you assume the forces on your free-body diagram as being positive when pushing (or compressing). You must look back at your free-body diagram to determine whether a computed force is tensile (T) or compressive (C).

6. **Calculate the corresponding stresses in each material.**

After you have the internal forces, you can compute the corresponding stresses (using negative values for the forces because of the (C) in the answer from Step 5) in each material:

$$\sigma_1 = \frac{P_1}{A_1} = \frac{(-10.65\text{ kip})}{(2\text{ in}^2)} = -5.33\text{ ksi} = 5.33\text{ ksi (C)}$$

$$\sigma_2 = \frac{P_2}{A_2} = \frac{(-7.35\text{ kip})}{(4\text{ in}^2)} = -1.84\text{ ksi} = 1.84\text{ ksi (C)}$$

7. **Calculate the corresponding deformations from Step 3 (or other values such as stresses and strains in the materials).**

After you know both internal forces, you can compute the deformation in Material 1 (which happens to be the same as the deformations in both Material 2 and the system's deformation.):

$$\Delta = \Delta_1 = 0.00083P_1\frac{\text{in}}{\text{kip}} = 0.00083(-10.65\text{ kip})\frac{\text{in}}{\text{kip}} = -0.0088\text{ in}$$

You can also verify that both materials deflect the same amount by substituting into the deformation equation for Material 2:

$$\Delta = \Delta_2 = 0.0012P_2\frac{\text{in}}{\text{kip}} = 0.0012(-7.35\text{ kip})\frac{\text{in}}{\text{kip}} = -0.0088\text{ in}$$

Both of these calculations produce the same numerical result for the displacement, which indicates that your compatibility equation worked correctly.

If you have more than two materials, you can expand the compatibility equation in Step 2 to include as many materials as you want — just remember that you also need to write the additional expressions in Step 3 and then relate all of the unknown forces in terms of one common force (it doesn't matter which you choose) in Step 4.

Flexure of multiple materials

Many flexural designs use multiple materials together to take advantage of particular material properties. For example, concrete is very strong in compression but very weak in tension. To utilize concrete as a material, designers often embed steel reinforcing bars (known as *rebar*) in regions of high tensile stresses to exploit the high tensile strength of steel.

Beams composed of multiple materials require a bit of manipulation before you can apply the flexural analysis principles I describe in Chapters 9 and 10. The biggest issue is tackling the assumption about homogenous material properties (see the earlier section "Clarifying assumptions for indeterminate methods"). Based on this assumption, if plane sections remain planar, you can assume that the normal strain along the cross section varies linearly with distance from the neutral axis (as shown in Figure 17-7a), which then means that the basic formulas for stress from flexural bending are also valid.

As I note in Chapter 9, the formula for the normal stress in flexural member (whose cross section lies in the XY plane) for a bending moment about the *x*-centroidal axis is

$$\sigma_{zz} = -\frac{M_x y}{I_{xx}}$$

However, you still need to take care of that pesky homogenous material assumption. As long as deformations remain small and within the elastic region, you can accomplish this task by creating a simple transformation relationship, which I discuss in the following section.

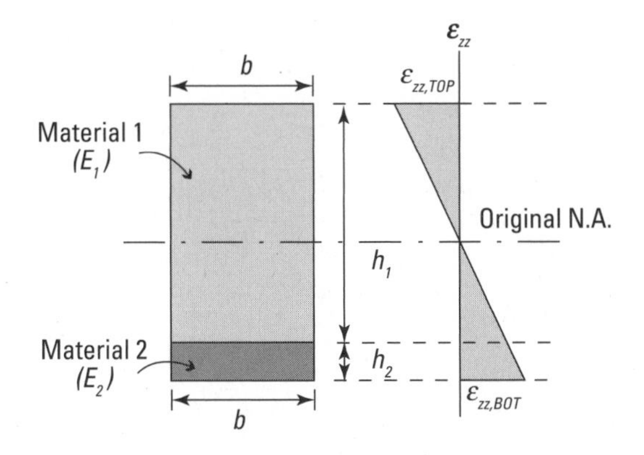

(a) Original System

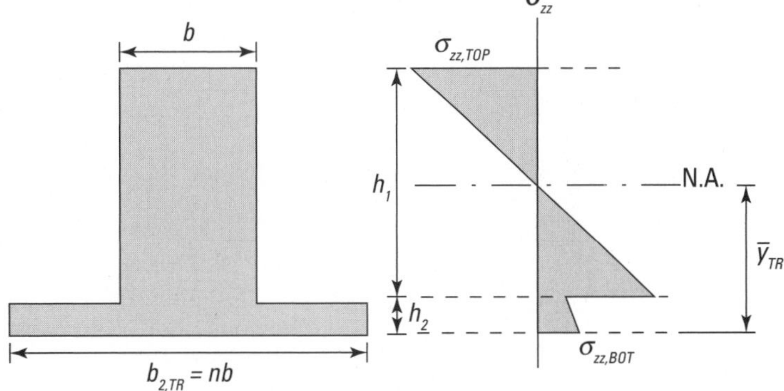

(b) Transformed Cross Section

Figure 17-7: Flexural members of multiple materials.

Transforming sections with a modulus ratio

The basic methodology for handling flexural members of multiple materials is to convert all materials within the cross section to equivalent cross sections of multiple materials by creating a modulus ratio n that is based on the Young's modulus of elasticity for each material:

$$n = \frac{E_2}{E_1}$$

where E_1 is the Young's modulus of elasticity of Material 1 and E_2 is the Young's modulus of elasticity of Material 2. With this ratio in hand, you then alter the cross-sectional area of Material 2 by multiplying the width of the section by the modulus ratio, a process that creates the *transformed cross section*. If the Young's modulus of elasticity of Material 2 is larger than the Young's modulus of elasticity of Material 1, the stress distribution in the cross section actually looks as shown in Figure 17-7b. After you have transformed the cross section, you then recalculate the cross section properties and continue with your analysis.

Following basic steps for working with transformed cross sections

To illustrate the methodology, consider the composite beam shown in Figure 17-7. Material 1 is aluminum (E_1 = 10,000 ksi), and Material 2 is steel (E_2 = 29,000 ksi). The dimensions for Material 1 are h_1 = 11.5 in high and b = 6 in wide. The dimensions for Material 2 are h_2 = 0.5 in high and b = 6 in wide. Suppose you want to determine the maximum stress at the bottom of the beam ($\sigma_{zz,BOT}$) and at the top of the beam ($\sigma_{zz,TOP}$) when subjected to a positive bending moment of M_x = +800 kip-in.

For a member of two materials, many textbooks and references use the convention of assigning the material with the larger stiffness (with the larger Young's modulus of elasticity) as Material 2, and the smaller stiffness (with the smaller Young's modulus of elasticity) as Material 1.

With this example in mind, consider the following process for analyzing flexural beams of two materials:

1. **Compute the modulus ratio for the materials.**

   To transform to an equivalent area of aluminum (Material 1), you need to compute the modulus ratio with respect to the steel (Material 2) as follows:

   $$n = \frac{E_2}{E_1} = \frac{(29{,}000 \text{ ksi})}{(10{,}000 \text{ ksi})} = 2.9$$

   Because this modulus ratio is greater than 1.0, the transformed width of Material 2 must be larger than its original width. Conversely, for a modulus ratio less than 1.0, the transformed width must be smaller than the original width.

2. **Construct a transformed cross section.**

   The height remains constant for each material, and the width of Material 1 remains unchanged. However, you multiply the width of Material 2 by the ratio of Step 1 such that

   $$b_{2,TR} = n \cdot b_2 = (2.9)\,(6 \text{ in}) = 17.4 \text{ in}$$

   The dimensions of Material 2 after you've transformed it are now h_2 = 0.5 in and $b_{2,TR}$ = 17.4 in.

3. **Compute the centroid of the transformed cross section.**

   Remember that you can find the centroid of the transformed a region by using the basic relationship

$$\bar{y}_{TR} = \frac{\sum y_i A_i}{\sum A_i}$$

   Using the principles in Chapter 4, you can compute the location of the neutral axis (at the centroid) of the transformed cross section:

$$\bar{y}_{TR} = \frac{\sum y_i A_i}{\sum A_i} = \frac{(6\text{ in})(11.5\text{ in})(6.25\text{ in}) + (17.4\text{ in})(0.5\text{ in})(0.25\text{ in})}{(6\text{ in})(11.5\text{ in}) + (17.4\text{ in})(0.5\text{ in})} = 5.58\text{ in}$$

4. **Compute the moment of inertia of the transformed cross section.**

   To compute the moment of inertia for a region about its own centroidal axis with respect to some other parallel axis (the centroid of the transformed section in this case), use the following formula:

$$I = I_O + Ad^2$$

   For the two rectangular regions of the transformed section about the transformed neutral axis:

$$I_{TR,NA} = \frac{1}{12}(17.4\text{ in})(0.5\text{ in})^3 + (17.4\text{ in})(0.5\text{ in})(0.25\text{ in} - 5.58\text{ in})^2$$
$$+ \frac{1}{12}(6\text{ in})(11.5\text{ in})^3 + (6\text{ in})(11.5\text{ in})(6.25\text{ in} - 5.58\text{ in})^2 = 1{,}039\text{ in}^4$$

   Remember that because you have multiple regions with different centroidal axes, you need to use the parallel axis theorem to compute the moment of inertia (see Chapter 5) as I've done here.

5. **Compute the stresses at the location of interest by using the section properties from Step 3 and Step 4.**

   At this point, you're ready to compute stresses at different locations. For this example, I choose to compute the stress at the top of the cross section (in the aluminum) by using the basic stress equation. Because you didn't transform this material, you just use the basic normal stress calculation:

$$\sigma_{zz,TOP,AL} = -\frac{M_x y_{TOP}}{I_{TR,NA}} = -\frac{(800\text{ kip}\cdot\text{in})(12\text{ in} - 5.58\text{ in})}{(1{,}039\text{ in}^4)} = -4.94\text{ ksi} = 4.94\text{ ksi (C)}$$

   To compute the stress in the steel, you need to perform a similar calculation, but now you must transform the stress back to the original material by incorporating the modulus ratio:

$$\sigma_{zz,BOT,ST} = -\frac{M_x y_{BOT}}{I_{TR,NA}}(n) = -\frac{(800\ \text{kip}\cdot\text{in})(0.25\ \text{in} - 5.58\ \text{in})}{(1{,}039\ \text{in}^4)}(2.9)$$

$$= +11.90\ \text{ksi} = 11.90\ \text{ksi (T)}$$

Torsion of multiple materials

To analyze torsion of a tube concentrically filled with a second material, you follow a similar set of rules as I cover for axial problems of multiple materials (see the earlier section "Axial bars of multiple materials"). The only difference is that instead of writing compatibility equations for the deformations of each material, you need to correlate the angles of twist for the two materials, which for concentric shafts (if one material doesn't slip relative to another) means they all have the same angle of twist: $\phi_1 = \phi_2 = \phi$.

Consider the 2-meter-long composite shaft of two concentric materials shown in Figure 17-8a. Material 1 is an inner core of brass (G_1 = 40 GPa), and Material 2 is an outer shell of steel (G_2 = 80 GPa) whose inner radius (r_1) is 40 mm and outer radius (r_2) is 80 mm. Assume the composite shaft is subjected to an applied torque of T = 200 kN-m.

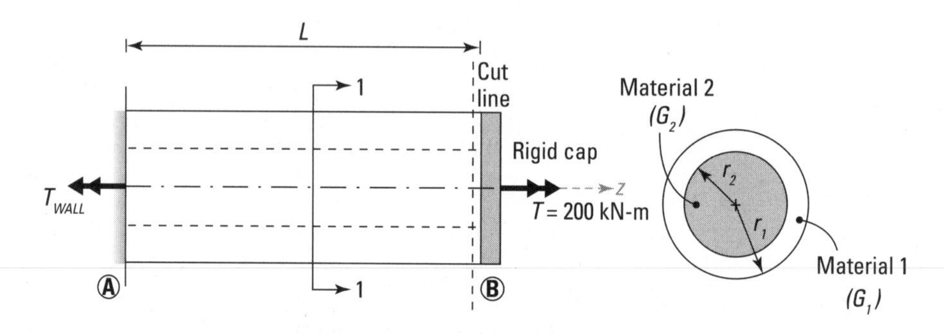

(a) Original System with Cross Section 1-1

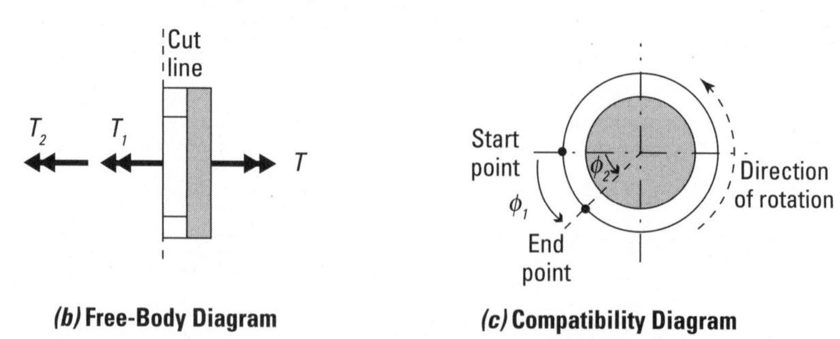

(b) Free-Body Diagram

(c) Compatibility Diagram ($\phi_1 = \phi_2 = \phi$)

Figure 17-8: Analyzing torsion of composite shafts.

You can use the following steps to analyze this and other indeterminate torsion systems of two materials:

1. **Cut a section through both materials including any known loads and write the torsion equilibrium equation for that piece.**

   Figure 17-8b shows you such a load cut; the following equation spells out the equilibrium equation.

   $$\sum F_z = 0 \Rightarrow -T_1 - T_2 + T = 0 \Rightarrow T_1 + T_2 = 200 \text{ kN}\cdot\text{m}$$

2. **Determine the compatibility requirements for both materials.**

   Because this problem involves concentric cross sections as shown in Figure 17-8c, you know that the angle of twist for each of the materials must be the same. Therefore, $\phi_1 = \phi_2 = \phi$.

3. **Write expressions for ϕ_1 and ϕ_2 in terms of their respective internal loads (T_1 and T_2) and their respective section properties.**

   For this example,

   $$\phi_1 = \frac{T_1 L_1}{J_1 G_1} = \frac{T_1(2 \text{ m})}{\frac{\pi}{2}(0.04 \text{ m})^4 (40\times 10^9 \text{ Pa})} = (1.243\times 10^{-5})T_1$$

   $$\phi_2 = \frac{T_2 L_2}{J_2 G_2} = \frac{T_2(2 \text{ m})}{\frac{\pi}{2}\left[(0.08 \text{ m})^4 - (0.04 \text{ m})^4\right](80\times 10^9 \text{ Pa})} = (4.144\times 10^{-7})T_2$$

   For composite shafts subjected to torsion, the lengths of the materials are usually the same: $L_1 = L_2$.

4. **Substitute the expressions from Step 3 into the compatibility equation of Step 2 and solve for one of the two internal forces in terms of the other.**

   For this example, I arbitrarily choose to solve for the torque in the brass (T_1) in terms of the torque in the steel (T_2).

   $$\frac{T_1 L_1}{J_1 G_1} = \frac{T_2 L_2}{J_2 G_2} \Rightarrow (1.243\times 10^{-5})T_1 = (4.144\times 10^{-7})T_2 \Rightarrow T_1 = 0.0333 T_2$$

5. **Substitute the equation of Step 4 into the equilibrium equation of Step 1 and solve for the unknown internal forces.**

   In this case, the unknown reaction is T_2.

   $$T_1 + T_2 = 200 \text{ kN}\cdot\text{m} \Rightarrow 0.0333T_2 + T_2 = 200 \text{ kN}\cdot\text{m}$$
   $$\Rightarrow T_2 = 193.5 \text{ kN}\cdot\text{m}$$

After you know T_2, you can substitute into the expression from Step 4 and solve for T_1.

$$T_1 + (193.5 \text{ kN} \cdot \text{m}) = 200 \text{ kN} \cdot \text{m} \Rightarrow T_1 = 6.5 \text{ kN} \cdot \text{m}$$

6. **Calculate the corresponding angle of twist for the combined section from Step 3 (or any other value, such as shear stress or shear strain in the composite shaft).**

   Suppose you want to compute the angle of twist of the combined shaft; you can then substitute these values back into the equations of Step 3:

$$\phi = \phi_1 = (1.243 \times 10^{-5})T_1 = (1.243 \times 10^{-5})(6.5 \text{ kN} \cdot \text{m}) = 8.08 \times 10^{-5} \text{ rad}$$

If you have more than two materials, you can expand the compatibility equation in Step 2 to include as many materials as you want — just remember that you also need to write the additional expressions in Step 3 and then relate all of the unknown forces in terms of one common torque (it doesn't matter which you choose) in Step 4.

Using Rigid Behavior to Develop Compatibility

Although you can analyze some indeterminate structures by relating the load in multiple materials to a single applied load on a cross section (as I discuss in the earlier section "Dealing with Multiple Materials"), that method isn't sufficient for some problems. If your system contains a rigid end cap or a rigid bar in it somewhere, you may actually be able to write a compatibility relationship based on multiple different locations.

These rigid members serve as the basis for your compatibility conditions. In this section, I show you how to analyze problems that contain rigid (or nondeformable) elements.

Rigid bar problems

A *rigid bar* is a type of lever that retains its basic straight shape while rotating about a pin or support location known as a fulcrum (see Figure 17-9a). The *fulcrum* serves as the basis for rotation of the rigid bar, and you measure all your length proportions from this location.

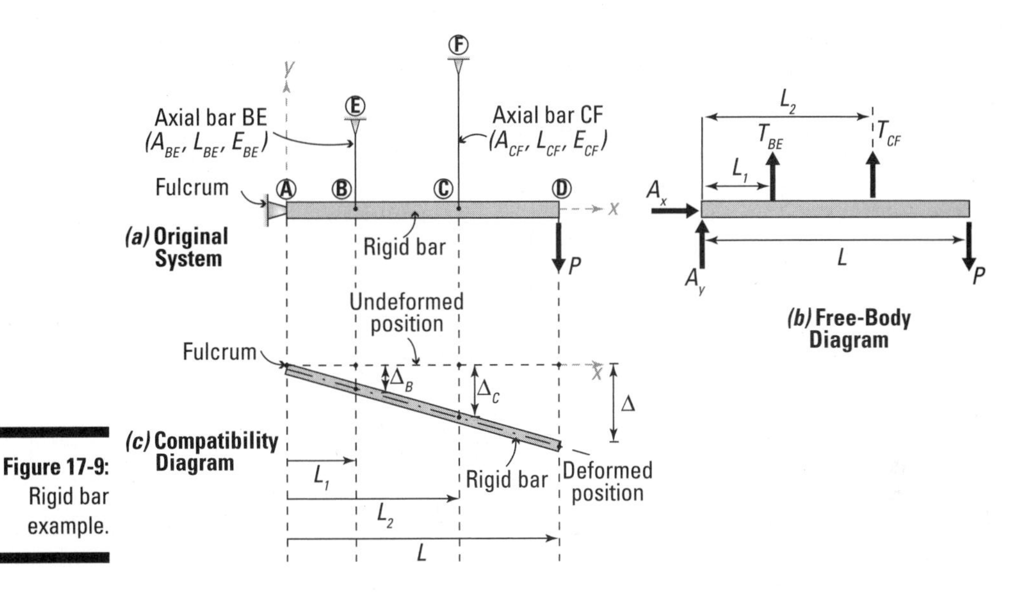

Figure 17-9: Rigid bar example.

Establishing proportions for compatibility

Rigid bars rotate about the fulcrum without deforming along their lengths — that is, a rigid bar remains straight even while rotating. With this information, you can then use proportions (or similar triangles) with respect to the fulcrum to establish your compatibility criteria:

$$\frac{\Delta_B}{L_1} = \frac{\Delta_C}{L_2} = \frac{\Delta}{L}$$

where the lengths have the relationships $L_1 = L_{AB}$, $L_2 = L_{AC}$, and $L = L_{AD}$ as shown in Figure 17-9. Δ_B is the vertical deflection at Point B; Δ_C is the vertical deflection at Point C; and Δ is the displacement at the tip of the rigid bar. You can calculate displacements at other locations as well by using similar triangles (or proportions) and the distance from the fulcrum to your location of interest along the rigid bar.

Following basic steps for solving rigid bar problems

Consider the rigid bar assembly in Figure 17-9. The bar is restrained by a axial bar *BE* (which has a cross-sectional area A_{BE} of 0.5 in^2, a length L_{BE} of 24 in, and a Young's modulus of elasticity E_{BE} = 29,000 ksi) and a second axial bar *CF* (which has cross-sectional area A_{CF} of 1.5 in^2, a length L_{CF} of 48 in, and a Young's modulus of elasticity E_{CF} of 29,000 ksi). Point B is a distance of $L_1 = L_{AB}$ = 30 in, Point C is a distance of $L_2 = L_{AC}$ = 40 in, and Point D is a distance of $L = L_{AD}$ = 60 in. Each of these dimensions is measured from the fulcrum point (or Point A in this example, as shown in Figure 17-9a) to the other points. A $P = 10$ kip load is applied at the end of the rigid bar (at Point D).

To analyze rigid bar systems such as the one in Figure 17-9, you need to follow a few basic steps:

1. **Draw the free-body diagram by isolating the rigid bar and including the location of the fulcrum.**

   Figure 17-9b shows the free-body diagram of this rigid bar. In this example, the fulcrum is located at Point A.

2. **Establish an equilibrium equation by summing moments about the fulcrum location on the free-body diagram of Step 1.**

   In this example, summing moments about Point A produces the following equilibrium equation:

   $$\sum M_A = 0 \Rightarrow T_{BE}(L_1) + T_{CF}(L_2) - P(L) = 0 \Rightarrow T_{BE}(30\text{ in}) + T_{CF}(40\text{ in}) = 600\text{ kip}\cdot\text{in}$$

   where T_{BE} and T_{CF} are the internal forces in their respective bars. This equation is the equilibrium equation for the rigid bar and provides a statics-based relationship between the internal forces in the axial bar and the applied external load.

   The reason you choose the fulcrum as your location for summing moments is to eliminate the unknown support reactions at that location. The only unknown forces that remain in the equilibrium equation are the internal forces from the axial bars and the applied point load on the rigid bar.

3. **Write the compatibility equations by establishing a proportion triangle with respect to the fulcrum and the displacement at the locations where the axial bars attach to the rigid bar.**

   If displacements remain very small, Δ_B and Δ_C remain approximately vertical, which allows you to use similar triangles (or proportions) to relate these displacements to each other. This proportion triangle is shown in Figure 17-9c. The compatibility relationship is expressed as

   $$\frac{\Delta_B}{L_1} = \frac{\Delta_C}{L_2} = \frac{\Delta}{L} \Rightarrow \frac{\Delta_B}{(30\text{ in})} = \frac{\Delta_C}{(40\text{ in})} = \frac{\Delta}{(60\text{ in})}$$

4. **Relate the deformations in the axial bars to the system displacements at the location where the axial bars are attached to the rigid bar.**

   In this example, you're dealing with the displacement Δ_B and Δ_C. Because *BE* and *CF* are axial bars, you know from statics that their internal forces must be entirely axial, and you know that the axial deformations in each bar can be directly related to their internal forces.

$$\Delta_B = \frac{T_{BE}L_{BE}}{A_{BE}E_{BE}} = \frac{(T_{BE})(24\text{ in})}{(0.5\text{ in}^2)(29{,}000\text{ ksi})} = 0.001655T_{BE}\ \frac{\text{in}}{\text{kip}}$$

$$\Delta_C = \frac{T_{CF}L_{CF}}{A_{CF}E_{CF}} = \frac{(T_{CF})(48\text{ in})}{(1.5\text{ in}^2)(29{,}000\text{ ksi})} = 0.001103T_{CF}\ \frac{\text{in}}{\text{kip}}$$

You can then substitute these relationships into the compatibility equation of Step 3 to produce another equation that relates the two forces, T_{BE} and T_{CF}:

$$\frac{0.001655T_{BE}}{(30\text{ in})} = \frac{0.001103T_{CF}}{(40\text{ in})} \Rightarrow T_{BE} = 0.50T_{CF}$$

In this example, I arbitrarily choose to solve for the axial force in *BE*, T_{BE}, in terms of the axial force in *CF*, T_{CF}. But you can actually solve for either force; it makes no difference in the end.

5. **Substitute the expressions from Step 4 into the equilibrium equation of Step 2 and solve for your chosen axial force.**

   In this case, that's T_{CF}. Use the following equation:

$$T_{BE}(30\text{ in}) + T_{CF}(40\text{ in}) = 600\text{ kip}\cdot\text{in} \Rightarrow (0.50T_{CF})(30\text{ in}) + T_{CF}(40\text{ in}) = 600\text{ kip}\cdot\text{in}$$
$$\Rightarrow T_{CF} = 10.91\text{ kip}$$

6. **Compute other axial forces from the equilibrium equation of Step 2.**

   Now that you know T_{CF}, you want to find T_{BE}.

$$T_{BE}(30\text{ in}) + (10.9\text{ kip})(40\text{ in}) = 600\text{ kip}\cdot\text{in} \Rightarrow T_{BE} = 5.47\text{ kip}$$

7. **Compute displacements at any location along the rigid bar by using the compatibility expression of Step 3.**

   For example, to compute the displacement at the tip of the rigid bar (or Point D),

$$\frac{(0.001655T_{BE})}{(30\text{ in})} = \frac{\Delta}{(60\text{ in})} \Rightarrow \Delta = \frac{(60\text{ in})}{(30\text{ in})}\left(0.001655(10.91\text{ kip})\frac{\text{in}}{\text{kip}}\right) = 0.036\text{ in}$$

Rigid end cap problems for axial and torsion cases

Another type of rigid object that you encounter from time to time in mechanics of materials is the rigid end cap. A *rigid end cap* is a solid object that connects multiple materials or multiple bars at a single location. Designers typically use rigid end caps to transmit a load to multiple members connected to it.

If you have a rigid end cap on an object of multiple materials, the rigid cap automatically tells you where to draw your free-body diagram and what your compatibility equation will look like. You just have to distinguish whether it's an axial application or a torsion application.

- **Axial rigid end caps:** You find the equilibrium equation by drawing a free-body diagram of the rigid end cap including internal forces of the materials connected to the rigid cap, and write your equilibrium equations. To determine compatibility, you relate the deformations of all materials attached to the cap to the displacement of the cap itself:

  $$\Delta_{CAP} = \Delta_1 = \Delta_2 = \ldots = \Delta_N$$

  where N is the number of different materials attached to each end cap. You then simply relate the deformations in each of the connecting materials as I show in the earlier section "Axial bars of multiple materials."

- **Torsional rigid end caps:** For torsional caps, you find the equilibrium equation by drawing an isolation box around the rigid end cap and summing moments (or more specifically, torques) for each of the materials acting on the cap. You determine compatibility by relating the cap's angle of twist to the angles of twist of all materials attached to the cap:

  $$\phi_{CAP} = \phi_1 = \phi_2 = \ldots = \phi_N$$

  where N is the number of different materials attached to each end cap. You then simply relate the angles of twist in each of the connected materials (see "Torsion of multiple materials" earlier in this chapter).

With rigid end caps for torsion applications, both the cap and all the materials must be concentric to the axis of rotation (or the longitudinal axis of the system).

Chapter 18

Buckling Up for Compression Members

In This Chapter

- Determining column properties and classifications
- Analyzing slender columns
- Including bending effects

Compression members are a type of axially loaded member in which the external forces are working to make the object shorter. Compression members appear in a wide assortment of applications, from the legs of the chair you're sitting on to members of trusses that span the rivers you cross as you travel down the highway.

Because compression members are axial members, you may think that you can readily apply the methods for axial stresses in Chapter 8 to your analysis of columns. However, the equations of axial stress (where stress is simply equal to force divided by area) are only valid up to a specific critical load. After this load, the behavior of members in compression changes, and the actual load that can be supported begins to decrease. It's this behavior, known as buckling, that you must consider in your analysis of members in compression.

In this chapter, I explain the basics of compression theory and show you how to determine a compression member's ultimate loads, which can be significantly lower than the loads you calculate from axial stress relationships. I also show you how the different cross-sectional properties of columns can affect the amount of load a column can carry. I start by illustrating how you can work with concentric axial compression loads and then conclude by showing you how to include additional bending moments.

Getting Acquainted with Columns

In the world of architecture, the word *column* may make you think of majestic Roman pillars from days long past or more-modern steel I-sections. Although the aesthetic beauty of a column may be an important characteristic if you're a tourist of ancient ruins, as a scientist or engineer, you're actually more interested in the type of column (how it's loaded) and its geometric properties. In this section, I explain the different types of columns and then show you how to compute an important cross-sectional property known as the slenderness ratio.

Typically, columns have a much longer axial (or longitudinal) dimension than their cross-sectional dimensions. Although this scenario isn't always the case, this general definition means that a column doesn't necessarily have to be vertical. Structures such as trusses often have a large number of compression members (or columns) that aren't oriented vertically.

Considering column types

Mechanics features two types of columns:

- A *concentric column* (see Figure 18-1a) is a member that is subjected to a compressive and *concentric axial load* (an axial load that acts through the centroid of a cross section).
- An *eccentric column* or *beam column* (see Figure 18-1b) is a member subjected to both a compressive axial load and a bending moment. Beam columns form when a column structure is subjected to an eccentric axial load, an applied end moment, or even external effects from transverse loads on the member or structure.

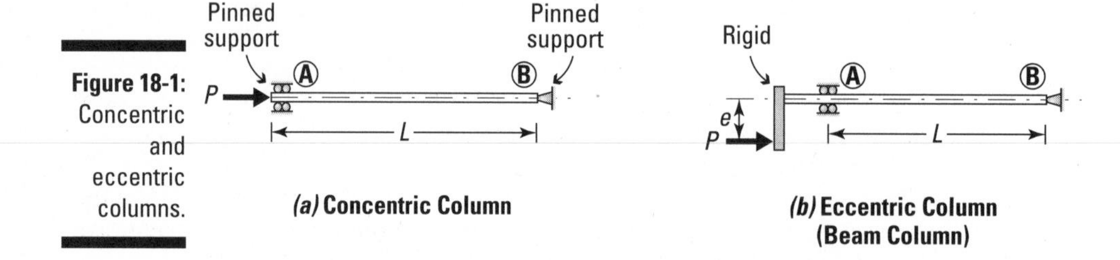

Figure 18-1: Concentric and eccentric columns.

Calculating a column's slenderness ratio

When you begin to analyze objects under applied loads, you must keep in mind that the amount of load an object can carry is limited by either a material failure or a structural instability. In a *material failure,* the stress in the object due to the applied loads reaches the limits that the material itself can sustain. A *structural instability,* on the other hand, is a failure of a member due to excessive deformations that change the basic static equilibrium conditions. To further complicate these matters, these deformations themselves are a function of the loads and member properties, making the study of structural instabilities into a potentially complex problem.

Columns are one of the most basic examples of a structural object that can illustrate both of these methods of failure. A very short column with a given cross section may be able to sustain a compressive load such that the material itself fails. However, the same cross section on a column of much longer length may only be able to support a fraction of the load carried by a shorter column. For this reason, you must come up with a way to classify a column in order to predict which mode of failure controls. That's where the slenderness ratio comes in.

The *slenderness ratio* is a measure of a column's ability to resist elastic instability or lateral displacements due to axial load, and it's a function of two parameters:

- **Effective length:** The *effective length* of a column is a parameter that incorporates the overall length (L) of the column with the type of support conditions. I explain more about these support reactions in the section "Incorporating support reactions into buckling calculations" later in this chapter.
- **Radius of gyration:** The *radius of gyration* (see Chapter 5) is a geometric property that relates the moment of inertia of a cross section about its given axis to the cross-sectional area. For symmetric cross sections, you usually compute the radius of gyration about the x- and y-centroidal axes, which respectively are given as

  $$r_x = \sqrt{\frac{I_{xx}}{A}} \quad \text{and} \quad r_y = \sqrt{\frac{I_{yy}}{A}}$$

  where I is the moment of inertia and A is the cross-sectional area. For unsymmetric sections, you compute the radius of gyration with respect to the principal axes of the cross section.

You express the slenderness ratio as KL/r, where KL is the effective length of the column, and r is the radius of gyration. Consider a column that has an effective length of 30 feet and radii of gyration of $r_x = 8.5$ inches and $r_y = 3.5$ inches. You can compute the slenderness ratios with respect to the x- and y-axes as follows:

$$\frac{KL}{r_x} = \frac{(30\text{ ft})\left(\frac{12\text{ in}}{1\text{ ft}}\right)}{(8.5\text{ in})} = 42.4 \quad \text{and} \quad \frac{KL}{r_y} = \frac{(30\text{ ft})\left(\frac{12\text{ in}}{1\text{ ft}}\right)}{(3.5\text{ in})} = 102.9$$

When computing the slenderness ratio for a column, you always choose the largest value, which for this example is 102.9.

Higher slenderness ratios typically result in lower column capacities. I explain more about how you can use the slenderness ratio to actually compute a column's capacity later in the chapter.

Classifying columns with slenderness ratios

After you know the slenderness ratio of a column (see the preceding section), you're ready to classify the type of column; this classification helps you decide how to proceed with analysis. Columns typically fall into one of three categories based on their slenderness ratios:

- **Short columns:** *Short columns* have a significantly large cross-sectional area and moment of inertia compared to their lengths, which results in a smaller slenderness ratio. A short steel column typically meets a slenderness ratio approximately less than 50.
- **Slender columns:** *Slender columns* are prone to *buckling* behavior (an instability that results from lateral displacements due to axial load). A slender steel column typically has a slenderness ratio greater than approximately 200. I explain buckling further in the section "Buckling Under Pressure: Analyzing Long, Slender Columns" a bit later in this chapter.
- **Intermediate columns:** *Intermediate columns* are columns that are classified as neither short columns nor slender columns. Their slenderness ratios fall somewhere between 50 and 200.

The slenderness ratio limits I describe in the preceding bullets are also actually a function of the type of material that the column is made from. For example, a slender steel column may have a slenderness ratio greater than 200, whereas a slender concrete column may have a slenderness ratio greater than 10.

Each of these types of columns exhibits a unique behavior, and you compute their overall strength in drastically different ways. I describe their various analysis techniques throughout this chapter.

Determining the Strength of Short Columns

Short columns (see the earlier section "Classifying columns with slenderness ratios") are the easiest class of column to work with because the stresses in the column cross section reach their material's yield point (see Chapter 14) before structural instability (or buckling) can occur.

You can compute the basic stress of a short column by simply calculating $\sigma = P/A$ as I show in Chapter 8. If a short column is subjected to a load of

P = 100 kip and has a cross-sectional area of A = 5.0 in^2, you find the axial normal stress in the column as follows:

$$\sigma = \frac{P}{A} = \frac{(100 \text{ kip})}{(5.0 \text{ in}^2)} = 20 \text{ ksi}$$

Thus, for this example, as long as the material of the column is capable of supporting a minimum stress of 20 ksi, the column is sufficient for these loads. However, if the column isn't classified as a short column, the maximum load that a column can support (known as the *critical load*) actually reduces based on the slenderness ratio. I explain this topic in more detail in the following section.

Buckling Under Pressure: Analyzing Long, Slender Columns

As long as a column stays straight, the principles of axial stress in Chapter 8 are still valid. However, columns don't always remain straight when subjected to a load, especially if that load is a compressive load.

For example, if you push on the ends of a piece of uncooked spaghetti, it begins to bend laterally. This lateral displacement from compression is actually an example of a structural instability known as *buckling,* and it results in a maximum load (known as the *critical buckling load*) that is less than the load that a short column of the same cross section can carry. In this section, I show you how to compute the critical buckling load for a slender column.

If a column displaces laterally, the axial loads actually become eccentric (or not in line with the centroid of all the cross sections along the member), which creates additional bending stresses (see Chapter 9).

Determining column capacity

To compute the *capacity* (or amount of load a column can carry) for a slender column, you need to take a few factors into consideration:

- **Slenderness ratio:** As I note in "Calculating a column's slenderness ratio" earlier in this chapter, the slenderness ratio compares the effective length of the column to one of the radii of gyration. The end support conditions also affect the slenderness ratio, and I discuss those conditions in more detail later in the chapter.
- **Young's modulus of elasticity:** The Young's modulus of elasticity is an important factor when determining a slender column's strength. For more on Young's modulus of elasticity, turn to Chapter 14.

- **Mode of buckling:** The *mode of buckling* refers to the number of half-sine shaped curves a column deforms into as it begins to buckle. The higher the mode number, the more force necessary to cause buckling.

To properly compute the capacity of a slender column, you have to first determine the type of failure or buckling that the column experiences — *elastic* or *inelastic* — a process that requires you to make some basic assumptions:

- **Concentric axial loads:** All loads in the columns of this section are concentric and axial, meaning that the column doesn't have any applied moment or eccentric axial loads.
- **Straight and prismatic columns:** All columns in this section must be straight and *prismatic* (or having a constant cross section). If a column isn't straight, you actually get a bending effect in the column, a result that violates the first assumption.

Computing the elastic buckling load

Elastic buckling is a type of buckling where the column's failure occurs below the material's yield stress. In 1757, Leonhard Euler developed an expression to estimate the critical elastic buckling load (known as the *Euler buckling load*) of a long slender column. In his derivation, Euler found that he could express the critical load as

$$P_{CR} = \frac{n^2\pi^2 EI}{L^2}$$

where n is the buckling mode number; E is Young's modulus of elasticity for the column material; I is the moment of inertia of the cross section with respect to the axis about which it's buckling; and L is the length of the column. The derivation of this formula assumes that the ends of the column are both *pinned supports,* which means they're free to rotate.

Defining the mode of buckling

The *buckling mode* is the shape that a column or compression element actually buckles into. Consider the columns of Figure 18-2, which show several different modes of buckling. In Figure 18-2a, you see the first mode of buckling (where $n = 1$), which is a simple bending example where all displacements are in the same direction (on the same side of the column).

If you brace the middle of the first column shown in Figure 18-2a and then push on the ends until buckling occurs, you actually end up with the buckled shape shown in Figure 18-2b, which is the second mode of buckling (or $n = 2$). As you may expect, you need significantly more effort — actually, four times as much — to cause a column to buckle into this shape. Similarly, Figure 18-2c shows the third mode of buckling (or $n = 3$) which takes nearly nine times the load to cause a column to buckle.

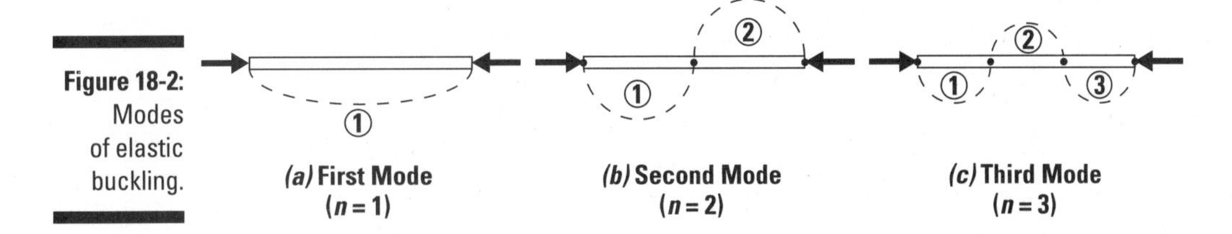

Figure 18-2: Modes of elastic buckling.

An infinite number of buckling modes actually exist, but after the first two or three, they become extremely rare under normal loading. However, vibrating loads can actually cause a column to buckle in these higher-order modes.

A column wants to buckle into its lowest mode shape whenever possible because that shape is usually the easiest to achieve. To increase a column's strength within the elastic buckling region, you simply need to brace the column to prevent it from buckling into a higher-mode shape.

Performing an elastic buckling calculation

With Euler's buckling load formula in hand, you can predict the load at which elastic buckling occurs for a given mode number. For example, consider a steel column with both ends pinned and subjected to the first mode of buckling (or $n = 1$). It has a length L of 40 feet, a Young's modulus of elasticity E of 29,000 ksi, and moments of inertia of $I_{xx} = 100\text{ in}^4$ and $I_{yy} = 33\text{ in}^4$.

For buckling in the y-direction in the first mode,

$$P_{CR} = \frac{n^2\pi^2 EI_{yy}}{L^2} = \frac{(1)^2\pi^2(29{,}000\text{ ksi})(33\text{ in}^4)}{\left[(40\text{ ft})\left(\frac{12\text{ in}}{1\text{ ft}}\right)\right]^2} = 41.0\text{ kip}$$

For buckling in the x-direction,

$$P_{CR} = \frac{n^2\pi^2 EI_{xx}}{L^2} = \frac{(1)^2\pi^2(29{,}000\text{ ksi})(100\text{ in}^4)}{\left[(40\text{ ft})\left(\frac{12\text{ in}}{1\text{ ft}}\right)\right]^2} = 124.2\text{ kip}$$

From these calculations, you can see that the axis about which buckling occurs plays a significant role in determining the critical buckling load. In this example, the elastic buckling load that causes buckling about the y-axis is about 30 percent of the load that causes buckling about the x-axis (or strong axis). In fact, this type of calculation is what lets an engineer know how to orient a column or in which directions to provide bracing.

Thus, for this example, the column does not fail due to buckling if the applied load is less than 41.0 kip — that is, the column doesn't experience a structural instability.

Computing elastic buckling stress

Although knowing the load that causes buckling is good information to have, you really need to determine the stress in the column at which buckling occurs in order to evaluate whether your applied axial load causes an axial stress in the column that exceeds the critical buckling stress. To compute the critical buckling stress, you can further modify the Euler buckling equations as follows to compute the critical stress as well as incorporate the effective length:

$$\sigma_{CR} = \frac{P}{A} = \frac{n^2\pi^2 E}{\left(\frac{KL}{r}\right)^2}$$

where K is a length correction factor that takes into account the type of support reactions other than the pinned-pinned ends of Euler's basic assumption, and r is the smallest radius of gyration of the cross section (for more on radius of gyration, turn to Chapter 5).

For the steel column of the previous example, if the largest slenderness ratio of the column is 300, you can classify this column as a slender column and use the following equation to compute its buckling stress:

$$\sigma_{CR} = \frac{n^2\pi^2 E}{\left(\frac{KL}{r_y}\right)^2} = \frac{(1)^2\pi^2(29{,}000\text{ ksi})}{(300)^2} = 3.17\text{ ksi}$$

This buckling stress of 3.17 ksi is actually a reasonably small number in the world of steel columns. An ASTM A36 column (which has a yield stress of 36 ksi) with this length and slenderness ratio would buckle well before it reached its yield stress (at about 10 percent of the actual material strength); that's a very inefficient design.

Incorporating support reactions into buckling calculations

Although columns with pinned ends (such as the ones I cover earlier in the chapter) cover a lot of scenarios in compression member design, not all compression members are pinned at both ends. The effect of end supports on a column can have a dramatic effect on the overall column capacity. For example, you need a larger load to cause a column with fixed ends to buckle than you need to buckle the same column with pinned ends, mostly because the moments at the fixed ends are fighting to keep the column from buckling. You account for these effects with an effective length correction factor (the K) in the slenderness ratio calculation.

Euler's buckling load equation already applies a correction factor of 1.0 to the length term (L) because the formulation assumes that the column is pinned at both ends (as shown in Figure 18-3a). However, if you restrain the rotation of the ends of the column such that they're actually fixed ends at both ends of the column (as shown in Figure 18-3b), the effective length factor becomes 0.5, or half of the pinned condition.

The case in Figure 18-3c is actually a column with one end fixed and the other end pinned. As you may expect, the buckling load for this case is somewhere between the buckling cases for both ends pinned and both ends fixed; in this case the correction factor is 0.7. Conversely, the column in Figure 18-3d shows a free end at one end and a fixed end at the other (which is similar to a flagpole situation). For this case, the effective length coefficient is actually 2.0, which is double the pinned-ends condition, meaning that the flagpole experiences buckling at significantly lower load values, mostly due to the lateral movement of the free end of the column, which causes an eccentric load.

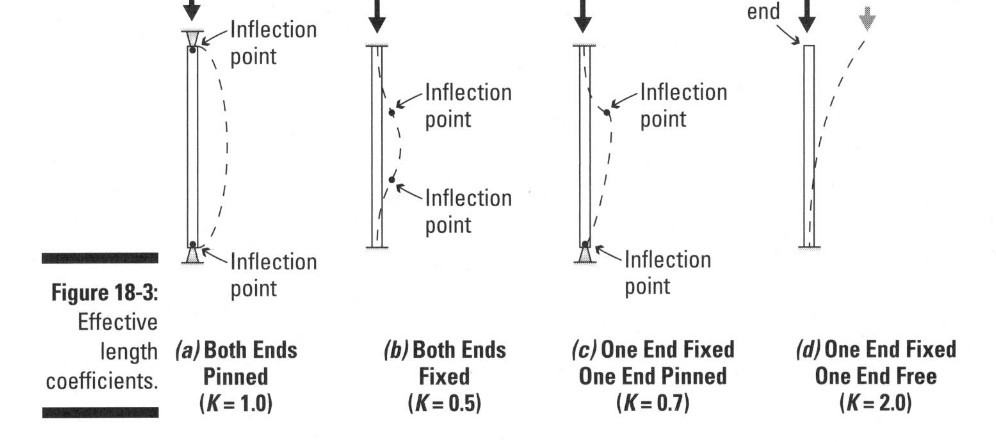

Figure 18-3: Effective length coefficients.

Consider the pinned column I describe earlier in the chapter, where $L = 40$ feet, $E = 29{,}000$ ksi, $I_{xx} = 100\ \text{in}^4$, and $I_{yy} = 33\ \text{in}^4$. If you modify this column so that its ends are fixed rather than pinned, the slenderness ratio changes from 300, which qualifies it as a slender column, to a much more manageable 150, which makes it an intermediate column and requires you to use a slightly different formulation of the Euler buckling equation, which I show in the following section.

Improving column capacity

When a column fails because of inefficient design, as a design engineer you may find it prudent to make a few modifications:

- **Increase the radius of gyration.** Increasing the size of the cross section generally increases the moment of inertia and the cross-sectional area, which typically results in a larger radius of gyration. If you increase the minimum radius of gyration, you end up reducing the slenderness ratio in that direction, which is often good for the capacity of a column.
- **Alter the mode number.** Modifying the structure by incorporating additional bracing or supports can tremendously increase capacity simply by altering the buckled shape.
- **Decrease the slenderness ratio in the controlling direction.** If the slenderness ratio is larger in a particular direction, you can reduce the effective length of the column by providing intermediate supports or providing additional material to reduce the radius of gyration in that direction.

  In some structures, a column isn't braced the same in all directions. You can actually end up having a column length in one direction be significantly smaller than in the other. In fact, the strong axis may actually be the limiting direction if the length and support reactions are unfavorable in that direction.
- **Modify the support reactions.** Restraining one or both ends of the member can reduce your K factor (and consequently your slenderness ratio) by as much as 70 percent if one end is fixed or a whopping 50 percent if both ends are fixed.

Working with Intermediate Columns

When a column falls into the intermediate column category (which basically means it's not a short column or a slender column in terms of its failure), it experiences *inelastic buckling,* a blend of those other two columns' behaviors. When a column experiences inelastic buckling, it doesn't fail by direct compression (which I discuss in Chapter 8) nor by elastic buckling (which I discuss earlier in the chapter). Instead, the material stresses at part of the cross section exceed the proportional limit at the time when buckling occurs.

Although you can actually handle inelastic buckling in multiple ways, perhaps the easiest method is to calculate the inelastic buckling load by using the *generalized Euler buckling formula:*

$$P_{CR} = \frac{n^2\pi^2 E_t I}{(KL)^2}$$

where this formula is the same as the elastic buckling load equation in the earlier section "Computing the elastic buckling load," except that you replace Young's modulus of elasticity (E) with the term E_t, or the tangent modulus of elasticity, which I discuss in Chapter 14.

Figure 18-4 shows the relationship among the three column classifications that I discuss earlier in this chapter with respect to the slenderness of the column.

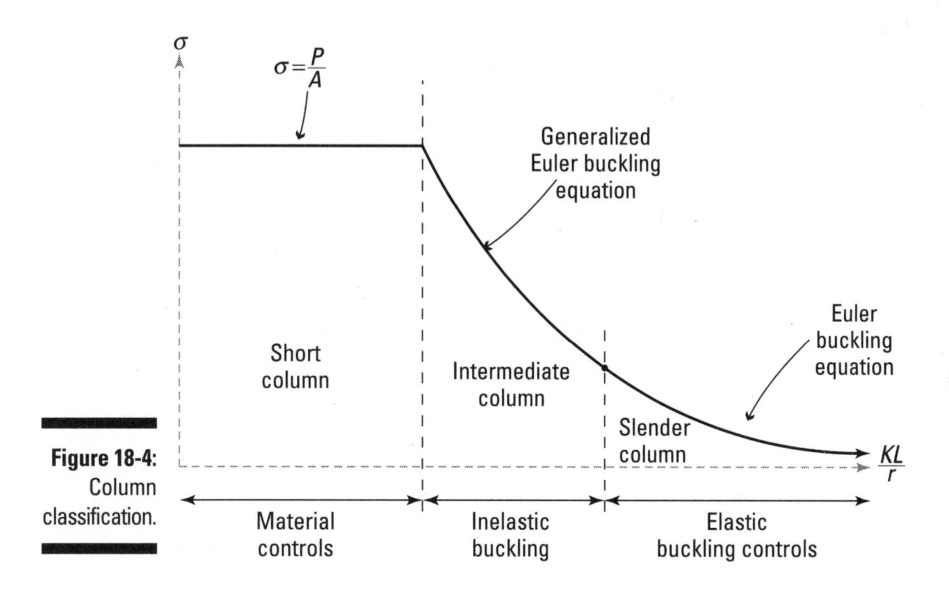

Figure 18-4: Column classification.

Incorporating Bending Effects

Throughout this chapter, I describe the critical buckling load calculations and assumptions that you need for working with columns that are straight and axially loaded concentrically. However, in many applications columns can be subjected to bending moments at the same time as they're axially loaded. Because the buckling phenomenon is related to the compressive normal stress in the cross section, buckling can occur even sooner if you add additional compressive stresses from bending, which means you must be sure to include their effects when you are analyzing columns.

Bending stresses produce both tension and compression stresses in the cross section. Although tension stresses theoretically help improve a column's performance against buckling, the compression stresses from bending also add to the compressive stresses from the axial loads, a situation that increases the likelihood of buckling.

To handle bending moments, you treat the moments on a cross section as an *eccentric axial load* (or a force applied at an eccentricity instead of being concentric), as I show in Figure 18-1 earlier in the chapter. You compute the maximum stress capacity of an eccentric column by

$$\sigma_{MAX} = \frac{P}{A}\left[1 + \frac{ec}{r^2}\sec\left(\frac{KL}{2r}\sqrt{\frac{P}{AE}}\right)\right]$$

which is also known as the *secant formula*. In this equation, P is the applied eccentric load; A is the cross-sectional area; e is the eccentricity of the load; c is the distance from the neutral axis to the extreme fiber in compression; r is the radius of gyration; E is the Young's modulus of elasticity of the column material; and KL is the effective length.

Remember that the $\sec(\theta)$ or secant operator is actually $1/\cos(\theta)$ and that this operation requires that you express the terms within the parenthesis in units of radians, so be sure to check the mode on your calculator! And then remember to switch it back when you're done.

For example, consider a steel column that is subjected to an axial load of $P = 100$ kip with an eccentricity of 4 inches. If the column has cross-sectional properties of a depth of 6 inches (which means the c is 3 inches), cross-sectional area A of 1.43in^2, radius of gyration r of 4.8inches, and an effective length KL of 480 inches, you can compute the maximum stress capacity of this column as follows:

$$\sigma_{MAX} = \frac{(100\text{ kip})}{(1.43\text{ in}^2)}\left[1+\frac{(4\text{ in})(3\text{ in})}{(4.8\text{ in})^2}\sec\left(\frac{(480\text{ in})}{2(4.8\text{ in})}\sqrt{\frac{(100\text{ kip})}{(1.43\text{ in}^2)(29{,}000\text{ ksi})}}\right)\right] = 22.8\text{ ksi}$$

For the case of the same column loaded concentrically (where the eccentricity $e = 0$), you can compute the maximum stress by substituting into the secant formula again:

$$\sigma_{MAX} = \frac{(100\text{ kip})}{(1.43\text{ in}^2)}[1+0] = 69.9\text{ ksi}$$

Thus, by simply increasing the eccentricity by 4 inches (which seems like a small distance), you actually reduce the stress to cause buckling from 69.9 ksi to 22.8 ksi (or approximately 32 percent of the stress from a concentric loading.)

Chapter 19

Designing for Required Section Properties

In This Chapter

- Establishing the basic principles of design
- Working with factors of safety and allowable stresses
- Calculating member section properties from known internal loads

Engineers are responsible for ensuring that the members of a structure or system are capable of supporting anticipated loads throughout the structure's life. In addition, engineers must make sure that the object performs its intended function and doesn't experience deflection, vibration, or cracking. On top of all that, they must also create the most economical and efficient design as possible.

The basis for design involves taking known loads and using your knowledge of statics to predict the external support reactions on a system. With these reaction loads known, you can compute internal forces and moments and then, if you know the sizes of the members, perform basic stress and strain analysis as I illustrate in Parts II and III.

However, one major problem arises with indeterminate structures: To compute the external support reactions, you have to know how much the object deforms under the applied loads. And before you can calculate deformation, you have to have some idea of the member's cross-sectional properties (such as area, moment of inertia, and radius of gyration, to name a few), which you can't possibly know without knowing the internal forces. So the issue you're faced with is, "Where do I start?"

Design requires you to make some initial guesses based on experience, equations, and sometimes even a bit of luck to determine the necessary cross-sectional properties. This initial guessing gives you a starting point for your design, which you must be sure to confirm with proper analysis. After you know the section properties, you can then compute stresses and strains and verify (or dispute) your assumptions.

The difficulty in design is that you can't compute the actual stresses when you're still trying to determine the required section properties of the member. However, you can use material properties to develop an upper limit on stresses and then use this limit to determine a required section property to ensure that actual stresses never exceed the limits of the material.

In this chapter, I show you how to calculate the capacity of a structural member and then how to compute a required area or moment of inertia to support a given internal force or moment. I also introduce factors of safety, which provide an extra degree of, well, safety to your structure.

Structural Adequacy: Adhering to Formal Guidelines and Design Codes

The goal of any engineer is to provide a safe, economical structure that performs its intended function for many years to come. A structure or object that meets all of these criteria is said to demonstrate characteristics of *structural adequacy*.

When you design an object, you can simplify the basis for structural adequacy into two basic criteria: strength and serviceability:

- **Strength:** *Strength* means that the structure is sufficient to support the anticipated design loads.
- **Serviceability:** *Serviceability* means that the structure can meet its intended function. For engineering structures, this criteria can mean that the object doesn't experience excessive deflections or vibrations (among other non-strength-related issues).

As an engineer or designer, you should keep these two criteria in the back of your mind at all times. Just because the beams of your living room floor may be adequate to support the loads of your house and your furniture (known as the service loads, which I discuss in the following section), you don't want the floor of your house to experience large amounts of deflection (especially if all your chairs are on wheels or casters).

For strength and serviceability considerations, engineers utilize *design codes* (material guidelines and standards of practice for design) to help them ensure minimum acceptable levels of performance. Literally hundreds of different codes help engineers and designers meet the requirements for strength and serviceability. These codes are based on experience and experimental research and are compiled by a wide range of organizations such as the following:

- **Fabrication organizations:** Manufacturing organizations such as the American Concrete Institute (ACI) and the American Institute of Steel Construction (AISC) provide recommendations based on research within their specific materials.
- **Professional engineering societies:** Professional groups such as the American Society of Civil Engineers (ASCE), the American Society of Mechanical Engineers (ASME), and the Society of Automotive Engineers (SAE) may establish guidelines and standards to help practicing engineers.
- **Government agencies:** Agencies and councils within your county, town, state, or nation may provide formal building code requirements. Many different codes exist, and knowing which specific code applies within a given project jurisdiction becomes important. Codes such as the International Building Code (IBC) and the Uniform Building Code (UBC) all provide guidelines for structural engineers.

An example of a potential structural building code requirement is the limitation of deflection to some pertinent ratio. For example, for many codes, the deflection in a floor beam is limited to

$$\Delta_{LL,MAX} \leq \frac{L}{360}$$

where L is the span of the beam and $\Delta_{LL,MAX}$ is the maximum allowable live load flexural deflection in the beam. For more on these deflection calculations, turn to Chapter 16.

Exploring Principles of the Design Process

When you start design, you need to keep a couple of simple concepts in mind: The member needs to be able to carry the anticipated loads as well as perform its intended task without deflecting or vibrating excessively. Other considerations — including dimensional requirements and, of course, cost — can also factor into your design decisions. These principles are important because you always want an object to be safe, functional, and economical. In this book, I assume that cost is no problem and stick to considering issues involving strength and serviceability calculations.

Explaining member strength and design loads

In design, you must first understand a bit of terminology regarding the loads that you expect to use in the design process:

- **Design loads:** *Design loads* are the loads that are applied to the object, including any factors of safety or other load increase. That is, design load = factor of safety × applied load. A *factor of safety* is a constant multiplier (usually larger than 1.0) meant to cover *overload conditions* (situations where the actual loads exceed the value that you normally expect) and other safety considerations. Different design methodologies require different values. For an allowable-stress-based design (such as what I describe in this chapter), you commonly see a factor of safety with a value of 2.0 or 3.0.

  These loads represent the maximum load (or the worst-case scenario) that you expect the object to ever experience. Design loads typically fall into two major categories:

  - **Factored loads:** *Factored loads* are loads you use to evaluate a structure for strength; they typically include factors of safety.
  - **Service loads:** *Service loads* are loads that act on an object or structure during normal usage. Engineers often use them to evaluate the serviceability criteria for deflections and vibrations (see the preceding section for more on serviceability). However, some design methods actually use service loads rather than factored loads.

    For the Load and Resistance Factor Design (LRFD) method, you categorize loads separately and apply individual multipliers: *dead loads* (permanent loads that never move) may have a multiplier of 1.2 or 1.4, while *live loads* (loads such as people, vehicles, and office furniture that move around during the life of a structure) may have a factor of 1.6. You then combine these individual factored loads into a single factored load combination that you use as the design load.

- **Member capacity:** *Member capacity* is a measure of the amount of load that you can expect a member to safely support. This value is often referred to as the *member strength* of an object.

The design loads are a function of the type of loading on the object. In many cases, an object can have multiple design loads. For example, if you take a pressure vessel (such as the ones I discuss in Chapter 8) and apply a torque to the ends of it (as I discuss in Chapter 11), the object has two different design loads: one for axial loads from the pressure vessel and another from twisting moments due to torsion.

Likewise, a member can have different capacities under different types of loading. For example, a long, slender column may have a large axial capacity but a very small flexural capacity. These combined effects are a serious consideration for engineers during the design process as I show in the "Interacting with Interaction Equations" section later in this chapter.

Creating a design criteria

In all cases, for an object to be considered adequate for a particular design, the whole design process basically boils down to a single equation. In short, you want to ensure that

capacity ≥ design loads

Capacity is usually expressed as a maximum load or maximum moment. However, working with stresses is typically more convenient because you can relate those stresses directly to material properties such as yield stress or ultimate stress (see Chapter 14). As you're working through the design process, remember that you also need to consider issues with instability (such as buckling of compression members in Chapter 18).

In most design methods, you relate the allowable design stress σ_{ALL} or τ_{ALL} to a relationship involving both the actual maximum stress that a member can support (σ_{MAX} or τ_{MAX}) and the factor of safety *F.S.*

$$\sigma_{ALL} = \frac{\sigma_{MAX}}{(F.S.)} \quad \text{and} \quad \tau_{ALL} = \frac{\tau_{MAX}}{(F.S.)}$$

In many design problems, you actually assign the yield strength or ultimate strength values to the actual value of the maximum stress (σ_{MAX}). For example, if the yield strength of a material is 50 ksi and you want a factor of safety of 3.0, you use an allowable design stress of σ_{ALL} = (50 ksi) ÷ (3.0) = 16.7 ksi. You then design your member's section properties to limit its allowable stress to 16.7 ksi under the applied design loads.

Although the maximum tensile yield strength is usually well defined for normal stresses, the shear yield stress τ_y isn't typically readily available. Experimental material tests indicate that the maximum shear stress is often about 50 to 60 percent of tensile yield stresses:

$$\tau_y = 0.5(\sigma_y)$$

where σ_y is the tensile yield stress of the material.

In some more-advanced design methodologies, applying an additional type of safety factor in the form of a reduction in capacity isn't uncommon. For example, in the design of bending members, some codes only allow you to use 90 percent of the actual capacity of a member, which results in a 0.9 reduction factor for the capacity.

Developing a Design Procedure

In general, the design process is usually a methodical approach that may require a few repetitions to determine the ideal section properties for a design. In every design method, you must be able to identify all the possible failure modes, determine loads for those failure modes, and then calculate the required section properties to prevent those failure modes. This section outlines a basic design procedure to determine required section properties.

Think of structural design as filling a bucket with water. The bucket represents the capacity of the structural object, and the water represents the load or stress applied. If you put too much water (design load) into your bucket (actual member capacity), the water overflows, creating a mess. Likewise, if you overload a structural member, something bad, such as a member breaking or deflecting too much, can (and usually does) happen, resulting in a completely different type of mess altogether.

Outlining a basic design procedure

Generally speaking, you can break down the procedure for designing an object into a few simple steps:

1. **Determine the design mode.**

   The *design mode* is the mode of failure, which can be either a limitation due to strength or a limitation due to a serviceability requirement such as deflection or vibration. I explain more about determining design requirements in the next section.

2. **Compute the maximum design load for strength or the allowable deformation/vibration for serviceability for the design mode of Step 1.**

   To help you find these maximum values, you almost always need to draw an internal force diagram, such as the shear and moment diagrams for flexural members (see Chapter 3) or torque diagrams for torsion in shafts.

3. **Apply factors of safety or load factors to compute design loads.**

   The factor of safety effectively increases the size of your bucket, ensuring that you can safely handle all the loads on your structure. The earlier section "Explaining member strength and design loads" gives you the lowdown on factors of safety.

4. **Determine the required section property to resist the design loads of Step 3.**

   In this step, you actually determine how the applied loads are related to the limiting stress of the material. I discuss more about these design requirements in the following section.

Determining design requirements from modes of failure

The hardest part of the design process is actually knowing which section property you need to calculate. This section property usually depends on the type of load on the structure. For simple cases, Table 19-1 shows you the section properties that you usually need to calculate depending on the type of loading on the object.

Table 19-1 Design Requirement Section Properties

Loading Type	*Possible Mode of Failure*	*Typical Design Requirement*
Axial tension	Axial normal stress	Area
Axial compression	Axial normal stress	Area
	Instability from buckling	Slenderness ratio
Direct shear	Shear stress	Area
Simple bending	Flexural normal stress	Moment of inertia
	Transverse shear stress	First moment of area, moment of inertia, thickness
Torsion	Shear stress	Torsion constant

For example, for a beam or frame subjected to bending, you can expect to have both normal and shear stresses from flexural loads. To handle the normal stresses, you typically need to compute the second moment of area (I), and for the shear stress from flexural shear, you need the first moment of area (Q) and the second moment of area (or the moment of inertia).

Structural objects can also be subjected to multiple simple effects simultaneously. For example, a column in a building is often subjected to an axial load as well as bending in multiple directions (such as the biaxial bending problems I discuss in Chapter 15). For these objects, you need to work with *interaction diagrams* that interrelate these different effects into a single capacity equation. Because one load uses up capacity, you can't use the same portion of the total capacity to resist additional loads. You can think of these effects as being two different fluids you pour into your bucket. After you put one fluid in, you decrease the available space in the bucket for the additional fluid.

I should point out that the design process involves considerably more than the basic calculations I present in the coming sections. A competent engineer must be familiar with the relevant design codes and standards that may require different or additional calculations from what I have presented here.

Designing Axial Members

Whether you're trying to determine the required size for a rope or designing a member in a truss, axial members are often one of the first types of design you learn. As with all types of design, designing for axial members has its roots in basic stress analysis and requires knowledge about both geometric and material properties, including Young's modulus of elasticity (E), of the material that you're planning to use.

Consider the truss assembly shown in Figure 19-1, which is subjected to a vertical load of 10 kip acting downward at Point C. The internal forces for Members AC and BC by using the equilibrium equations from statics (see Chapter 3) and are indicated on their respective members in the diagram. For this example, I assume that the factor of safety is 3.0 and that the bars are solid, square cross sections (having equal dimensions for width and height) made from steel (E = 29,000 ksi) with a maximum yield stress of 50 ksi.

You can find the allowable stress of the material based on the maximum yield stress and the factor of safety as I show in the earlier section "Creating a design criteria." For this example, the allowable stress σ_{ALL} = (50 ksi) ÷ (3.0) = 16.7 ksi.

This example actually has two different axially loaded members, one in tension and one in compression. The following sections show you how to deal with each of these members.

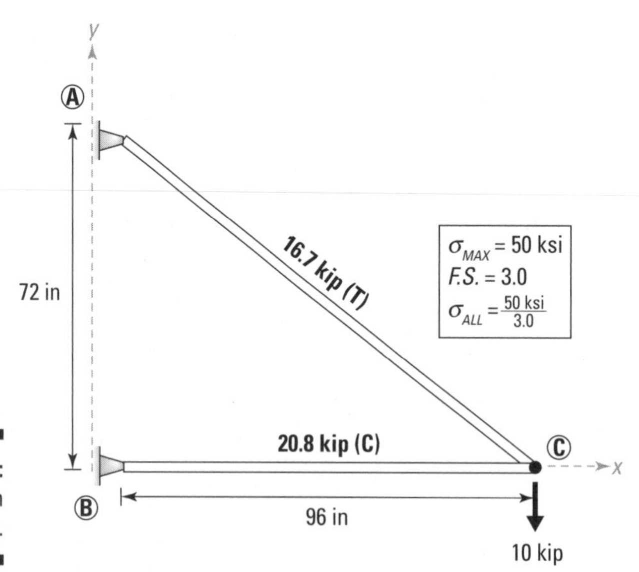

Figure 19-1: Axial design example.

Calculating for simple tension members

For tension members, you can rearrange the basic axial stress equation from Chapter 8 a bit to find a required cross-sectional area A_{REQ}:

$$A_{REQ} = \frac{P}{\sigma_{ALL}}$$

where P is the design axial load and σ_{ALL} is the allowable stress of the material (which also takes into account the safety factor). For this example, you can determine the required cross-sectional area as follows:

$$A_{REQ} = \frac{P}{\sigma_{ALL}} = \frac{(16.7 \text{ kip})}{(16.7 \text{ ksi})} = 1.0 \text{ in}^2$$

For a square bar, $A = b^2$, so a bar with cross-sectional dimensions of 1 in x 1 in would be sufficient for this tension member *AC*.

Guessing a column classification for compression loads

Working with compression members is a bit more complex than working with tension members (see the preceding section). When you analyze a column (or compression member), you must consider *buckling,* a function of the *effective length* (the length of the column multiplied by a factor describing the end support conditions) and the *radius of gyration* (a cross-sectional property based on the moment of inertia and the cross-sectional area). Flip to Chapter 18 for more information on columns and compression members.

Unfortunately, without knowing a member's size, you can't possibly know which type of buckling the member experiences. And without knowing which type of buckling to design for, you can't determine the critical loads on a column. At this point, all you can do is guess a type of buckling, perform your calculations based on this assumption, and then verify your assumption after you've selected a size.

To get started for the example in Figure 19-1, I assume that the column of Member BC is a short column and determine a minimum required area based on the assumption that short columns experience a material failure (that is, the cross section is able to reach its allowable stress limit before buckling occurs). You start with the same basic stress calculation you use for tension members:

$$A_{REQ} = \frac{P}{\sigma_{ALL}} = \frac{(-20.8 \text{ kip})}{(-16.7 \text{ ksi})} = 1.24 \text{ in}^2$$

which makes the bar dimensions $b = h = 1.11$ in^2. For this bar, the radius of gyration is 0.320, which makes the slenderness ratio for this member a whopping $KL/r = 299$! As I note in Chapter 18, the slenderness ratio limit for slender steel columns is 200, which means my assumption that Member BC is a short column was incorrect. I actually need to design this member as a slender column subjected to elastic buckling effects.

For this example, Member BC experiences the first mode of buckling ($n = 1$), and has an effective length of $KL = (1.0)(96\text{ in}) = 96$ in. You can determine the critical buckling load P_{CR} of a slender column from the Euler buckling equation (see Chapter 18):

$$P_{CR} = \frac{n^2\pi^2 EI}{(KL)^2}$$

In this equation, the one value that remains unknown is the moment of inertia of the member; if you don't know this value, you can't determine whether the applied load will cause the column to buckle. But you can determine the minimum moment of inertia for the column by setting the design load P_{DESIGN} equal to the critical buckling load P_{CR} and incorporating the factor of safety *F.S.* as follows:

$$P_{DESIGN} \le \left(\frac{P_{CR}}{F.S.}\right) \quad \text{or} \quad (F.S.)(P_{DESIGN}) \le P_{CR}$$

Substituting this equation into the Euler buckling equation and solving for the moment of inertia,

$$I_{MIN} = \frac{(KL)^2 (F.S.) P_{DESIGN}}{n^2\pi^2 E}$$

For this example, you can calculate the minimum required moment of inertia with the following equation:

$$I_{MIN} = \frac{(KL)^2 (F.S.) P_{DESIGN}}{n^2\pi^2 E} = \frac{(96\text{ in})^2 (3.0)(20.8\text{ kip})}{(1)^2 \pi^2 (29{,}000\text{ ksi})} = 2.01\text{ in}^4$$

At this point, you're ready to choose a member size from design tables or compute the dimensions from the basic moment of inertia formulas. For a square cross section (where $b = h$), the moment of inertia is given as $I = bh^3/12 = b^4/12 = 2.01$ in^4. Solving this expression for the dimensions, $b = h = 2.22$ in, a figure that's substantially larger than when you initially assumed the member was a short column.

When working with rectangular (and square) sections, you have to be very careful about assigning the proper dimensions to the variables *b* and *h*, because these dimensions help determine the orientation of the member with respect to the applied loads. For square tubes, however, the variable assignment doesn't make any difference because both dimensions have the same value.

Because of the effects of buckling on your design, you need a square bar with more than double the dimensions of your original guess!

As you gain experience, you start to develop a sense (sometimes referred to as *engineering judgment*) for columns such as Member BC. Because the length is very long (96 inches) for a reasonably small load (which makes the required area very small), classifying this column as slender is a fairly safe bet. However, until you're comfortable coming to this conclusion, confirming any guesses or intuition you choose to rely on with calculations is always a good idea.

Designing Flexural Members

Flexural members such as the ones in Chapter 9 are among the most common structural items you design. You encounter these objects in a wide variety of applications, the simplest being horizontal beams in roofs or floors of buildings.

The first step to analyzing any flexural member is to perform a static analysis, determine support reactions, and then sketch the shear and moment diagrams to help you locate the maximum shear force and maximum moment along the length of the beam. (Chapter 3 covers all these tasks.) Without shear and moment diagrams, you have no way of knowing where the maximum internal shear force and moment occur.

Consider the steel beam shown Figure 19-2 (with cross section lying in the XY plane) that is subjected to both a point load and a uniformly distributed load. For this beam, the steel's maximum normal stress at yielding is given as 36 ksi, and its maximum shear stress is 24 ksi. I assume a factor of safety of $F.S. = 1.5$, making $\sigma_{ALL} = 36 \text{ ksi} \div 1.5 = 24 \text{ ksi}$ and $\tau_{ALL} = 24 \text{ ksi} \div 1.5 = 14.4 \text{ ksi}$. Figure 19-2 provides the already-developed support reactions and shear and moment diagrams (which you may recognize from Chapter 3).

For the example in Figure 19-2, the shear and moment diagrams clearly show that the maximum positive moment in this beam is $M_{MAX} = 84$ kip-ft and occurs at Point C, while the maximum shear force occurs at Point B and has a value of $V_{MAX} = 31$ kip.

Planning for bending moments with the elastic section modulus

As I indicate in Chapter 9, you can calculate flexural bending stresses from the basic formula

$$\left|\sigma_{zz,ALL}\right| = \frac{M_x y}{I_{xx}} = \frac{M_x}{S_x}$$

In the second part of the equation, S_x is known as the *elastic section modulus* with respect to the *x*-axis. As you may recall, you can also find an S_y, which is the elastic section modulus with respect to the *y*-axis. For more on actually computing the elastic section modulus, turn to Chapter 9.

The formula for the elastic section modulus requires both the cross section's moment of inertia and the distance from the neutral axis to the *extreme fiber* (the outer edge) of the cross section, both of which are independent of each other and dependent on the dimensions of the cross section. For the beam of the Figure 19-2 design example, you can compute the required elastic section modulus as follows:

$$S_{x,REQ} = \left|\frac{M_x}{\sigma_{zz,ALL}}\right| = \frac{(84\text{ kip-ft})\left(\frac{12\text{ in}}{1\text{ ft}}\right)}{(24\text{ ksi})} = 42.0\text{ in}^3$$

Working with section property tables

In design, you often don't know the dimensions of the beam at the beginning of the design process. However, by utilizing the elastic section modulus, you can compute a single value and use design tables to look up a specified value. Table 19-2 shows an example of a typical entry in a steel manual for an I-beam (shown in Figure 19-3). Figure 19-3 shows the location of dimensions b_f, t_f, and t_w for the standard I-shaped cross section.

Table 19-2 Sample Section Property Table

1 Section	2 Area (in²)	3 Depth (in)	4 b_f (in)	5 t_f (in)	6 t_w (in)	7 I_{xx} (in⁴)	8 r_x (in)	9 S_x (in³)
W12x35	10.3	12.5	6.56	0.520	0.30	285	5.25	45.6
W10x45	13.3	10.1	8.02	0.495	0.310	248	4.33	49.1
W8x48	14.1	8.5	8.11	0.685	0.400	184	3.61	43.3

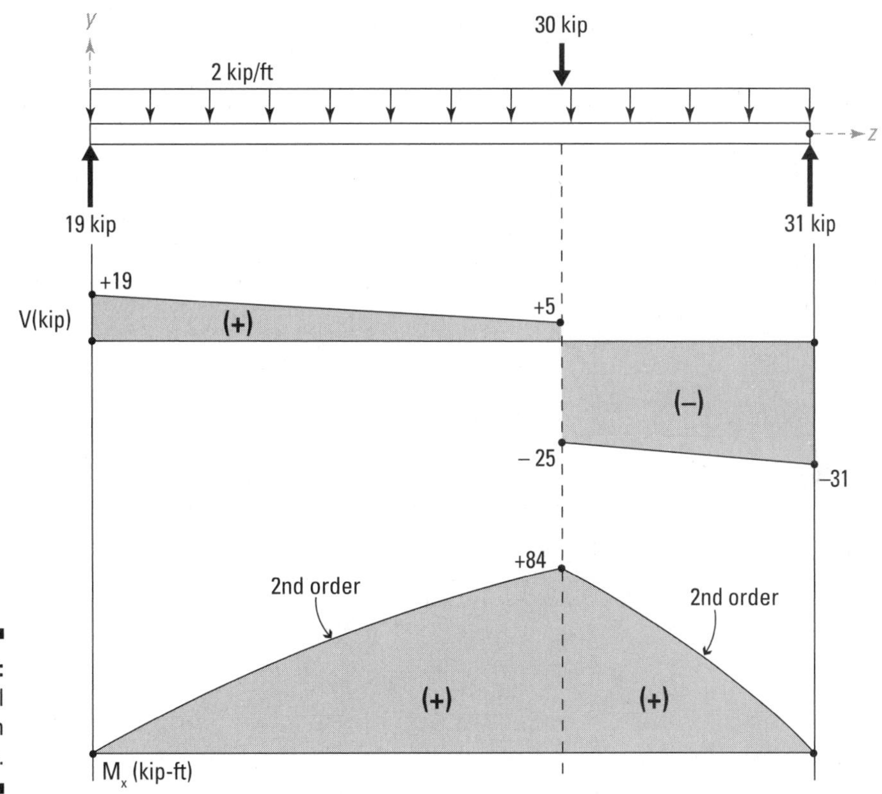

Figure 19-2: Flexural design example.

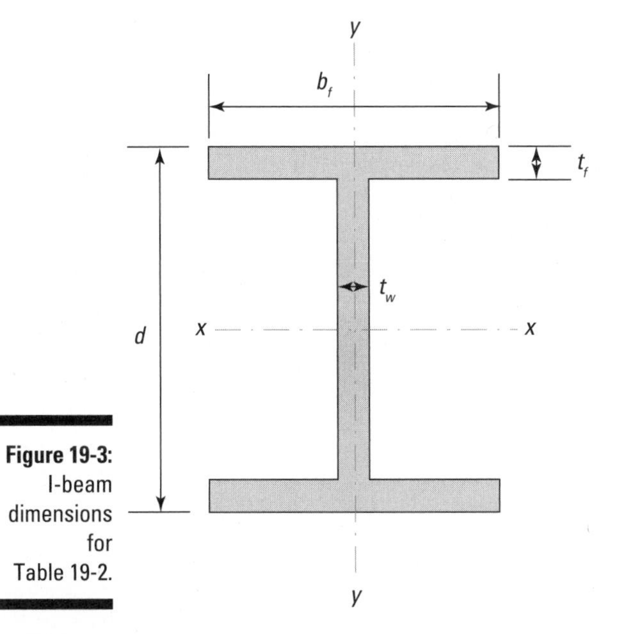

Figure 19-3: I-beam dimensions for Table 19-2.

Column 1 is the designation of the section you're working with. Most manufacturers fabricate their structural members to conform to a specified set of dimensions within prescribed tolerances. In this table, the *W* prefix indicates that the member conforms to a wide-flange design specification (which happens to include most I-shaped sections). The remaining columns 2 through 9 are computed section properties for a given shape. Talk about a timesaver! If you use a standard shape, you don't even have to do those repetitious section property calculations I show in Part I.

Turn to the back of your mechanics of materials textbook, and you may actually find a short table in an appendix that contains similar information for different cross sections.

With these data values, you can perform a wide variety of stress calculations relatively quickly. Most design tables typically contain additional values, including elastic section properties about the *y*-axis (which I don't list in Table 19-2), as well as additional shapes such as hollow tubes, round pipe, and channels (to name a few).

Looking at Table 19-2, you can see that each of those three shapes provides a sufficiently large elastic modulus and would be a suitable candidate to support the moments from the applied loads. However, designing flexural members involves many more considerations; you also need to design for flexural shear, connection details, and many other situations that I don't necessarily cover in this book.

Selecting a most-efficient section

So how do you know which section to choose from a list of suitable options? The short answer is usually "Choose the most-efficient section," which basically means you should select the shape that costs the least. Considerations for choosing a most-efficient section include the following:

- **Most economical (smallest cross-sectional area):** In most materials, the member that weighs the least is often the most economical because weight is often directly proportional to the cost. And the member that weighs the least usually has the smallest cross-sectional area (which you can see in column 2 of Table 19-2).
- **Physical dimension requirements:** Physical space restrictions can play a role as well. If the beam you choose is 20 inches deep and your structure only has room for a 12-inch beam, you have a problem. For these situations, you need to keep an eye on the overall depth (column 3 of the table) of the member you select.

- **Serviceability:** In beam design, you also want to keep an eye on the minimum moment of inertia (I_{xx}). If you choose a beam with a very small moment of inertia, the deflections of your beam will be larger for a given loading. Using the deflection calculations in Chapter 16 and the appropriate design codes from "Structural Adequacy: Adhering to Formal Guidelines and Design Codes" earlier in this chapter, you may find that the moment of inertia actually controls, which means you then need to look to the value in column 7.

Accounting for flexural shear

Typically, design of flexural members (such as beams) is controlled by bending moments, so this category is a good place to start. After a design is complete, you can then check that the section remains adequate for shear effects as well. The shear capacity is usually performed as a follow-up check after you've designed the section. Depending on the code you use, your design requires different formulas and safety factors. In some codes, you're allowed to use an average shear stress calculation (which I explain in Chapter 6):

$$\tau_{AVG} = \frac{V}{A}$$

where V is the internal shear force and A is the area on which the shear is acting. Rearranging the average shear stress equation,

$$A_{REQ} = \frac{V_{MAX}}{\tau_{ALL}} = \frac{(31\text{ kip})}{(14.4\text{ ksi})} = 2.15\text{ in}^2$$

Looking at column 2 of Table 19-2, you can see that each of the three chosen sections should provide sufficient shear capacity because their cross-sectional areas are significantly larger than the minimum required area you just calculated. If you find that your shear capacity is insufficient, you can always go back and increase the member size to satisfy shear requirements; doing so typically results in a design that's still adequate for the flexural effects as well.

I should point out that many design codes often don't allow you to count the entire cross-sectional area for shear capacity when the cross section has flanges or legs. In these cases, you may only be allowed to use the area of the web, $A_{WEB} = (t_w)(h)$, where t_w is the web thickness shown in column 6 at the neutral axis and h is the height of the beam or web — usually the value in column 3. (The *web* is a part of a cross section that connects other structural elements such as the vertical region of the I-beam cross section.) For Figure 19-3, the area of the web, A_{WEB}, of a W10x45 is still (10.1 in)(0.31 in) = 3.13 in^2, which is still more than the A_{REQ} that you calculated earlier in this section.

After you've selected a preliminary member size, you need to make sure to go back and add an additional load to account for the member's self weight. In many common structures, a beam can require more than 100 pounds per foot (and actually more for really large beams) just for self weight.

Designing for Torsion and Power

One of the most important design problems related to torsion is the transmission of power from a motor or engine through a rotating shaft. Spinning shafts apply torque based on the speed of their rotation and the power being transmitted. The more slowly a shaft spins, the higher the torque is for a given power output.

To ensure that the shaft has sufficient dimensions, you typically work with the *J/c* ratio, which is a variation of the elastic section modulus for flexure that I describe in the earlier section "Designing for bending moments with the elastic section modulus:"

$$\frac{J}{c} \geq \frac{\text{power}}{2\pi(\text{rpm})\tau_{ALL}}$$

where *J* is the torsion constant (see Chapter 11); *c* is the outer radius of the shaft; *power* is the power (in watts) the shaft is transmitting; *rpm* is the speed of the rotating shaft in revolutions per minute; and τ_{ALL} is the maximum allowable shear stress in the shaft (as I define in the earlier section "Creating a design criteria").

For example, consider a rotating shaft used to transmit 700 kW at a speed of 90 rpm. If the shaft is steel with a tensile yield strength of σ_Y = 250 MPa and a factor of safety of *F.S.* = 1.5, you can then establish the dimensional requirements *(J/c)* for the shaft. First, you need to determine the allowable shear stress for the shaft material:

$$\tau_{ALL} = \frac{\tau_{MAX}}{(F.S.)} = \frac{0.5(\sigma_Y)}{(F.S.)} = \frac{0.5(250 \times 10^6 \text{ Pa})}{1.5} = 83.33 \times 10^6 \text{ Pa}$$

Then you can determine the J/c ratio:

$$\frac{J}{c} \geq \frac{(800{,}000 \text{ watt})}{2\pi\left(90 \ \frac{\text{rev}}{\text{min}}\right)\left(\frac{1 \text{ min}}{60 \text{ sec}}\right)\left(83.33\times 10^{6} \text{ Pa}\right)} = 0.00102 \text{ m}^3$$

After you have this ratio, you can then compute viable alternatives for the shaft's dimensions. For a solid shaft,

$$\frac{J}{c} = \frac{\pi}{2}c^3 = 0.00102 \text{ m}^3 \Rightarrow c = 0.0866 \text{ m} = 86.6 \text{ mm}$$

A radius of 86.6 millimeters corresponds to a shaft diameter of 173.2 millimeters.

Similarly, if you want to use a hollow shaft, you can use the expression

$$\frac{J}{c} = \frac{\pi\left(r_o^4 - r_i^4\right)}{2r_o} = 0.00102 \text{ m}^3$$

From this calculation, you can easily create a spreadsheet table to determine satisfactory inner (r_i) and outer (r_o) radii to transmit this torque. For static shafts subjected to applied torque, you simply need to follow the examples in Chapter 11.

Interacting with Interaction Equations

For simple designs, where objects are only subjected to a single load type, the design process is usually fairly simple. The basic relationship that you must satisfy is

$$\text{applied loads} \leq \text{member capacity}$$

where the applied loads are the design loads for the object, and the member capacity is the strength of the object in resistance to the load. If you rearrange this basic expression, you can create an inequality that looks something like the following:

$$\frac{\text{applied loads}}{\text{member capacity}} \leq 1.0$$

If you think of the 1.0 as being a 100 percent of the strength of the member, or the size of your bucket, as I describe in the earlier section "Developing a Design Procedure," you can extend this bucket analogy to problems involving multiple load types. This relationship is known as an *interaction equation,* and design codes may require you to check many different types of interaction. One of the most popular interaction equations is the one for the interaction of axial load and bending moment (in one direction):

$$\left(\frac{P}{P_{MAX}}\right)+\left(\frac{M}{M_{MAX}}\right)\leq 1.0$$

If you use up the capacity of a member on the axial load, you may not have enough capacity to resist the moments.

For example, consider a cross section that has an axial capacity of $P_{MAX} = 100$ kN and a moment capacity of $M_{MAX} = 400$ kN-m. If you subject the member to an axial load of $P = 30$ kN, you can find the maximum applied moment by solving the interaction equation for the applied load M:

$$M\leq\left(1.0-\left(\frac{P}{P_{MAX}}\right)\right)\left(M_{MAX}\right)=\left(1-\left(\frac{30\text{ kN}}{100\text{ kN}}\right)\right)\cdot\left(400\text{ kN}\cdot\text{m}\right)=280\text{ kN}\cdot\text{m}$$

For this example, the maximum moment that can be applied to the section can only be 280 kN-m.

Chapter 20

Introducing Energy Methods

In This Chapter

- Explaining the law of conservation of energy in mechanics of materials
- Working with strain energy
- Developing relationships for impact

The equilibrium methods that I describe earlier in this book take advantage of the equations of statics and the relationship between stress and strain of the object. These methods work very well as long as the basic assumptions in Chapter 14 remain valid.

However, many objects in engineering may not obey all those basic assumptions. For these cases, you need to use a more-generalized approach such as the *energy methods* I show in this chapter, which are all about calculating energy and its effect on stress-strain behavior of elastic objects.

In this chapter, I demonstrate the basics of computing internal strain energy and provide a bit of a physics refresher to help you with computing external work calculations. I conclude the chapter by showing how you can use these internal energy calculations to compute internal forces and stresses caused by impact of one object onto another.

The study of energy methods I present in this chapter is a basic overview of energy methods focusing on linear systems; you can apply these ideas to simple applications such as calculating stresses and strains due to the impact of one object onto another. Analysis techniques involving energy methods can become quite involved in extreme cases, so for more advanced applications, check out an advanced mechanics or physics textbook.

Obeying the Law of Conservation of Energy

In physics, the classic definition of *energy* is the ability of an object to perform work. Energy comes from a wide variety of sources, such as motion, chemical reactions, heat, light, and so on. The units of energy are the Joule (J), which is equivalent to 1 N-m, in SI units and the pound-foot (lb-ft) in U.S. customary units.

Now, before you go out and strap a solar panel on your house, I'm not talking about the energy that powers the appliances in your home. Engineers of structural objects are typically concerned with two specific types of energy:

- **Kinetic energy:** *Kinetic energy* is the energy of the mass of a system due to its motion, and it remains constant for a constant speed. Only a change in velocity can cause a change in kinetic energy. In general, the expression used to calculate kinetic energy (U_K) is

  $$U_K = \frac{1}{2}mv^2$$

  where m is the mass of a system and v is the velocity of the mass.
- **Potential energy:** *Potential energy* is a measure of the energy stored in an object. For example, if you hold a weight in the air, the potential energy is the gravitational force acting on the mass multiplied by the height through which the object is capable of traveling before reaching equilibrium. In equation form, you calculate the potential energy for such a weight as

  $$U_p = mgh$$

  where m is the mass of the suspended object; g is the acceleration due to gravity acting on the object; and h is the height of suspension.

The *law of conservation of energy* is a significant principle in the application of physics. This law states that over time, the total energy of a system remains unchanged (or is conserved). The law of conservation of energy is the inspiration for the saying "Energy can neither be created nor destroyed." Energy can change forms only.

In physics, the law of conservation of energy says that if you release a suspended object above the ground and allow gravity to grab hold of it, its potential energy transforms into kinetic energy, at which point the object increases in velocity as it falls toward the ground.

When you start working with energy methods, you must be careful to keep your different energy forms separated into two main categories:

- **Internal energy:** *Internal energy* is the energy of an object that develops through stresses and strains in reaction to outside sources (or external energy sources). Mechanics of materials calls this internal energy *strain energy,* and I discuss that topic in more detail in the following section.
- **External energy:** *External energy* is the energy put into an object either by impacting the object directly through motion (which is classified as a kinetic form of energy) or through external potential energy or thermal energy sources.

Energy methods in mechanics of materials use the law of conservation of energy by obeying the principle that any energy put into an object or system from an external energy source is transformed into the internal energy stored inside the object in any of several different forms (such as deformation, changes in velocity, and heat emission).

You can express the law of conservation of energy in mechanics of materials by the following:

$$U_{EXTERNAL} = U_{INTERNAL}$$

where $U_{EXTERNAL}$ is the system's external energy and $U_{INTERNAL}$ measures the system's internal energy. For example, if you bend a paper clip back and forth quickly and then touch the bend, you may notice that the clip is warm (or even hot) to the touch. The external energy you put into the clip from the bending process was transformed into internal energy from the deformation of the clip (the bending displacement) and energy from internal friction (the heat).

Both the energy from bending and the energy from heat contribute to the total internal energy of the system. However, in most basic mechanics of materials applications, the portion of the internal energy contributed from the deformations is often significantly larger than the energy from internal friction. For this reason, neglecting the internal energy contributed from internal friction is fairly common practice.

Working with Internal and External Energy

The law of conservation of energy tells you that the external applied energy must be equal to internal strain energy. The next challenge in your application of energy methods is knowing how to compute these basic energy values. In this section, I show you several different forms of internal strain energy calculations and then explain how you relate them to different types of external work energy calculations.

Finding the internal strain energy

Strain energy is a measure of the internal potential energy of an object. If you squeeze a kitchen sponge in the palm of your hand, the energy you use to compress the sponge is the external work applied to the sponge. This energy is stored inside the sponge in the form of *internal strain energy* (energy due to stresses and strains). As long as the sponge has remained within the elastic region of its stress-strain relationship (see Chapter 14), this stored internal energy source is completely released when you open your hand because the stored energy is transformed into the energy that causes the object to return to its original shape. In equation form, the external work W you applied to the sponge is stored as an internal energy U within the object as follows:

$$W = U = AL\int_0^{\varepsilon_1} \sigma \cdot d\varepsilon = (\text{volume}) \cdot \int_0^{\sigma_1} \frac{\sigma}{E} \cdot d\sigma$$

where volume is simply the geometric volume of the object; σ is the internal stress; ε is the corresponding strain; and E is the value of Young's modulus of elasticity for the material.

I should also point out that energy is always a scalar quantity and doesn't have a direction associated with it. So if you have an internal strain energy in conjunction with a chemical reaction energy, you can simply add these two effects together to compute a total internal energy. But I'll leave the chemical energy to the chemists and focus simply on mechanical strain energy here.

If you want to work with a system of multiple objects, you need only compute the strain energy of each individual part and then add them together to get the total energy of the system:

$$U_{INTERNAL,TOTAL} = \sum_{i=1}^{n} (U_{INTERNAL})_i$$

where n is the number of members in the system. Depending on the type of loading, stresses and strains can be quite different (see Part II). So your analysis of objects by using strain energy methods must begin with identifying the type of load being applied: axial, torsion, and flexure. You can then use this information to choose the appropriate strain energy calculations that I describe in the coming sections.

Strain energy under axial loading

You can calculate the internal strain energy of an axially loaded bar from the following equation:

$$U = AL\left(\frac{\sigma^2}{2E}\right) = \frac{P^2 L}{2AE} = \frac{AE\Delta^2}{2L}$$

Consider the system of two axial bars AB and AC that have the length and cross section dimensions shown in Figure 20-1.

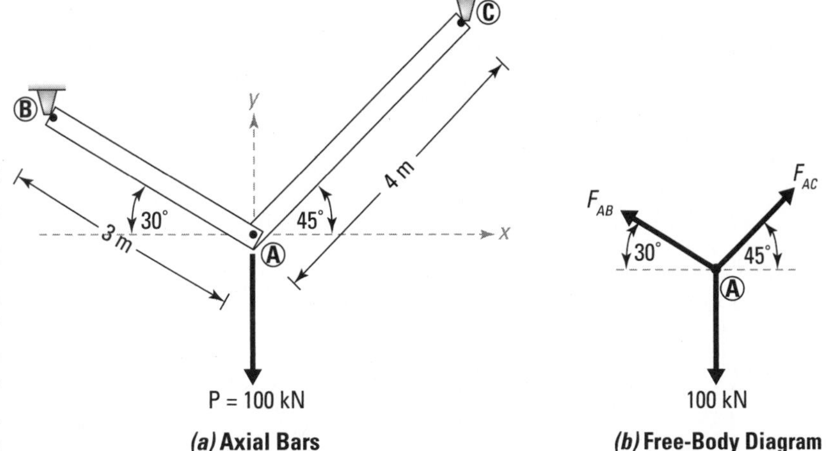

Figure 20-1: Computing internal strain energy for a system of axially loaded bars.

If a load of P = 100 kN is applied to the end of this system, you can use your basic statics skills to deduce that the internal force F_{AB} in bar AB is 75.8 kN (T) and that the internal force F_{AC} in bar AC is 92.8 kN (T). With the statics out of the way, you can then compute the internal energy of the system:

$$U=\frac{(F_{AB})^2 L_{AB}}{2A_{AB}E_{AB}}+\frac{(F_{AC})^2 L_{AC}}{2A_{AC}E_{AC}}=\frac{(+75,800\text{ N})^2(3\text{ m})}{2(0.01\text{ m}^2)(70\times10^9\text{ Pa})}+\frac{(+92,800\text{ N})^2(4\text{ m})}{2(0.05\text{ m}^2)(70\times10^9\text{ Pa})}$$

$$\Rightarrow U=(12.3\text{ N}\cdot\text{m})+(4.9\text{ N}\cdot\text{m})=17.2\text{ N}\cdot\text{m}$$

Strain energy in beams

The internal strain energy of a beam is a bit more complex in that it requires using a bit of calculus to evaluate the internal energy:

$$U=U_B+U_V=\int_0^L\frac{M^2}{2EI}dz+\int_0^L\frac{\alpha V^2}{2AG}dz$$

where U_B is the internal strain energy due to bending and U_V is the internal strain energy due to shear. In these integrals, M and V are the generalized internal moment and internal shear equations, respectively, for the beam loading; E is the Young's modulus of elasticity; G is the shear modulus of elasticity; I is the moment of inertia of the beam; and A is the cross-sectional area of the beam. The first term in the summation represents the internal strain energy due to flexural bending, and the second term represents the internal strain energy due to flexural shear. In the flexural shear portion, you can see a numerical constant α that you can evaluate with yet another integral:

$$\alpha = \frac{A}{I^2} \int \frac{Q^2}{b^2} dA$$

where Q is the first moment of area of the cross section about the neutral axis and b is the width of the cross section.

In engineering, energy due to flexural shear, or the U_V term in the earlier equation, is often neglected. Calculating this term is actually cumbersome, and the result only ends up accounting for somewhere between 1 and 3 percent of the total energy of a member in flexure. However, this omission only works if the beam dimensions are such that flexural stresses control. If a beam is very deep and very short, you need to include the shear deformations (and hence the flexural shear energy contribution) in your calculations.

Consider the steel beam (E = 29,000 ksi) shown in Figure 20-2 that has a span of 10 ft and a moment of inertia of I = 100 in^4. The beam is subjected to a uniformly distributed load of w = 3 kip/ft.

You can write the generalized equation for the moment as follows:

$$M = 15z - \frac{3}{2}z^2$$

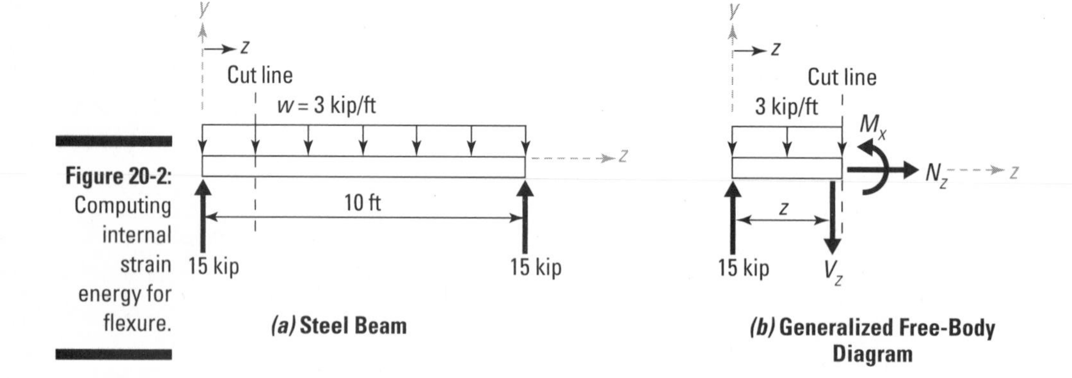

Figure 20-2: Computing internal strain energy for flexure.

Considering the energy from flexural shear as negligible, you can compute the internal strain energy of this beam due to the bending moments as follows:

$$U \approx U_B = \int_0^L \frac{M^2}{2EI} dz = \int_0^{10} \frac{\left(15z - \frac{3}{2}z^2\right)^2}{2EI} dz = \frac{30,000}{2(29,000 \text{ ksi})(100 \text{ in}^4)} \left(\frac{144 \text{ in}^2}{1 \text{ ft}^2}\right) = 0.745 \text{ lb} \cdot \text{ft}$$

Strain energy in circular shafts subjected to torsion

You can compute the internal strain energy for a circular shaft subjected to torsion by using the following equation:

$$U = \frac{T^2 L}{2GJ} = \frac{GJ\phi^2}{2L}$$

where T is the internal torque; G is the shear modulus of elasticity; J is the torsion constant for the shaft; L is the length over which the torque is constant; and ϕ represents the angle of twist over the length. For more on torsion calculations, flip to Chapter 11.

Setting the internal strain energy equal to the external work energy

As I discuss earlier in the chapter, the premise behind working with energy methods in mechanics of materials is that the internal strain energy (see the preceding section) must be equal to the external work energy.

Depending on the type of applied loading, you can compute the external work energy by using the following basic relationships:

- **Axial external work:** For an axial load, the external work done by a point load P that moves through a displacement Δ is given by

  $$W = \frac{1}{2}P\Delta$$

- **Bending external work:** For a bending effect, the external work done by a concentrated moment M that moves through a given rotation θ is given by

  $$W = \frac{1}{2}M\theta$$

  For a bending affect resulting from a concentrated load P that moves through a given displacement Δ is given by

  $$W = \frac{1}{2}P\Delta$$

This relationship is the same relationship in the equation work energy of an axial bar. This correlation illustrates the analogy that structural systems can be considered to behave as types of springs, because the work energy calculation becomes directly affected by the stiffness of the member under load.

- **Torsional external work:** For a torsional effect, the external work done by a concentrated torque T that moves through an angle of twist ϕ is given by

  $$W = \frac{1}{2}T\phi$$

The units on both the applied moment (M) and the applied torque (T) are already expressed in either N-m (for SI) or lb-ft (for U.S. customary), which happen to match the units of energy already. If the angle of rotation (θ) and the angle of twist (ϕ) are both expressed in radians (or a unitless measure), the units on the external energy also work out.

After you have the internal strain energy computed (see the preceding section), you can compute the deflection by setting the internal strain energy equal to the external work. Consider the axial system from Figure 20-1 earlier in the chapter. If the applied load P = 100 kN moves through a vertical deflection Δ_V, you determine the external work as follows:

$$U_{EXTERNAL} = \frac{1}{2}P\Delta_y = (0.5)(100\text{ kN})\Delta_y = (50\text{ kN})\Delta_y$$

Setting the external work equal to the internal work ($U_{INTERNAL}$ = 16.9 N-m), you can rearrange the expression and solve directly for the vertical displacement in the direction of the load.

$$(50{,}000\text{ N})\Delta_y = 17.2\text{ kN}\cdot\text{m} \Rightarrow \Delta_y = 3.44\times10^{-4}\text{ m} = 0.344\text{ mm}$$

Remember to be mindful of the units in your calculations. For example, if an applied load is expressed in kN but your energy calculations are in J or N-m, you have to make sure the units on the external energy calculations agree with the units on the internal energy calculations.

A very popular method of structural analysis known as Virtual Work uses the equations in this section. In fact, the method of Virtual Work uses an imaginary *unit load* (which is a *unit force* [1 N or 1 lb] for linear displacements and a *unit moment* [1 N-m or 1 lb-ft] for rotation calculations) in the direction of interest to help predict the displacements and deformation in that direction.

Brace Yourself: Figuring Stresses and Displacements from Impact

The energy methods of this chapter truly shine in the determination of stresses and displacements from impact. *Impact* occurs when one object strikes another with a dynamic force. As you may imagine, the impact causes external energy to be transmitted to the object, which then deforms (or experiences stresses and strains) in response. The calculations of internal strain energy remain the same as in the earlier "Finding the internal strain energy" section; the challenge here is to be able to determine the external work done by the moving load. But first, you must make a few assumptions:

- Displacements are directly proportional to the load applied.
- Material behaves elastically (see Chapter 14), meaning it doesn't deform permanently due to the impact loads.
- No energy is lost because of localized deformation (or damage) at the point of impact.
- All impact-related energy is transferred completely to the impacted object.

Determining impact from kinetic energy

When a force from a mass is applied *statically* (without acceleration or velocity) to an object, the resulting energy is less than the energy that results when the same force hits (or impacts on) the same object with a constant velocity. This energy increase is due to the contribution of kinetic energy (which I discuss in the earlier section "Obeying the Law of Conservation of Energy"). The relationship between a static load and deflection is given by the basic relationship

load = stiffness × deformation

All structural objects can be classified as some type of spring, meaning that when a load is applied, the amount it deforms is a function of the geometric dimensions of the object and the material from which the object is made.

To illustrate the effect of impact, consider the simple spring with stiffness k as shown in Figure 20-3. A small mass m moving at a constant velocity v impacts the end of the spring.

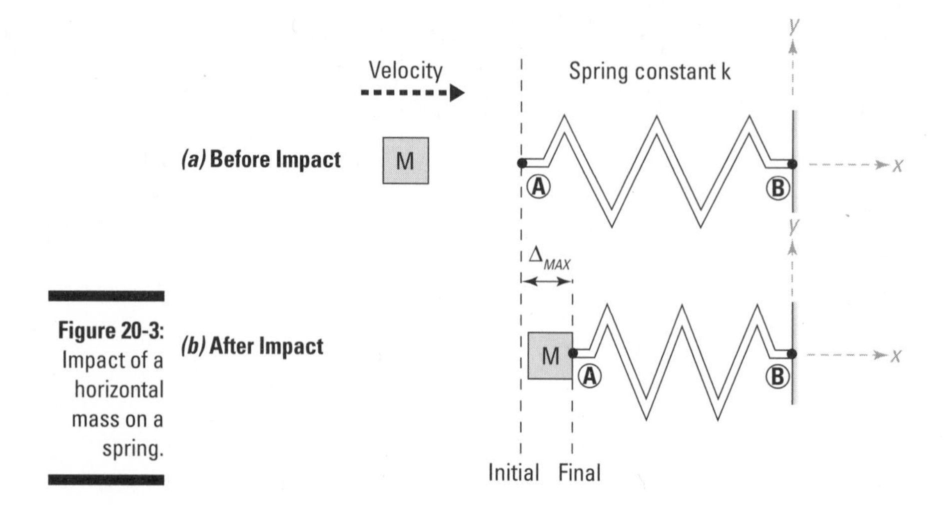

Figure 20-3: Impact of a horizontal mass on a spring.

If you use the analogy that all members can be considered a type of spring, the stiffness k of an axial member is equal to the numeric constant AE/L of the member and the stiffness of a bending member is a function of some constant value and an EI/L^3 term.

In this example, the internal strain energy from the spring is the same as the axial member; you compute it as follows:

$$U_{INTERNAL} = \frac{(P_{MAX})^2 L}{2AE}$$

The external work is actually equal to the kinetic energy of the mass:

$$U_{EXTERNAL} = \frac{1}{2} mv^2$$

You can then set these two expressions equal and solve for the internal force P_{MAX} that results from impact:

$$P_{MAX} = \sqrt{\frac{mv^2 AE}{L}}$$

After you have this force computed, you can then calculate the stress that results as I show in Chapters 6 and 8.

Determining energy relationships through vertical impact factors

You calculate vertical impact much the same as you do horizontal impact (which I cover in the preceding section). If an object starts at rest and is dropped from a height h, the external work is equal to the potential energy of the mass m as shown in Figure 20-4.

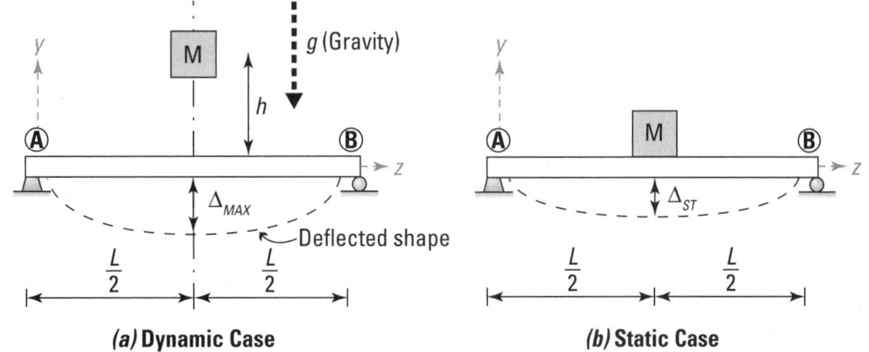

Figure 20-4: Vertical impact factors.

For the dynamic cases shown in Figure 20-4a, you can express the external energy done as

$$U_{EXTERNAL} = mg(h + \Delta_{MAX}) = 0$$

where Δ_{MAX} is the displacement due to impact.

Setting the external energy equal to the total bending external work $W_{EXTERNAL}$ for the beam (as I describe in the section "Setting the internal strain energy equal to the external work energy") gives the expression

$$W_{EXTERNAL} = \frac{1}{2}P_{MAX}\Delta_{MAX} = \frac{1}{2}k(\Delta_{MAX})^2$$

where k is the stiffness of the member at the location of impact.

Setting the external energy equal to the external work , you get the following energy expression:

$$U_{EXTERNAL} = W_{EXTERNAL} \Rightarrow mg(h + \Delta_{MAX}) = \frac{1}{2}k(\Delta_{MAX})^2$$

From Figure 20-4b, you express the static displacement Δ_{ST} as follows if the load (mg) is acting at the point of impact:

$$\Delta_{ST} = \frac{mg}{k} \Rightarrow k = \frac{mg}{\Delta_{ST}}$$

Substituting this relationship into the conservation of energy equation in the earlier section "Obeying the Law of Conservation of Energy," you can apply a bit of algebra and then the quadratic formula to show that

$$\Delta_{MAX} = \Delta_{ST} + \sqrt{(\Delta_{ST})^2 + 2(\Delta_{ST})h} = \left(1 + \sqrt{1 + \frac{2h}{\Delta_{ST}}}\right)\Delta_{ST}$$

$$\Rightarrow \Delta_{MAX} = (I.F.)\Delta_{ST}$$

where *I.F.* is an *impact factor* for the vertical load being dropped on the beam:

$$I.F. = \frac{\Delta_{MAX}}{\Delta_{ST}} = 1 + \sqrt{1 + \frac{2h}{\Delta_{ST}}}$$

For a single mass applied at the midspan of a beam, you can use the principles of Chapter 16 to show that

$$\Delta_{ST} = \frac{mgL^3}{48EI}$$

which means that you can express the impact factor for a mass m dropped from a height h onto the middle of a beam of span L as

$$I.F. = \frac{\Delta_{MAX}}{\Delta_{ST}} = 1 + \sqrt{1 + \frac{2h(48EI)}{mgL}}$$

The impact factor is directly related to the beam's stiffness parameters (E, I, and L), the mass, and the height you're dropping the mass from.

All you need to do is compute the static displacement of an equivalent weight, and you can compute the impact factor. Because the impact factor is a ratio of the impact displacement over the static displacement, you can apply Hooke's law (see Chapter 14) to show that the stresses can be affected by the same impact factor:

$$\sigma_{MAX} = (I.F.)\sigma_{ST}$$

Depending on the height and mass of the weight being dropped, impact can significantly increase the stresses in an object.

Part V
The Part of Tens

In this part . . .

In this part, I provide a basic list of ideas and concepts to watch out for when applying mechanics of materials to the world around you. I also provide a list of ten useful tips for solving mechanics of materials problems.

Chapter 21

Ten Mechanics of Materials Pitfalls to Avoid

In This Chapter

- Using the correct aspects of mechanics of materials
- Noting important steps and formulas

Mastering mechanics of materials is no easy feat; with all the formulas, constants, and variables, you may feel like you'll buckle before your structure does. To help you sidestep potential troubles, I've gathered the ten most common mistakes I see mechanics of materials students make. Although this chapter doesn't cover all aspects of the topic, it does provide a good checklist of things to keep in mind when working with mechanics of materials.

Failing to Watch Your Units

One of the biggest unit mix-ups comes in dealing with the megapascal (MPa), which equals 1 million N/m^2 and is a common SI unit for material properties such as Young's modulus of elasticity and the shear modulus of elasticity. The problem occurs because force is often given in SI units of Newton (N) or kiloNewton (kN), and you may also encounter cross-sectional dimensions measured in millimeters or area values expressed in mm^2. In order to match these units, you either need to convert the meters portion of the MPa to millimeters or convert the area back to m^2.

Not Determining Internal Forces First

This one should seem fairly obvious; after all stresses and strains are measures of internal behaviors within an object, so you have to start there. If you're interested in computing the normal stresses on an element, you need to consider both internal axial forces as well as internal bending moments. If you want to work with shear stresses, you need to determine the internal shear from flexure and any torsional effects on the object. Internal force diagrams (which I cover in Chapter 3) prove very handy when working with stresses and strains.

Choosing the Wrong Section Property

Sometimes the hardest part about working with stresses is remembering which section property you need to use. When you're working with normal stresses (such as from internal axial loads or bearing and contact pressures), you need to use the cross-sectional area of the member or the contact region. When you're calculating normal stresses from bending, you need a moment of inertia (or an elastic section modulus that uses that value). For shear stresses from torsion, you need a polar moment of inertia; for shear stress from flexural shear, you need both the first and second moments of area.

Forgetting to Check for Symmetry in Bending Members

To compute bending stresses in a symmetric cross section, you need only compute the principal moments of inertia of the cross section; these moments conveniently happen to be the same as the moments of inertia about the centroidal axes. However, some professors like to be a little sneaky by putting a nonsymmetric problem on an exam. When this situation happens, remember that you also need to compute a product moment of inertia.

The basic equations for computing normal stresses due to bending require that the cross section have at least one axis of symmetry, which physically means that the product moment of inertia is equal to zero. In the event that a section is unsymmetrical, the product moment is never zero, which means you need to then compute the principal moments of inertia and their orientation angles. These figures then help you to determine the orientation of the neutral axes of the cross section.

Carelessly Combining Stresses and Strains

Remember that you can combine multiple stress effects by using the principle of superposition from Chapter 15. Normal stresses can combine with other normal stresses, and shear stresses can combine with other shear stresses. Watch the signs on the stresses: Tension normal stresses are usually assumed as positive, and compression normal stresses are usually assumed as negative. Shear stresses involving torsion have one sign on one edge of the cross section and the opposite sign on the other. Similarly, you can combine normal strains together and shear strains together.

Ignoring Generalized Hooke's Law in Three Dimensions

Make this formula your friend. Instead of memorizing multiple formulas for the relationship between stress and strain, remember that the generalized Hooke's law always works, even for uniaxial and plane stress situations. You just need to identify which of the normal stresses have a zero value; if no stress is acting in a given direction, the stress has to be zero. You can even analyze plane stress elements because the out-of-plane stress component is automatically zero. For more on Hooke's law, turn to Chapter 14.

Classifying Columns Incorrectly

When considering compression members (columns), don't forget that the length of the member plays a key role in computing the capacity of the member. If the column is classified as a short column, it follows the same normal stress calculations you use for computing normal stresses from axial loads. However, if the slenderness ratio becomes too big, buckling becomes a consideration, and you need to look to the Euler buckling equation, which includes the slenderness ratio. For more on columns, turn to Chapter 18.

Overlooking that Principal Normal Stresses Have No Shear

A principal stress element can have up to two normal stresses, but shear stress never acts on the principal element at that orientation. Conversely, a principal shear stress element not only has the shear stress, but it also has the same value of normal stress acting on all sides of the element.

Neglecting to Test the Principal Angle after You Calculate It

If you're using equations (instead of Mohr's circle) to compute transformed stresses and strains, the principal angle that the equation gives you from the basic formula is only one of two possible answers. Although you may be able to compute both of the principal stresses (or strains), you don't know which of these values is associated with the principal angle that you calculate from the formulas. To determine this piece of information, simply substitute the principal angle and the current state of stress (or strain) back into the transformation equations, and the transformed value that is associated with the angle pops right out.

If you're using Mohr's circle, this step isn't necessary. You can tell which principal value the angle is associated with by simply examining the circle directly. Mohr's circle even tells you which way you need to rotate the element to get to a particular principal value. See Chapters 7 and 13 for more on Mohr's circles.

Falling Victim to Tricky Issues with Mohr's Circle

If you choose to utilize the graphical method of Mohr's circle rather than transformation equations, keep the following pointers in mind:

- **As you start plotting the stress and strain coordinates on Mohr's circle, don't forget that both the positive shear stress and positive shear strain values always plot below the horizontal axis.** Although doing so may upset your middle-school math teacher, this change guarantees that the direction you rotate on Mohr's circle agrees with the direction you rotate your transformed elements.
- **The angle you measure on the circle is always twice the angle you would use with the transformation equations.** The procedure for constructing Mohr's circle is the same for both stress and strain. So if you're rotating an element from one orientation to another, multiply the angle by two; if you're calculating an angle from geometry on Mohr's circle, divide that angle by two.
- **The shear strain must be plotted at half its value.** The Mohr's circle for strain (see Chapter 13) has one additional tricky spot: You divide the shear strain that plots on the vertical axis by two. So if you measure a shear strain of $+500\mu$, you actually need to change the vertical coordinate to $+250\mu$.

Chapter 22

Ten Tips to Solving Mechanics of Materials Problems

In This Chapter

- Applying statics to your mechanics problems
- Thinking in terms of failure
- Working with stresses and strains in design

So your professor walks in, throws a strange problem on the board, and informs you that today is your day to do analysis work. Armed with your knowledge of mechanics of materials (and possibly with this book), you set out on the mechanics of materials path laid before you.

The biggest problem you face is that no two problems (especially design problems) are ever the same. You have to determine how to quickly and effectively tackle a mechanics of materials problem. Fortunately for you, though, the actual equations of mechanics of materials never change; the method in which you need to apply them does. In this chapter, I present a general list of ten basic steps that you can use to tackle even the most ornery of mechanics of materials problems.

Do Your Statics

I'm sorry to say you just can't avoid statics. Without statics, mechanics of materials problems are pretty much impossible (unless of course you have all the internal forces already given to you, which in design is rarely the case). For the sake of argument in this chapter, assume you aren't that lucky.

The first step in tackling a mechanics of materials problem is to work through the statics involved in the problem, which means you need to draw free-body diagrams. These diagrams include exposed internal forces, external forces, support reaction forces, and self weight. You can then use the equations of equilibrium to possibly compute the magnitudes of unknown support reactions.

Expose Internal Forces

After you have your basic statics done and the support reactions computed (see the preceding section), you need to determine internal forces by cutting the object to expose the internal forces (such as axial and shear forces and internal moments) required to balance one of the cut pieces of your structure. If you later want to calculate stresses or strains at a particular point (see the later section "Compute Strains and Deformations for Your Stress Elements"), you need to expose the internal forces at that point first. When you've included the internal forces on your free-body diagram, you can then apply the equations of equilibrium again to compute the unknown internal forces (as I outline in Chapter 3).

If you repeat this process at multiple locations (or write the generalized equations instead), you can create internal force diagrams such as shear and moment diagrams, axial force diagrams, or torque diagrams. All these diagrams help you quickly locate the maximum internal forces within a member.

Identify How the Object Can Break

One of the hardest parts of mechanics of materials is determining exactly which stress or strain you need to calculate in the first place. One trick I like to use is to picture how an object can possibly break. To identify these situations, I usually start at one end of the assembly and mentally slice the object perpendicular to the longitudinal axis (if one exists), and then I confirm that I've accounted for that particular break scenario in my previous calculations. If I haven't, I know I need to take care of the stress calculations at that location as well. I then repeat this process throughout the remainder of the object or system.

If you see a hole or change in geometry, you know you need a slice at that location for sure because at those locations you can get localized increases in stress that exceed the average stresses in other sections.

Consider two plates bolted together by a single bolt. If you pull on the ends of each plate with a constant force, the assembly can break apart in a number of ways. For example, if the bolt breaks, the assembly will definitely fall apart. To check for this failure, you use an average shear stress calculation that involves the internal force of the bolt acting over its cross-sectional area. Another possibility is that the stress in the plate itself may become too much, causing the plate to fail. In that scenario, you need to check two normal stresses: the average normal stress of the gross cross-sectional area and the maximum normal stress acting adjacent to the hole on the net cross section. After you know these stresses, the formulas basically become a number-crunching process.

If the internal load is acting parallel to the cross section, you need a shear stress, so turn to Chapters 10 and 11. If the internal load is acting perpendicular to the cross section, you have a normal stress calculation in your future and may want to check out Chapters 8 and 9 for help with those. Just remember to watch out for members in compression, because buckling (see Chapter 18) is always a concern.

Compute Appropriate Section Properties

When you know the stress calculation you need to perform, the section property you need is generally fairly apparent from the corresponding formula. Most section properties require the location of the centroid (or neutral axis) as a reference, so you may as well do that first. After you have the centroid located and all the internal forces calculated at your location of interest, keep the following considerations in mind.

- **Axial loads:** Beware of the presence of holes or openings because you need to compute a net area at those locations. If your axial load is a compression load, you also need to compute the radius of gyration (which you use to compute the slenderness ratio) in order to consider buckling as a possibility.
- **Bending moments:** Bending moments always need a moment of inertia (which I discuss in Chapter 5). Make sure you correctly determine which of the cross section's neutral axes is being displaced and then calculate the moment of inertia about that axis. If both axes are moving (as in the cases of biaxial bending problems), you need to compute both moments of inertia, which for symmetric sections are the principal moments of inertia.

  For unsymmetrical sections, you also need the product moment of inertia so that you can calculate the principal moments of inertia.
- **Flexural shear:** For flexural shear, you need to compute the first and second moments of area (see Chapter 5) for your location of interest.
- **Torsion:** Torsion problems always require you to compute the torsion constant. For circular shapes, this value is the same as the polar moment of inertia (see Chapter 5). For other shapes (including rectangles), you need to use the approach I illustrate in Chapter 11.

Sketch Combined Stress Elements

As long as your loading meets the material assumptions I outline in Chapter 14, you can combine stresses from simple cases to create a combined load case. So don't be scared of those pressure vessels with torsion and bending loads on them.

Just remember that you can only add normal stresses to other normal stresses, and you can only add shear stresses to other shear stresses. After you have these values determined, you can sketch the combined stress element, which I discuss in Chapter 15.

Transform Those Stresses!

Maximum stresses don't always line up with the stresses you calculate from the loading. In fact, you can compute these maximum stresses (the principal normal stresses and maximum shear stresses) by using either Mohr's circle for stress or the stress transformation equations. While you're at it, you should also determine which orientation these stresses occur at within the object. This step becomes especially important if you're investigating welds in metal or fiber stresses in wood construction because the orientation of those stresses plays a critical role in the design process.

The majority of failure theories (which you find out about in an advanced mechanics class) need the basic principal stress values in order to be able to perform the theories' advanced calculations. I show you how to calculate those principal stress values in Chapter 7.

Have Your Material Properties Handy

Make sure that you have the proper material properties for the object you're investigating. These material properties can have a tremendous effect on design calculations. At the very least, you usually need Young's modulus of elasticity and Poisson's ratio for each material within the object (see Chapter 14). You should also note the yield stress and ultimate stress if you're doing any sort of design work. Although Young's modulus of elasticity and Poisson's ratio are usually fairly constant across different types of the same material, the yield stress and ultimate stress can occur at uniquely different values. For example, steel comes in a wide variety of compositions, such as A36 (which has a yield stress of approximately 36 ksi) or A992 (which has a yield stress beginning at 50 ksi).

Apply Factors of Safety and Local Code Requirements

If you can compute stresses from the applied loads and you know the stress your object's material can withstand, you're on your way to making an informed design decision. But remember, you can't just jump straight into design with only these stresses; you need to incorporate a factor of safety to turn your actual stresses into allowable stresses. To compute allowable stresses, you need to know only the correct factor of safety (which is often specified in your local design codes). A factor of safety less than 1.0 is usually an unsafe design. Depending on your design and local codes, allowable factors of safety are often closer to 2.0 and are even higher in special circumstances.

Compute Strains and Deformations for Your Stress Elements

After you have your stresses computed, you're ready to put Hooke's law (see Chapter 14) to work to help you estimate the strains your object may be experiencing. After all, a structure that experiences too much deformation may not be able to perform its intended function, even if it's capable of supporting the intended loads. In systems that are statically indeterminate, the deformations are what actually let you compute the support reactions, as I describe in Chapter 16.

Design for Deflections

When designing flexural members (and particularly beams), remember that in addition to designing for stresses, you also need to design for deflections — which means you need to use the appropriate moment of inertia for your problem. Deflections can often control the minimum moment of inertia, the maximum span length of a flexural member, or even the type of material you use (specifically the Young's modulus of elasticity), even when the stresses are acceptable.

Index

• Symbols & Numerics •

• A •

• B •

• E •

• F •

• G •

• H •

• J •

• K •

• L •

• N •

• O •

• P •

• T •

• U •